Current Research in

EMBRYOLOGY

Current Research in

EMBRYOLOGY

Sabine Globig

*Associate Professor of Biology, Hazard Community
and Technical College, Kentucky, U.S.A.*

Apple Academic Press

Current Research in Embryology

First Published in the Canada, 2011
Apple Academic Press Inc.
3333 Mistwell Crescent
Oakville, ON L6L 0A2
Tel. : (888) 241-2035
Fax: (866) 222-9549
E-mail: info@appleacademicpress.com
www.appleacademicpress.com

The full-color tables, figures, diagrams, and images in this book may be viewed at www.appleacademicpress.com

First issued in paperback 2021

ISBN 13: 978-1-77463-223-9 (pbk)
ISBN 13: 978-1-926692-81-4 (hbk)

Sabine Globig

Cover Design: Psqua

Library and Archives Canada Cataloguing in Publication Data
CIP Data on file with the Library and Archives Canada

CONTENTS

Introduction 9

1. What Makes us Human? A Biased View from the Perspective 13
   of Comparative Embryology and Mouse Genetics
   André M. Goffinet

2. Maternal Diabetes Alters Transcriptional Programs in the 26
   Developing Embryo
   Gabriela Pavlinkova, J. Michael Salbaum and Claudia Kappen

3. Increased Expression of Heat Shock Protein 105 in Rat Uterus 47
   of Early Pregnancy and Its Significance in Embryo Implantation
   Jin-Xiang Yuan, Li-Juan Xiao, Cui-Ling Lu, Xue-Sen Zhang,
   Tao Liu, Min Chen, Zhao-Yuan Hu, Fei Gao and Yi-Xun Liu

4. Ovarian Hyperstimulation Syndrome and Prophylactic 66
   Human Embryo Cryopreservation: Analysis of Reproductive
   Outcome Following Thawed Embryo Transfer
   Eric Scott Sills, Laura J. McLoughlin, Marc G. Genton,
   David J. Walsh, Graham D. Coull and Anthony P. H. Walsh

5. Neural Differentiation of Embryonic Stem Cells In Vitro: 76
   A Road Map to Neurogenesis in the Embryo
   Elsa Abranches, Margarida Silva, Laurent Pradier, Herbert Schulz,
   Oliver Hummel, Domingos Henrique and Evguenia Bekman

6. Defining Human Embryo Phenotypes by Cohort-Specific   105
    Prognostic Factors

    Sunny H. Jun, Bokyung Choi, Lora Shahine, Lynn M. Westphal,
    Barry Behr, Renee A. Reijo Pera, Wing H. Wong and Mylene W. M. Yao

7. Search for the Genes Involved in Oocyte Maturation and   121
    Early Embryo Development in the Hen

    Sebastien Elis, Florence Batellier, Isabelle Couty, Sandrine Balzergue,
    Marie-Laure Martin-Magniette, Philippe Monget, Elisabeth Blesbois and
    Marina S. Govoroun

8. Ectopic Pregnancy Rates with Day 3 Versus Day 5 Embryo   152
    Transfer: A Retrospective Analysis

    Amin A. Milki and Sunny H. Jun

9. Transcriptome Analysis of Mouse Stem Cells and Early Embryos   160

    Alexei A. Sharov, Yulan Piao, Ryo Matoba, Dawood B. Dudekula,
    Yong Qian, Vincent VanBuren, Geppino Falco, Patrick R. Martin,
    Carole A. Stagg, Uwem C. Bassey, Yuxia Wang, Mark G. Carter,
    Toshio Hamatani, Kazuhiro Aiba, Hidenori Akutsu, Lioudmila Sharova,
    Tetsuya S. Tanaka, Wendy L. Kimber, Toshiyuki Yoshikawa,
    Saied A. Jaradat, Serafino Pantano, Ramaiah Nagaraja,
    Kenneth R. Boheler, Dennis Taub, Richard J. Hodes, Dan L. Longo,
    David Schlessinger, Jonathan Keller, Emily Klotz, Garnett Kelsoe,
    Akihiro Umezawa, Angelo L. Vescovi, Janet Rossant, Tilo Kunath,
    Brigid L. M. Hogan, Anna Curci, Michele D'Urso, Janet Kelso,
    Winston Hide and Minoru S. H. Ko

10. Transcriptional Profiling Reveals Barcode-Like Toxicogenomic   181
    Responses in the Zebrafish Embryo

    Lixin Yang, Jules R. Kemadjou, Christian Zinsmeister,
    Matthias Bauer, Jessica Legradi, Ferenc Müller, Michael Pankratz,
    Jens Jäkel and Uwe Strähle

11. Release of sICAM-1 in Oocytes and In Vitro Fertilized   207
    Human Embryos

    Monica Borgatti, Roberta Rizzo, Maria Beatrice Dal Canto,
    Daniela Fumagalli, Mario Mignini Renzini, Rubens Fadini,
    Marina Stignani, Olavio Roberto Baricordi and Roberto Gambari

12. Three-Dimensional Analysis of Vascular Development in the   222
    Mouse Embryo

    Johnathon R. Walls, Leigh Coultas, Janet Rossant and
    R. Mark Henkelman

13. Nucleologenesis and Embryonic Genome Activation are
    Defective in Interspecies Cloned Embryos Between Bovine
    Ooplasm and Rhesus Monkey Somatic Cells   254

    Bong-Seok Song, Sang-Hee Lee, Sun-Uk Kim, Ji-Su Kim,
    Jung Sun Park, Cheol-Hee Kim, Kyu-Tae Chang, Yong-Mahn Han,
    Kyung-Kwang Lee, Dong-Seok Lee and Deog-Bon Koo

14. Expression of Transmembrane Carbonic Anhydrases, CAIX
    and CAXII, in Human Development   278

    Shu-Yuan Liao, Michael I. Lerman and Eric J. Stanbridge

    Index   305

INTRODUCTION

Embryology (literally, the study of embryos) is the branch of biological science that deals with the formation and early development of an individual organism, from fertilization of the egg (ovum) to birth. There are generally considered to be four types of embryological study: descriptive embryology, comparative and evolutionary embryology, experimental embryology, and teratology. Descriptive embryology is a branch of microscopic anatomy dealing with the structure of embryos and fetuses. Comparative embryology compares the development of embryos of two or more species. The observed similarities and differences may be used in taxonomic and phylogenic studies. Experimental embryology tests hypotheses about development using genetic and embryo manipulations, and addresses questions about mechanisms underlying observed developmental changes, including whether these changes are autonomous or due to environmental conditions. Teratology is an interdisciplinary field of embryology and pathology that focuses on abnormal development and congenital malformations.

Embryology also includes the study of the necessary accessory tissues; in mammals, this usually includes the placenta, fetal membranes, and umbilicus. While the field of human embryology generally includes fertilization, there is really no consensus as to when fertilization occurs: some sources adhere to the traditional idea that penetration of the ovum by the sperm comprises fertilization, while others insist that it does not occur until fusion of the two nuclei has taken place. Similarly, the definition of the term "embryo" has also proven to be an area of dispute: most consider the zygote (fertilized ovum) to be distinct from an embryo, but whether an embryo exists after one or several mitotic divisions of the zygote, or only after implantation in the uterine wall, remains an area of disagreement. It is generally agreed, however, that in humans, the zygote initially develops into a ball of cells, the morula, then a larger blastula, and at around day 5 after fertilization, a hollow blastocyst of about 70–100 cells begins to form, with an inner cell mass (ICM), or embryoblast (which subsequently forms the embryo), and an outer layer of cells, or trophoblast (which later forms the embryologic placenta). The blastocyst then embeds in the endometrial lining of the uterine wall during the second week of development, if development proceeds. From the third through the eighth week, known as the embryonic period, the embryoblast continues to develop and differentiate, clearly becoming an embryo, and the trophoblast joins with endometrial tissue to develop into the placenta. From the 9th week until birth (about the 38th week), the developing organism is termed a fetus.

Embryology, however, is not limited to human, or even mammalian, embryology. Much of today's knowledge in the field has arisen from research into the development of simpler organisms, including invertebrates. Roundworms, fruit flies, and sea urchins, as well as zebrafish, frogs, chickens, and mice, have all served as research models that led to the discovery of fundamental principles and mechanisms in the regulation of development in mammals, including humans. Gene research and biotechnology advances have provided the tools to delve deeper into the mechanisms involved. And the benefits of a deeper understanding of developmental processes include advances in many areas: human fertility, congenital diseases and defects, premature births and infant mortality, animal husbandry, veterinary medicine, and cancer research.

Currently, embryology has become an important research area for studying the genetic control of the development process and its link to cell signaling, its importance for the study of certain diseases and mutations, and in links to stem cell research. The findings of developmental biology can help to understand developmental abnormalities such as the chromosomal abnormalities that cause Down syndrome. Mutations in human regulatory genes underlie the pathogenesis of a wide spectrum of human congenital diseases, and genetic screening for mutations in many of these human genes is under way in hundreds of testing laboratories throughout the world. Many of these regulatory genes are tumor suppressor genes, further emphasizing their clinical significance. And the pathogenetic mechanisms of many drugs and foods now provide a basis for the prevention of many birth defects.

The latest research in embryology includes stem cell research, reproductive medicine (in vitro fertilization, fetal tissue transplantation), organ and tissue transplantation, animal cloning successes, and research into programmed cell aging and death and its role in normal development and aging versus tumor growth. The study of human embryo development has recently taken on added significance due to these advances, which have opened the door for genetic engineering and tissue and gene therapy. Together, these embryological advances have raised a number of ethical, religious, and legal issues that all of us must confront, and which have already been raised in the media and the political arena.

— Professor Sabine Globig

What Makes us Human? A Biased View from the Perspective of Comparative Embryology and Mouse Genetics

André M. Goffinet

ABSTRACT

For a neurobiologist, the core of human nature is the human cerebral cortex, especially the prefrontal areas, and the question "what makes us human?" translates into studies of the development and evolution of the human cerebral cortex, a clear oversimplification. In this comment, after pointing out this oversimplification, I would like to show that it is impossible to understand our cerebral cortex if we focus too narrowly on it. Like other organs, our cortex evolved from that in stem amniotes, and it still bears marks of that ancestry. More comparative studies of brain development are clearly needed if we want to understand our brain in its historical context. Similarly, comparative genomics is a superb tool to help us understand evolution, but again, studies should not be limited to mammals or to comparisons between human and chimpanzee, and more resources should be invested in investigation of many vertebrate phyla. Finally, the most widely used rodent models for studies of cortical development are of obvious interest but they cannot be considered models of a "stem cortex" from which the human type evolved. It remains of paramount importance to study cortical development directly in other species, particularly in primate models, and, whenever ethically justifiable, in human.

Report

What makes us human?

A reader: "Gosh, who does he thinks he is, he who claims to answer that question?"

Indeed, so complex is the question that one may legitimately wonder whether it is worth asking it. A check on the internet as of this year (2006) returns different websites, from the reputed Smithsonian Institution to some that sound rather like crackpots. Most religions will tell us that Man was created by God and that our human condition will forever remain a mystery. Starting from a different viewpoint, information theory – and common sense – teach us that understanding something requires more analytical power than the object under investigation itself, thus leading to a similar conclusion. As a scientist, however, I am drawn almost inexorably to think about our "humanness" as a scientific question: even though there is no global answer, it is a question about which we can at least formulate ideas and hypotheses that can be checked by observation (e.g. fossil record) or experiment. This short commentary is aimed to those who share this endeavour.

Let me begin with two preambles. First, even the standard response "what makes us human is our brain" has obvious limitations. Imagine a creature with a human brain in the body of an ape: would she/he feel human? Would we regard her/him as human? Probably not: Being human is indeed a status that relies on, and integrates many parameters: our brain and body, of course, but also our history as a person and as a group, up to the social setting in which we live [1]. Arguably, such remarks apply to the definition of all biological species, but they are particularly striking for ours: it is not anthropomorphic to say that we have reached a unique point of sophistication in our relationships to each other and to other creatures. This remark made, I shall leave socio-biology aside – after all it is not my field – and concentrate on the point of view of brain development and evolution, about which I feel a bit less ignorant.

The other preliminary point that I wish to discuss briefly is the assumption "What makes our brain human, is our cerebral cortex, and particularly prefrontal lobes". Although I basically agree with this assertion as a first approximation, it should be noted that the evolution and development of cerebral cortical performances did not occur in isolation. Rather, the process was contingent upon the development of the nervous system and the rest of the body as a whole. To mention the most celebrated example: It is plain obvious that language is a defining trait of our species; language requires evolutionary acquisition of specific features in the larynx, and of neurological coordination of laryngeal muscles and other structures that are not related to the cerebral cortex, yet are unique to our species. The same remark can be made about the acquisition of hand skills and there are several other examples. Organisms evolve as entities. For obvious operational reasons, we mostly study the evolution of parts. But we should keep constantly in mind that, when it comes to interpretation of data, it makes little sense to discuss parts without considering the whole.

The two nuances above being made, I think most of us will agree that one of the main differences between us and other animals lays in our cognitive abilities. Every dog owner knows that animals have emotions or something close to it. They can be sad or joyous, they have their temper. They remember – sometimes at least – what they are taught. Chimpanzees can be trained, with much patience and care, to read and learn elementary symbols. There are countless anecdotes and observations indicating that elephants look depressed when loosing a mate, and may show some signs of mourning, such as returning to visit the site of death. Clearly, memory, intelligence, emotions, consciousness or even a moral sense of right and wrong are not absolutely unique to humans [2]. A recent report even claims that mice show "empathy"[3]. But only in our species have cognitive abilities reached a unique level of sophistication. As far as we know, even when compared to apes, we are the only creature who invented language, his own written rules, who runs his social life with a moral code of right and wrong, allowing us to propose that there have been some (several?) quantum leap(s) during the evolution of our brain, although there is much uncertainty about the stages in our evolution when these leaps occurred.

If we differ from animals in general, and from apes in particular, by our cognitive capacity, in which our cerebral cortex plays a key part, I would reformulate the question "What makes us human?" and ask: "What is so special about our brain and cerebral cortex"?

Man has a large brain. Some animals have larger ones, but, mostly, man has the largest brain weight when allometric correction is made for body size. Animals with the largest brains include cetaceans and elephants that we regard generally as intelligent. Interestingly, chimpanzee and gorilla score average for their brain size relative to body weight [4].

Studies of cranial endocasts indicate quite clearly that Neanderthal had larger cranial capacity and presumably larger brains than us. Brain size, although of obvious importance, is not everything. Cetaceans, even small ones such as Dolphins, have huge brains, with large and foliated cortical surfaces. Their temporal lobes are larger than ours, and this could be related to their fantastic spatial memory. Yet their cognitive skills, although far from negligible, cannot be compared to ours. The neocortex in cetaceans retains many features of its ancestral character such as a relatively low number of granular interneurons, and a relatively simple neuronal differentiation (dendrite trees). Phylogenetic isolation may have resulted in development of the nervous system chiefly by increase in nerve cell numbers (associated with great cortical expansion), by quantitative expansion without substantial architectonic evolution [5]. Neuronal numbers may not systematically vary linearly with cortical size or surface. In contrast to cetaceans, human neurons are characterized by a most elaborate architectonic organization, by a high proportion of some neuronal types, such as Cajal's "double bouquet" cell [6], and by its exquisite connectivity. Again, such parameters do not necessarily correlate with neuronal numbers and are obviously inaccessible to measurements of endocasts.

Genes that control development are preferential targets of the evolutionary process, and I would argue that the identification and study of mechanisms that regulate cortical development in different species are central to understanding our cortex. For example, developmental studies over the last ten years or so have clearly shown that the two main neuronal populations of the mammalian cortex, namely excitatory glutamatergic pyramidal cells and GABAergic interneurons are generated in different sectors of ventricular zones and migrate to the cortex along different routes. As schematized in Fig 1, the main body of cortical neurons form glutamatergic cells that migrate radially, whereas GABAergic cells generated in the ganglionic eminence, primarily the medial part (MGE), migrate to the cortex tangentially [7-9]. Whereas this developmental pattern is well established in rodents, it is also described in chick [10] and it may thus be a general feature of all amniotes. On the other hand, in the human brain, in addition to the MGE, GABAergic interneurons are also generated in the cortical VZ and migrate radially to the cortex [11]. Clearly, even though this has already been extensively studied, more works need to be done on the origin of cortical interneurons in different species.

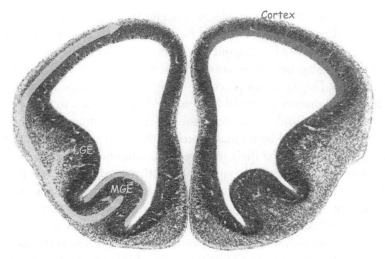

Figure 1. Coronal section in the forebrain of an embryonic mouse at 12.5 days of gestation (preplate stage), showing the lateral and medial ganglionic eminences (LGE, MGE) from which GABAergic interneurons migrate to the cortical anlage (left, yellow). Glutamatergic neurons destined for the cortex are generated locally in the cortical ventricular zone and migrate radially (right, red). Courtesy of V. Pachnis.

One would think that solid data are available in the literature and can help us answer most of the comparative questions stated above? Not at all: our knowledge of comparative brain anatomy – not to mention comparative brain development – remains rudimentary [9]. This field has been neglected by grant agencies – and by most scientists – for several decades, being pursued only by a few dedicated colleagues. Yet, if we accept that the question of human origins and humanness are relevant, then I would argue that systematic comparative studies of brain anatomy and development, using state of the art techniques, are urgently needed and should be appropriately supported. A few years ago, a Human Brain Project was proposed, but I no not believe we should focus too narrowly on the human, and even on the mammalian cortex, which will be best understood in its evolutionary context. "Nothing in biology makes sense except in the light of evolution" (Dobzansky, [12]).

It is our brain that makes us human, and our brain develops under control of our genetic makeup. Hence the saying: "Our "humanness" is in our genes". This leads us to believe that, by comparing DNA sequences, we might be able to pick up human specific features and explain the whole thing. There is currently a lot of publicity about the comparison between the human and chimpanzee genomes. It is widely thought that the subtle genetic differences thus defined will point to key genetic determinants of the human species. Although these studies are fascinating and necessary, and will undoubtedly yield considerable insight, I believe this reasoning is somewhat simplistic. There is rarely an evident correlation between a genetic difference and the resulting phenotypic effects: a change that would be considered almost irrelevant may very well be most important, and vice versa. The genetic program (mostly DNA sequences) is not the phenotype, as the latter results from running the genetic program during development, and is the product of epigenetic history. During development, the epigenetic landscape unfolds in a highly non linear and massively parallel fashion, making it difficult to understand relationships a posteriori, and usually impossible to predict outcome from basic principles. Our brain evolved by natural selection, that is the survival of phenotypes (hence genomes) with the highest rate of reproduction and best suited to changing environments. The evolutionary history of the vertebrate brain is poorly understood because brain tissue does not fossilize, but also because research in comparative developmental neurobiology is still in its infancy as outlined above. The cortex may have been absent in early vertebrates: It is reduced to a periventricular layer in anamniotic vertebrates. The cortex increased in size and organization in stem amniotes, the ancestors of living reptiles, birds and mammals [9]. It gained prominence in synapsids, the lineage leading to mammals, and evolved explosively in primates. Evolution from our common ancestor with apes is only the latest major radiation in this ongoing process, and this latest step and recent brain evolution did not evolve from scratch. "Evolution is a tinkerer" [13] and can only build on a prior structure. If we want to understand brain evolution, we need to tackle DNA sequences, like the fossil record, in their historical perspective and avoid focusing narrowly. I very much doubt that we can understand the genetic control of the human brain without addressing basic genetic mechanisms in living animals that belong to as many branches of the tree as possible. It might even be possible to rescue enough DNA from fossil material, although I doubt its quality will be sufficient. Examples of what I think are elegant approaches are the work of the Haussler group on ultraconserved elements present in the genome of human and several other vertebrates [14], that led to identification of 'human accelerated regions', HAR1, and of a novel RNA gene (HAR1F) that is expressed specifically in Cajal-Retzius neurons in the developing human neocortex [15], and a recent study of human lineage-specific gene amplification [16]. But this is only an encouraging beginning and a lot remains to be done.

After the sequencing of the human and mouse genomes, obvious biomedical priorities, I thought that sequencing centers would pursue with amphioxus, a fish, a frog, a turtle, a snake or lizard, Sphenodon, chick, a crocodilian, etc... But this is not what happened! Rather, genome sequencing is under way or almost completed for several horses, cattle, dogs, cats and other pets. As often, economic considerations and non scientific arguments prevail. But this can be changed and the price and speed of sequencing allows a wider perspective. If we believe that the questions of human origins and humanness are relevant, then concerted sequencing efforts to investigate as many branches of the tree as possible should be funded and undertaken actively. By comparing sequences on a global phylogenetic scale, we might be able to identify some of the unique, subtle genetic changes that are specific and essential to the evolution of the primate lineage and to the development of the human cortex.

The rapid increase of brain size and complexity during recent evolution in the primate lineage is well known and widely discussed elsewhere. To many, such changes in brain size in such a short time appear difficult to explain by natural selection. I believe this view is wrongly based on a sort of subconscious postulate that evolution works more or less linearly, namely that small changes in phenotypes reflect minor changes in the DNA, whereas large changes in phenotypes require huge modifications in DNA. But why should it be so, when high non linearity is in fact the rule rather than the exception. Like development, evolution works in a most non linear way. Whether this non linearity is chaotic will remain forever unknown, as we cannot rewind the tape and we have no way to produce evolution experimentally, except for a few very small and limited cases. But the point is that evolution is highly non linear, and I think some recent experimental observations can be interpreted in this frame of mind. Let me give a few examples.

The process of increased cortical surface by foliation is generally considered essential during evolution towards the human cortex. As I hinted to above, this was not the sole mechanism used by evolution to increase cortical performance, but few will doubt its importance. The process is often assumed to be complex, because it appears as such to our investigations. Yet, cortical folding can vary widely within closely related lineages, as can be appreciated by consulting the superb website "Comparative Mammalian Brain Collections" [17]. For example, in monotremes, Echidna has an elaborate, highly foliated cortex, whereas Platypus is almost lissencephalic [18]. Similar examples can be found in other phyla, including primates, some of which are almost lissencephalic. Furthermore, a mixture of brain hypertrophy with variable levels of gyrated cortex is artificially accomplished in mice by elegant, yet relatively simple manipulations [19], such as germline inactivation of caspases 3 or 9 [20,21], increased expression of beta-catenin in transgenic mice [22,23], incubation of embryonic cortex in vitro in the presence of lysophosphatidic acid [24,25], or manipulation of ephrin/Eph signaling [26]. Intriguingly, mice with inactivation of the phosphatase PTEN [27], and with brain-specific inactivation of alpha-catenin [28] have brain hypertrophy, but no or very little increased cortical foliation, showing that brain size, cell numbers and foliation are not always correlated. The cerebellum of Mormyrid fishes, with its large size and extensive foliation [4], provides yet another example indicating that huge increase in surface of a cortex can probably evolve or be produced quite easily.

These observations suggest that the production of a gyrated surface does not need extensive genetic changes, and could have evolved in any phylum. But it did not evolve very often, and I see at least two reasons for this, that can be tested experimentally or by comparative studies. First, acquiring a large foliated cortex may be relatively easy, but it may be more difficult to make this large cortex work efficiently. Transgenic mice with increased cortex are often not viable and, although this remains to be studied in detail, the resulting large cortex does probably not work well. This illustrates the point made above, that evolution of the brain does not occur in isolation but in the context of the whole organism. An increase in cortical size must be accompanied with balanced growth of the mesodermal components that support and vascularize the brain. Also, the increase in cortical surface and the organization of many adjacent radial cortical columns may increase neuronal excitability and susceptibility to seizures. A consequence of a highly geometrical arrangement of radial cortical columns is to facilitate modifications of the membrane potential by field effects ("ephaptic" interactions), largely believed to be involved in the oscillations of electrocortical rhythms and in generation of seizures [29]. This quasi crystalline arrangement presumably has advantages in terms of computational power, but also comes at a price, as ephaptic excitation facilitates the tangential spreading of activity and decreases the threshold for aberrant epileptic discharges. The second reason, not in contradiction with the first, may be more important. Namely, the acquisition of a large brain and particularly of a large foliated cortex may not be an evolutionary advantage per se. Most species that are hugely successful in the evolutionary sense do not have a large brain or high cognitive power. Large brain size and increased computational power presumably proved evolutionary useful quite recently, in early Homo, and the reason why this parameter was selected positively at some point in our phylogenetic history remains unknown. Perhaps, like several traits, increased cortical surface was just "tried" at some point in the primate lineage and, once the track had been taken, there was little choice but to keep the option, because it would have cost much more to turn back than to continue. The option finally paid off, as humans are obviously a very successful species, at least at this moment and provided we do not end it all ourselves by burning or exploding the planet.

Contrary to common belief, stem mammals probably did not have a rodent-like forebrain. Rodents are highly evolved animals that are not directly related to stem mammals, from which the lineage leading to primates is in fact more directly derived [30]. Although this is not proven, I consider it likely that stem mammals had a relatively unspecialized, basic cortex, possibly with some foliation, from which highly specialized lissencephalic cortices evolved in lineages such as rodent, whereas other lineages kept some foliation and even increased it in some branches, most notably ours. It is generally accepted that evolution works more easily on relatively undifferentiated forms [31] and that neoteny of primates was a contributing factor to their rapid evolution. If this view is correct, then the mouse cortex is not a model of the "primitive" cortex of stem mammals, and inferences from mouse data in terms of cortical evolution should be made with appropriate caution.

Another illustration that the mouse, with all its advantages, should not be considered the sole model for cortical development and evolution concerns the role of Cajal-Reztius cells, early neurons in the cortical marginal zone that degenerate massively around birth [32]. Studies in reeler and other mutant mice and human genetic studies clearly demonstrate that Reelin secreted by Cajal-Retzius cells is absolutely required for normal cortical development in mice and for foliation of the human cortex [33]. Yet, in mice, genetic ablation of most of Cajal-Retzius cells does not perturb cortical development much, indicating a large redundancy [34]. Although this remains to be studied further, it seems that there is a huge excess of Cajal-Retzius cells and of Reelin in rodents, and that this is not the case in the human, where a provision of Reelin seems to be provided over an extended period of time in the marginal zone [35].

As discussed above, the unique cognitive power of the human brain seems to be due to the evolutionary acquisition of multiple factors such as high neuronal number, large foliated cortex, optimal architectonic organization, complex neuronal types, highly organized and elaborate connections. The evolution of the human brain has proceeded at amazing speed. We have reached a stage where our cognitive capacity increases more through technological innovation, and cultural evolution vastly outpaces biological evolution. However, I see no reason why brain evolution should stop at the present human level and it is impossible to escape the question of its future. Without going into science fiction, I would like to mention one feature of the cerebral cortex that receives scant attention – it is in fact generally ignored – and that I find very intriguing, namely the subpial granular layer (SGL). The SGL, is a transient contingent of cells that are apparently generated from a basal region, close to the hilus of the sylvian fissure and the paleoventricle, and migrate tangentially in the subpial cortical marginal zone during mid- and late gestation. The SGL is much more developed in human than other mammals, to the point that it is sometimes considered human-specific, even though a diminutive SGL has now been described in other mammals. Initially described more than a century ago by Ranke, the SGL was examined in some detail by Brun [36], who concluded that its cells differentiate into glia and/or probably die after entering the cortical ribbon radially. A more recent study indicated that the SGL cells are likely neuronal and enter the cortex radially [37], but remained inconclusive about their fate. By analogy with the development of the external and internal granular layers of the cerebellum, a reasonable hypothesis could be that subpial neurons contribute to the cortical neuronal population but that this went undetected because it represents a minority of neurons. The SGL might provide a source of interneurons in addition to the main contingent that originates from the ganglionic eminences. Like in the cerebellum, an increase in SGL cells could result in an increase in surface and folding of the cerebral cortical surface and, who knows, result in a cerebral cortex with increased computational power. Even though the SGL is best studied in man, the idea that it could play a role during cortical development and evolution is not pure speculation. Some diminutive SGL is present in mammalian models and techniques are available to define better the cellular constitution of the SGL and the fate of its cells after they enter the cortex, to identify their repertoire of gene expression, the transcription factors implicated in their differentiation.

I am convinced that we should not limit ourselves to the analysis of the mouse and non mammalian species, but that we should also study actively brain development in primates, and particularly in human, using state of the art techniques. Of course, ethical considerations are of the utmost importance and must be taken into account. But studies of brain development in primates and man, using the whole arsenal of modern technology are a unique way to trace novel, original avenues and to address scientifically the question of the evolution of our cerebral cortex, and of our biological nature.

Conclusion

The human cerebral cortex is at the core of human nature. Our cortex evolved from that in stem amniotes and cannot be understood if excluded from this evolutionary context. If we want to understand better our cerebral cortex, more efforts should be invested in comparative studies of embryonic development, using state of the art technologies. Genomic sequencing efforts should be directed at all branches of the vertebrate tree rather than focused narrowly on mammals. Finally, the rodent cortex is not a perfect model of the stem mammalian cortex and specific studies of primate and human cortices are necessary. In addition to its fundamental interest, an improved scientific knowledge of human nature will help us define better our place and thus our rights and our duty in relation to our environment and to ourselves. This is after all the ultimate ecological challenge!

Acknowledgements

I wish to thank Neil Smalheiser for inviting me to contribute to this series of comments on "What makes us human?" I also thank Gundela Meyer, as well as Catherine Lambert de Rouvroit and Fadel Tissir, for numerous stimulating discussions about cortical evolution.

References

1. Landes DS: The Wealth and Poverty of Nations. New York, W.W.Norton & Co; 1999.

2. de Waal F: Good natured. Harvard University Press; 1996.

3. Langford DJ, Crager SE, Shehzad Z, Smith SB, Sotocinal SG, Levenstadt JS, Chanda ML, Levitin DJ, Mogil JS: Social modulation of pain as evidence for empathy in mice. Science 2006, 312:1967–1970.

4. Butler AB, Hodos W: Comparative Vertebrate Neuroanatomy. New York, Wiley-Liss; 1996.

5. Morgane PJ, Glezer II, Jacobs MS: Comparative and evolutionary anatomy of the visual cortex of the dolphin. In Comparative Structure and Evolution of Cerebral Cortex, Part II. Edited by: Jones EG and Peters A., Springer Verlag; 1990:215–262.

6. Yanez IB, Munoz A, Contreras J, Gonzalez J, Rodriguez-Veiga E, DeFelipe J: Double bouquet cell in the human cerebral cortex and a comparison with other mammals. J Comp Neurol 2005, 486:344–360.

7. Metin C, Baudoin JP, Rakic S, Parnavelas JG: Cell and molecular mechanisms involved in the migration of cortical interneurons. Eur J Neurosci 2006, 23:894–900.

8. Fragkouli A, Hearn C, Errington M, Cooke S, Grigoriou M, Bliss T, Stylianopoulou F, Pachnis V: Loss of forebrain cholinergic neurons and impairment in spatial learning and memory in LHX7-deficient mice. Eur J Neurosci 2005, 21:2923–2938.

9. Molnar Z, Metin C, Stoykova A, Tarabykin V, Price DJ, Francis F, Meyer G, Dehay C, Kennedy H: Comparative aspects of cerebral cortical development. Eur J Neurosci 2006, 23:921–934.

10. Cobos I, Puelles L, Martinez S: The avian telencephalic subpallium originates inhibitory neurons that invade tangentially the pallium (dorsal ventricular ridge and cortical areas). Dev Biol 2001, 239:30–45.

11. Letinic K, Zoncu R, Rakic P: Origin of GABAergic neurons in the human neocortex. Nature 2002, 417:645–649.

12. Dobzansky T: Nothing in biology makes sense except in the light of evolution. The American Biology Teacher 1973, 35:125–129.

13. Jacob F: Evolution and Tinkering. Science 1977, 196:1161–1166.

14. Bejerano G, Pheasant M, Makunin I, Stephen S, Kent WJ, Mattick JS, Haussler D: Ultraconserved elements in the human genome. Science 2004, 304:1321–1325.

15. Pollard KS, Salama SR, Lambert N, Lambot MA, Coppens S, Pedersen JS, Katzman S, King B, Onodera C, Siepel A, Kern AD, Dehay C, Igel H, Ares M Jr., Vanderhaeghen P, Haussler D: An RNA gene expressed during cortical development evolved rapidly in humans. Nature 2006.

16. Popesco MC, Maclaren EJ, Hopkins J, Dumas L, Cox M, Meltesen L, Mc-Gavran L, Wyckoff GJ, Sikela JM: Human lineage-specific amplification, selection, and neuronal expression of DUF1220 domains. Science 2006, 313:1304–1307.

17. Comparative-Mammalian-Brain-Collections   http://www.brainmuseum.org/index.html

18. Rowe M: Organisation of the cerebral cortex in monotremes and marsupials. In Cerebral Cortex, Vol 8B: Comparative structure and Evolution of Cerebral Cortex, Part II. Edited by: Jones EG and Peters A. New York, Plenum; 1990:263–334.

19. Rakic P: Neuroscience. Genetic control of cortical convolutions. Science 2004, 303:1983–1984.

20. Kuida K, Zheng TS, Na S, Kuan C, Yang D, Karasuyama H, Rakic P, Flavell RA: Decreased apoptosis in the brain and premature lethality in CPP32-deficient mice. Nature 1996, 384:368–372.

21. Kuida K, Haydar TF, Kuan CY, Gu Y, Taya C, Karasuyama H, Su MS, Rakic P, Flavell RA: Reduced apoptosis and cytochrome c-mediated caspase activation in mice lacking caspase 9. Cell 1998, 94:325–337.

22. Chenn A, Walsh CA: Regulation of cerebral cortical size by control of cell cycle exit in neural precursors. Science 2002, 297:365–369.

23. Chenn A, Walsh CA: Increased neuronal production, enlarged forebrains and cytoarchitectural distortions in beta-catenin overexpressing transgenic mice. Cereb Cortex 2003, 13:599–606.

24. Price DJ: Lipids make smooth brains gyrate. Trends Neurosci 2004, 27:362–364.

25. Kingsbury MA, Rehen SK, Contos JJ, Higgins CM, Chun J: Non-proliferative effects of lysophosphatidic acid enhance cortical growth and folding. Nat Neurosci 2003, 6:1292–1299.

26. Depaepe V, Suarez-Gonzalez N, Dufour A, Passante L, Gorski JA, Jones KR, Ledent C, Vanderhaeghen P: Ephrin signalling controls brain size by regulating apoptosis of neural progenitors. Nature 2005, 435:1244–1250.

27. Yue Q, Groszer M, Gil JS, Berk AJ, Messing A, Wu H, Liu X: PTEN deletion in Bergmann glia leads to premature differentiation and affects laminar organization. Development 2005, 132:3281–3291.

28. Lien WH, Klezovitch O, Fernandez TE, Delrow J, Vasioukhin V: alphaE-catenin controls cerebral cortical size by regulating the hedgehog signaling pathway. Science 2006, 311:1609–1612.

29. McCormick DA, Contreras D: On the cellular and network bases of epileptic seizures. Annu Rev Physiol 2001, 63:815–846.

30. Colbert EH, Morales M, Minkoff EC: Colbert's Evolution of the Vertebrates. 5th edition. Wiley-Liss; 2001:576.

31. Raff RA: The Shape of Life. Chicago University Press; 1996.

32. Meyer G, Cabrera Socorro A, Perez Garcia CG, Martinez Millan L, Walker N, Caput D: Developmental roles of p73 in Cajal-Retzius cells and cortical patterning. J Neurosci 2004, 24:9878–9887.

33. Tissir F, Goffinet AM: Reelin and brain development. Nat Rev Neurosci 2003, 4:496–505.

34. Yoshida M, Assimacopoulos S, Jones KR, Grove EA: Massive loss of Cajal-Retzius cells does not disrupt neocortical layer order. Development 2006, 133:537–545.

35. Meyer G, Goffinet AM: Prenatal development of reelin-immunoreactive neurons in the human neocortex. J Comp Neurol 1998, 397:29–40.

36. Brun A: The subpial granular layer of the foetal cerebral cortex in man. Its ontogeny and significance in congenital cortical malformations. Acta Pathol Microbiol Scand 1965, Suppl 179:3–98.

37. Gadisseux JF, Goffinet AM, Lyon G, Evrard P: The human transient subpial granular layer: an optical, immunohistochemical, and ultrastructural analysis. J Comp Neurol 1992, 324:94–114.

Maternal Diabetes Alters Transcriptional Programs in the Developing Embryo

Gabriela Pavlinkova, J. Michael Salbaum and Claudia Kappen

ABSTRACT

Background

Maternal diabetes is a well-known risk factor for birth defects, such as heart defects and neural tube defects. The causative molecular mechanisms in the developing embryo are currently unknown, and the pathogenesis of developmental abnormalities during diabetic pregnancy is not well understood. We hypothesized that the developmental defects are due to alterations in critical developmental pathways, possibly as a result of altered gene expression. We here report results from gene expression profiling of exposed embryos from a mouse diabetes model.

Results

In comparison to normal embryos at mid-gestation, we find significantly altered gene expression levels in diabetes-exposed embryos. Independent validation of altered expression was obtained by quantitative Real Time Polymerase Chain Reaction. Sequence motifs in the promoters of diabetes-affected genes suggest potential binding of transcription factors that are involved in responses to oxidative stress and/or to hypoxia, two conditions known to be associated with diabetic pregnancies. Functional annotation shows that a sixth of the de-regulated genes have known developmental phenotypes in mouse mutants. Over 30% of the genes we have identified encode transcription factors and chromatin modifying proteins or components of signaling pathways that impinge on transcription.

Conclusion

Exposure to maternal diabetes during pregnancy alters transcriptional profiles in the developing embryo. The enrichment, within the set of de-regulated genes, of those encoding transcriptional regulatory molecules provides support for the hypothesis that maternal diabetes affects specific developmental programs.

Background

Maternal diabetes disturbs embryonic development and can cause diabetic embryopathy, with cardiovascular malformations, neural tube defects and caudal dysgenesis as the most characteristic congenital malformations [1,2]. Diabetes-induced dysmorphologies have been ascribed to increased apoptosis [3,4], perturbation of prostaglandin synthesis and metabolism [5-7], deficiency in membrane lipids [8-11], and generally altered metabolism in the embryo [12]. Several studies have associated oxidative stress with the maternal diabetic condition, and the administration of anti-oxidants reduced the incidence of developmental defects in experimental models of intrauterine exposure to diabetes [5,13-19]. However, it is unclear how systemic metabolic disease results in particular developmental defects that are restricted to specific tissues in diabetic embryopathy [20].

Growing evidence suggests that maternal diabetes alters expression of developmental genes in the embryo, resulting in abnormal morphogenesis. Decreased expression of Pax3, a gene involved in neural tube defects [21,22], was found in embryos from diabetic mouse dams at gestation day 8.5, with neural tube defects evident by day 10.5 [4]. Pax3 deregulation, presumably through oxidative stress [23], is also associated with heart defects that involve neural crest cell derivatives [24]. We recently showed that Wnt signaling is affected in diabetes-exposed mouse embryos [25]. These findings support the notion that diabetic pregnancy leads to altered expression of molecules that play key roles in patterning and development of embryonic tissues, implicating altered transcriptional regulation as a potential pathogenic mechanism in diabetic embryopathy. In order to identify genes and pathways affected by maternal diabetes, we performed gene expression profiling of diabetes-exposed mouse embryos using oligonucleotide microarrays.

Results and Discussion

Animal Model of Diabetic Embryopathy

Mouse embryos were isolated from diabetic or control dams at embryonic day 10.5 (E10.5) because at this stage neural tube defects are easily detectable. The frequency of NTDs in diabetes-exposed embryos was approximately 17% (16/96 diabetes-exposed embryos) compared to 0% in normal pregnancies (0/220). Except where noted, no malformed embryos were used for gene expression studies. We found no significant differences in litter size or number of resorbed embryos between diabetic (n = 11) and non-diabetic pregnancies (n = 10; P = 0.07). We also did not detect any developmental delay in apparently unaffected diabetes-exposed embryos; their morphological appearance, i.e. features of brain development, limb development, and somite numbers were commensurate with developmental age (Kruger et al., manuscript in preparation). All experiments used the FVB inbred strain. We here report the results from two independent expression profiling experiments (Figure 1), using individual embryo samples in Experiment I and and a pooling strategy in Experiment II.

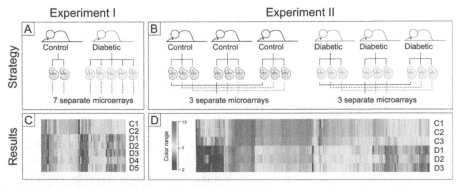

Figure 1. Experimental approaches to determine gene expression profiles of normal and diabetes-exposed embryos. Panels A and B depict the two independent microarray experiments. Panels C and D depict expression profiles where each colored vertical line represents the expression signal for one gene and row represent individual embryos (Exp. I) or samples (Exp. II). Red represents increased expression, blue reflects decreased expression, and intermediate colors represent minor changes (the color range was chosen along an arbitrary scale). Using a hierarchical clustering algorithm (with euclidean distance metric and centroid linkage rule, as implemented in GeneSpring), these graphic representations shows that expression profiles for embryos exposed to maternal diabetes differ considerably from control unexposed embryo profiles.

Gene Expression Profiling in Embryos Exposed to Diabetes

In Experiment I, we surveyed the expression profiles of 2 control and 5 diabetes-exposed embryos. In an initial comparison of expression levels between control and diabetes-exposed embryos, 302 probe sets passed the "fold-change>2" criterion, and their expression profiles were visualized using hierarchical clustering (Figure 1, Panel C). Control embryos exhibited profiles similar to each other, and differences to the expression profiles of individual diabetes-exposed embryos are visually obvious. These results support the hypothesis that maternal diabetes affects gene expression in the exposed embryos.

This survey covered about 14000 genes, and after application of the analysis criteria (see Methods for details), we identified 126 genes (~1% of the total) with expression levels that were changed in diabetes-exposed embryos by more than 2-fold relative to controls (Table 1). An additional 378 genes displayed expression differences between 1.5- and 2-fold (data not shown). The majority (83%) of the 126 genes we identified were expressed at lower levels in diabetes-exposed compared to control embryos, and this was reflected in the larger dataset as well (72.8% of the additional 378 genes were decreased in expression). This decreased gene expression was not the result of developmental delay, since the morphology of diabetes-exposed embryos was stage-appropriate.

Table 1. Genes affected by maternal diabetes classified by cellular function.

Functional category	# of genes	% of total	Gene Symbol
Transcription factors	15	12	Bcl11a, Cited4, Creb1, Crsp2, Hif1a, Klf9, Lin28, Nsd1, Rb1cc1, Rnf14, Zfa, Zfp60, Zfp294, Zfp385, 2610020O08Rik
DNA-binding/chromatin	8	6	Atrx, Baz1b, Exod1, Hist1h2bc, Hmga1, Msl31, Setdb1, Top2b
Signal transduction	15	12	Ap1g1, Arid4b, Gad1, Grb10, Mapk10, Phip, Pik3c2a, Pkia, Ptp4a3, Ptprs, Rabgap1, Rp2h, Stam2, Ywhag, Zcsl3
Cell surface, incl. receptors	13	10	Agtr2, Aplp2, Cxadr, Efnb2, Epha3, Ghr, Gpr65, Il6st, Itgav, Pdgfra, Ptprk, Sema3a, Tgfbr1
Extracellular matrix/adhesion	9	7	Adam10, Ctse, Hs6st2, Ndst1, Ogt, Pcdh18, Pxn, Sel1h, Twsg1
Cytoskeleton/microtubules	10	8	Dcx, Dnm1l, Epb4.1l2, Gmfb, Kif11, Mtap2, Sncg, Tubb2b, Tubb2c, Vcl
RNA-binding	7	6	Arl5a, Dcp1a, Pabpc1, Rnpc2, Rod1, Sfrs2, Syncrip
Transporter/channels	9	7	Abcb7, Aqr, Cacna2d1, Mbtps1, Slc2a1, Slc16a3, Slc25a22, Stx17, Tm9sf3
Metabolism/enzymes	6	5	Aldh18a1, Blvrb, Gmpr, Pfkl, Tnks2, Upp1
Lipid metabolism	6	5	Etnk1, Hdlbp, Scd2, Sgpl1, Sptlc1, Stom
Metal-ion homeostasis	3	2	LOC669660, Mt2, Tfrc
Protein catabolism	9	7	Arih2, Nedd4, Supt16h, Usp7, Usp12, Eif3s10, Gopc, Lin7c, Vps35
Cell cycle/apoptosis	3	2	Api5, Birc4, Kras
Other	6	5	Dysf, Ivns1abp, Pelp1, Plekha5, Trim2, Trim44
Unknown	7	6	Heatr1, BC067396, 1300007C21Rik, 6330503C03Rik, 6330527O06Rik, 6330578E17Rik, LOC640370
total	126	100	

Validation of Microarray by Quantitative Real-Time PCR

Changes in gene expression detected by microarray were validated by quantitative real time PCR (Q-RT-PCR) for selected genes with potential relevance to diabetes or embryonic development (Table 2). Embryo samples were from different pregnancies than those employed for the microarray studies, and only embryos were used that appeared morphologically normal. Of all genes assayed, 16 exhibited differential expression in the Q-RT-PCR assay, confirming the microarray results in independent embryo samples. Three genes exhibited no differences, and seven genes were differentially expressed in both assays ($P < 0.05$); however, the change occurred in opposite directions. The discrepancies were traced back to (i) annotation problems: Hmga1, Lin28, Phip, (ii) different length of 3'UTR sequences where location of the microarray probe would not query all transcripts arising from the respective genes: Sema3a, Rod1, Slc2a1, or (iii) alternative splicing: Ogt, Tfrc [26], Creb1 [27]. The Q-RT-PCR assays were designed to specifically amplify a region of the transcript different from that covered by the microarray probe in order to obtain an independent measurement. For the majority of genes we tested, the independent assay confirms the initial finding that expression levels are altered in embryos exposed to maternal diabetes.

Table 2. Validation of microarray results by quantitative RT-PCR.

| Gene Symbol | Microarray | | qRT-PCR | |
	Fold change	t-test (p-value)	Fold change	t-test (p-value)
Adam10*	-2.37	0.0024	-1.56	0.035
Api5*	-2.30	0.0004	-1.33	0.007
Atrx	-2.15	0.0016	-1.33	0.046
Bazlb*	-2.38	0.0012	-1.58	0.017
Cxadr*	-2.00	0.0034	-1.38	0.0002
Dcx*	-3.05	0.0001	-2.20	<0.0001
Efnb2*	-2.04	0.0437	-1.72	0.010
Hif1a	-2.52	0.0041	-1.49	0.011
Il6st	-2.28	0.0198	-1.77	0.039
Mt2	3.45	0.0053	2.16	0.027
Mtap2	-3.02	0.0000	-1.98	0.0002
Pcdh18*	-5.77	0.0017	-1.35	<0.0001
Pdgfra l	-2.01	0.0100	-1.42	0.018
Tgfβ r1	-3.26	0.0002	-1.89	0.010
Twsg1	-4.91	0.0002	-1.89	0.005
Vcl	-2.37	0.0020	-2.03	0.002
Hmga1	2.13	0.0060	-2.14	<0.0001
Lin28*	2.00	0.0182	-1.30	0.032
Ogt	2.52	0.0310	-1.51	0.019
Slc2a1	2.15	0.0252	-1.68	0.012
Phip	-2.76	0.0006	1.49	0.011
Rod1*	-2.32	0.0004	1.20	0.041
Sema3a	-2.05	0.0003	1.74	0.024
Creb1	-3.33	0.0007	1.05	ns
Cited4*	2.11	0.0026	-1.10	ns
Tfrc*	-2.20	0.0023	1.11	ns

We cannot formally exclude altered mRNA stability as a factor causing the observed changes in mRNA levels, but it would be difficult to explain how the stability of relatively few transcripts could be altered in a selective fashion. Rather, we find that many transcription factor genes are down-regulated in their expression in diabetes-exposed embryos, and this trend is also reflected in the group of genes with 1.5- to 2-fold differences in expression; most likely therefore, the observed lower levels of gene expression are due to diminished or deregulated transcription.

Molecular Classification of Genes Altered in Response to Maternal Diabetes

The biological roles of many products encoded by the 126 diabetes-affected genes are known (Table 1 and Figure 2, Panel A). Most intriguingly, the largest functional category is comprised by transcription factors (15/126) and DNA-binding molecules known to affect transcriptional regulation (8/126), with both categories together comprising 18% of the identified genes. Thus, relative to the 7% of genes in the mouse genome that encode transcription factors [28], we found transcription regulatory genes highly overrepresented among our diabetes-affected genes. This is also reflected in enrichment of this category in DAVID annotation, http://david.abcc.ncifcrf.gov/. A combined 22% of the genes encode cell surface receptors (10%) and signal transduction molecules (12%) that ultimately converge on transcription. The transcription factor Pax3, which was previously identified as affected by maternal diabetes [4] was not represented on the microarray, but was changed as expected when assayed by Q-RT-PCR [25]. These results indicate that transcription factors and signaling molecules are prominent targets for perturbation by maternal diabetes, and that altered transcriptional regulation plays a major role in the response of embryos to intrauterine exposure to diabetic conditions.

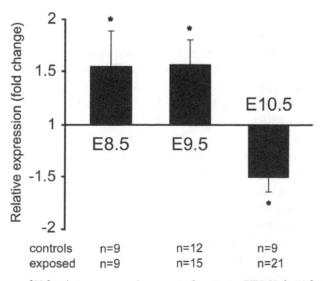

Figure 2. Expression of Hif1α during mouse embryogenesis. Quantitative RT-PCR for Hif1α at various stages of development normalized to expression levels of Pole4. n = number of individual embryos tested.

Transcription Regulation

If maternal diabetes deregulates cohorts of genes in the developing embryo through shared pathways of transcriptional regulation, one would expect (i) occurrence of common transcription factor binding sites (TFBS) in regulatory regions associated with multiple genes, and (ii) overrepresentation of such TFBS relative to other genes in the genome. We therefore analyzed the promoter sequences (5kb upstream of the transcription start) of our diabetes-affected genes for the presence of TFBS. As expected for genes with mostly broad expression patterns [29], we found a diverse set of conserved motifs in the upstream regions of these genes. Intriguingly, there was prominent over-representation of binding sites for the transcription factors FOXO1 and FOXO4 (-Log(p) = 12.416 and -Log(p) = 10.544, respectively), which are known to be involved in the response to oxidative stress, and for HIF1 (-Log(p) = 10.219), which is involved in the response to hypoxia. Out of 109 genes for which results were returned, FOXO1 and FOXO4 sites were enriched in 68% of the genes. NRF2 motifs were found in 76 promoters, further supporting the notion that oxidative stress may be involved in the response of diabetes-exposed embryos. HIF1 motifs were enriched in promoter regions of 22 genes (20.2%). These TFBS occur in combinations with sites for other transcription factors with a known role in responses to hypoxia, such as ATF4, E2F1 and E2F4, EGR1, ETS1, IRF1, NfkappaB, SOX9, SP3, and XBP1. Given the proposed role of oxidative stress and hypoxia in the pathogenesis of diabetic embryopathy [5,6,14,17,18,30], it is striking that 97% of the genes affected by maternal diabetes carry in their upstream regions potential binding sites for transcription factors that are involved in responses to oxidative stress and hypoxia.

Both conditions have been reported to be associated with diabetic pregnancy [14,31], and would predict activation of hypoxia-regulated pathways in the embryonic response to diabetic conditions. Paradoxically, the expression of Hif1α, a key regulator of embryonic responses to hypoxia [32], was reduced in our diabetes-exposed embryos at E10.5. In a post-hoc analysis of our microarray data with specific focus on HIF1 target genes [33], 22 HIF1 targets showed altered expression (fold-change >1.5) and passed one of the t-tests; 9 genes passed both statistical filters. Twenty HIF1-regulated genes exhibited increased expression in diabetes-exposed embryos, possibly reflecting an embryonic response to increased hypoxia. Further support for this idea comes from Hif1α message levels that are increased in diabetes-exposed embryos at E8.5 and E9.5 (Figure 3), and this increased Hif1α expression could be responsible for the increased expression of HIF target genes at E10.5. Together, these results implicate oxidative stress and hypoxia pathways in deregulated gene expression in diabetes-exposed embryos and identify the molecular targets of these pathways.

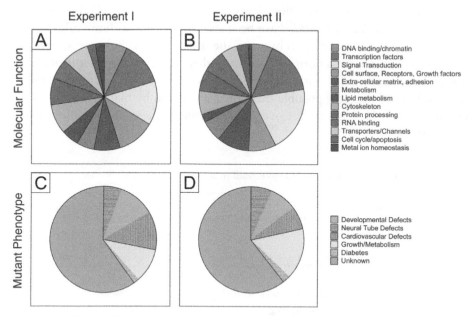

Figure 3. Classification of diabetes-affected genes by molecular function and function in vivo. Panel A depicts the representation of molecular classes of encoded products for the diabetes-affected genes identified in Experiment I. Panel B depicts the representation of molecular classes for the diabetes-affected genes identified in Experiment II. Genes encoding products with unknown function were omitted. Panel C depicts in vivo phenotypes based upon MGI annotation for genes identified in Experiment I; Panel D depicts in vivo phenotypes for genes in Experiment II. Only knockout phenotypes (null and conditional) were included.

Functional Roles of Genes Deregulated by Maternal Diabetes

Our identification of genes whose expression is affected by exposure to maternal diabetes suggests that those genes could be involved in the developmental defects in diabetic pregnancies. This notion is supported by qualitative expression information (MGI) for 69 of these genes, with 62 detected in the embryonic CNS, and 35 in the embryonic cardiovascular system. With both CNS and heart frequently being affected in diabetic embryopathy, genes with abnormal expression in these tissues might contribute to pathogenesis of birth defects.

Functional data also support this hypothesis: most remarkably, about one fourth of the diabetes-affected genes are known genes for which a functional role in embryonic development was established experimentally: in knockout mutant mice, 35 of the genes we identified have been shown, by genetics, to be required for mouse embryonic development (Table 3 and Figure 3, Panel C). This implies that these genes, under conditions of maternal diabetes but in the absence of genetic alterations, are subject to gene-environment interactions and respond to the intra-uterine environment of a diabetic pregnancy. Deficiencies in 15 of our genes (Agtr2, Cxadr, Dysf, Hif1a, Il6st, Itgav, Pdgfra, Pxn, Sema3a, Tgfbr1, Vcl, Adam10, Epha3, Efnb2, and Sfrs2) have been shown to cause cardiovascular defects or disease in mouse models. In humans, the risk for congenital heart disease is 2.8 times higher in infants born to diabetic mothers compared to the offspring of non-diabetic mothers [2] with higher odds ratios for specific malformations [34,35]. In light of this, it is possible that down-regulation by maternal diabetes of one or more of the genes we discovered could contribute to abnormalities of the heart [36].

Table 3. In vivo function of genes affected in diabetes exposed embryos.

Function	#	GeneSymbol
metabolic/growth defect	15	Ap1gl, Aplp2, Ghr, Grb10, Hmga1, Mapk10, Mbtps1, Mtap2, Nedd4, Ptprs, Scd2, Tnks2, Top2b, Upp1, Zfp385
diabetes	2	Ghr, Hmga1
embryonic	35	Abcb7, Adam10, Agtr2, Ap1gl, Aplp2, Bcl11a, Creb1, Cxadr, Dcx, Efnb2, Epha3, Gad1, Grb10, Hif1a, Il6st, Itgav, Kras, Mbtps1, Ndst1, Nedd4, Nsd1, Ogt, Pdgfra, Ptprs, Pxn, Scd2, Sema3a, Setdb1, Sfrs2, Slc2a1/Glut1AS[a], Tfrc, Tgfbr1, Top2b, Twsg1, Vcl
cardiovascular	15	Adam10, Agtr2, Cxadr, Dysf, Efnb2, Epha3, Hif1a, Il6st, Itgav, Pdgfra, Pxn, Sema3a, Sfrs2, Tgfbr1, Vcl
neural tube defects	7	Adam 10, Hif1a, Pdgfra, Tfrc, Tgfbr, Twsg1, Vcl

Information was obtained from GO annotations and hand curated with input from MGI and PubMed sources. All results are from knockout models, except for: [a] Glut1 antisense transgenic mice. References are provided in Additional file 4.

Central nervous system malformations occur in about 5% of children born to diabetic mothers [35], which represents an up to 15-fold higher risk of over pregnancies unaffected by diabetes. Intriguingly, we found seven genes affected by maternal diabetes that previously have been associated with neural tube defects (NTD): Hif1a, Pdgfra, Twsg1, Adam10, Tgfbr1, Tfrc, and Vcl. Thus, deregulated expression of these genes in diabetes-exposed embryos might predispose embryos to neural tube defects. We also analyzed our dataset from Experiment I for differences between the 5 embryos that were exposed to maternal diabetes, of which two exhibited defective closure of the neural tube. In this comparison, we identified only two genes (Etnk1, Gmfb) that passed the criteria filter of >1.5-fold change and both statistical significance tests. Both genes were detected in the initial diabetes-exposed versus normal comparison; we did not discover any genes that were uniquely altered in NTD embryos, implying that differences between NTD-affected and -unaffected individuals with regard to gene expression are mostly quantitative. Our results are consistent with the hypothesis that diabetes of the mother alters expression of specific known heart-defect and neural-tube-defect genes, and that these genes may be responsible for the birth defects in diabetic pregnancies.

Table 4. Wnt-pathway genes affected by maternal diabetes.

	Control Mean	(+/-SD)	Exp. II Mean	(+/-SD)	p-value	fold-change	Gene Symbol	Gene Title
1450007_at	602	(± 141)	76	(± 23)	0.0031	-7.93	1500003O03Rik	similar to EF-hand Ca2+ binding protein p22
1450056_at	423	(± 43)	103	(± 42)	0.0008	-4.12	Apc	adenomatosis polyposis coli
1455231_s_at	326	(± 71)	97	(± 3)	0.0052	-3.36	Apc2	adenomatosis polyposis coli 2
1426966_at	1257	(± 52)	530	(± 52)	6.6^{-5}	-2.37	Axin1	axin 1
1444031_at	17	(± 5)	47	(± 6)	0.0028	2.80	Camk2d	calcium/calmodulin-dependent protein kinase II, delta
1417176_at	3354	(± 833)	1102	(± 385)	0.0132	-3.04	Csnk1e	casein kinase 1, epsilon
1422887_a_at	3473	(± 351)	1625	(± 222)	0.0015	-2.14	Ctbp2	C-terminal binding protein 2
1430533_a_at	3643	(± 841)	333	(± 352)	0.0033	-10.95	Ctnnb1	beta-catenin
1458662_at	167	(± 35)	75	(± 18)	0.0152	-2.23	Daam1	dishevelled associated activator of morphogenesis 1
1450978_at	965	(± 42)	430	(± 40)	9.0^{-5}	-2.24	Dvl1	dishevelled homolog 1
1417207_at	993	(± 91)	327	(± 46)	0.0004	-3.04	Dvl2	dishevelled homolog 2
1455220_at	181	(± 12)	76	(± 12)	0.0004	-2.39	Frat2	frequently rearranged in advanced T-cell lymphomas 2
1437284_at	2115	(± 66)	965	(± 32)	1.1^{-5}	-2.19	Fzd1	frizzled homolog 1
1418532_at	1720	(± 389)	551	(± 193)	0.0096	-3.12	Fzd2	frizzled homolog 2
1450044_at	2329	(± 163)	884	(± 174)	0.0005	-2.63	Fzd7	frizzled homolog 7
1423348_at	419	(± 33)	188	(± 35)	0.0011	-2.23	Fzd8	frizzled homolog 8
1427529_at	136	(± 24)	57	(± 21)	0.0125	-2.40	Fzd9	frizzled homolog 9
1455689_at	485	(± 124)	204	(± 22)	0.0182	-2.37	Fzd10	frizzled homolog 10
1451020_at	434	(± 144)	156	(± 28)	0.0303	-2.79	Gsk3b	glycogen synthase kinase 3 beta
1417409_at	1585	(± 117)	664	(± 114)	0.0006	-2.39	Jun	Jun oncogene
1425795_a_at	1069	(± 122)	423	(± 92)	0.0018	-2.53	Map3k7	mitogen activated protein kinase kinase kinase 7
1452497_a_at	192	(± 85)	39	(± 4)	0.0353	-4.97	Nfatc3	nuclear factor of activated T-cells, calcineurin-dependent 3
1423379_at	497	(± 87)	81	(± 9)	0.0012	-6.17	Nfatc4	nuclear factor of activated T-cells, calcineurin-dependent 4
1419466_at	353	(± 31)	166	(± 5)	0.0005	-2.13	Nkd2	naked cuticle homolog 2
1448661_at	385	(± 42)	189	(± 47)	0.0057	-2.04	Plcb3	phospholipase C, beta 3
1439797_at	306	(± 67)	100	(± 27)	0.0078	-3.05	Ppard	peroxisome proliferator activator receptor delta
1426401_at	1094	(± 85)	525	(± 41)	0.0005	-2.08	Ppp3ca	protein phosphatase 3, catalytic subunit, alpha isoform
1427468_at	1103	(± 170)	207	(± 169)	0.0029	-5.32	Ppp3cb	protein phosphatase 3, catalytic subunit, beta isoform
1450368_a_at	186	(± 53)	83	(± 16)	0.0321	-2.23	Ppp3r1	protein phosphatase 3, regulatory subunit B, alpha isoform (calcineurin B, type I)
1452878_at	366	(± 35)	172	(± 32)	0.0021	-2.13	Prkce	protein kinase C, epsilon
1448695_at	242	(± 84)	70	(± 28)	0.0284	-3.44	Prkci	protein kinase C, iota
1424287_at	190	(± 55)	34	(± 12)	0.0084	-5.53	Prkx	protein kinase, X-linked
1451358_a_at	2310	(± 218)	1154	(± 250)	0.0038	-2.00	Racgap1	Rac GTPase-activating protein 1
1416577_a_at	5908	(± 412)	11936	(± 1884)	0.0056	2.02	Rbx1	ring-box 1
1423444_at	1347	(± 351)	657	(± 105)	0.0311	-2.05	Rock1	Rho-associated coiled-coil containing protein kinase 1
1425465_a_at	611	(± 128)	194	(± 73)	0.0080	-3.15	Senp2	SUMO/sentrin specific peptidase 2
1416594_at	722	(± 105)	300	(± 44)	0.0030	-2.41	Sfrp1	secreted frizzled-related protein 1
1422485_at	2828	(± 354)	1189	(± 425)	0.0068	-2.38	Smad4	MAD homolog 4
1434644_at	780	(± 274)	29	(± 15)	0.0090	-27.22	Tbl1x	transducin (beta)-like 1 X-linked
1429427_s_at	221	(± 5)	460	(± 97)	0.0132	2.08	Tcf7l2	transcription factor 7-like 2, T-cell specific, HMG-box
1455592_at	1336	(± 506)	306	(± 101)	0.0258	-4.37	Vangl2	vang-like 2 (van Gogh homolog)
1448818_at	212	(± 68)	82	(± 16)	0.0326	-2.60	Wnt5a	wingless-related MMTV integration site 5A
1420892_at	645	(± 120)	213	(± 15)	0.0035	-3.03	Wnt7b	wingless-related MMTV integration site 7B

Confirmation of Major Findings by a Separate Profiling Experiment

Our initial survey employed embryo samples that came from two pregnant females: one STZ-treated diabetic and a control untreated dam. This presents the theoretical possibility that any differences between progeny of the two dams reflect differences between pregnancies in addition to diabetic state. Also, we used individual embryo samples, and this approach is likely to incur substantial variability in the data and thus understimation of molecular changes. To address both concerns, we conducted Experiment II, in which equal amounts of RNA was pooled from three embryos of same gestational age into one sample, with each embryo derived from a different dam; for the diabetic as well as the control condition, we prepared three such pools, respectively (Figure 1, Panel B). Taking advantage of technical advances, these samples were processed and hybridized to the Affymetrix Mouse 430 2.0 chip, which surveys 39000 transcripts. Using the same analysis criteria as before, we identified 2231 transcripts of which 276 (12.37%) showed increased levels of expression in the diabetes-exposed samples, and 1955 (87.63%) exhibited decreased expression compared to the controls (Figure 1, Panel D). Thus, we confirm the general trend in the results from Experiment I. Of the differentially expressed transcripts, 179 lacked identifying features, such as a name, RefSeq or ENSEMBLE IDs, or Unigene Mm. cluster number; they also lacked any annotation information. This left us with 2052 annotatable genes. Classification by molecular functions revealed a distribution of molecular properties (Figure 3, Panel B) highly similar to that in Experiment I (Figure 3, Panel A). Again, genes encoding transcription factor and DNA-binding regulatory molecules were significantly enriched, accounting for 16.3% of the deregulated genes; strong enrichment was confirmed by DAVID annotation. Annotation for function in vivo identified 1836 gene entries in MGI; for 1095 of those, phenotype information was not available. However, 747 genes were associated with documented phenotypes in mouse mutants, of which 388 are developmental phenotypes by virtue of embryonic, neonatal, or perinatal death of homozygous mutant offspring. Again, the distribution of particular phenotypes in Experiment II (Figure 3, Panel D) was very similar to that of Experiment I (Figure 3, Panel C). Metabolic abnormalities were reported for mutants of 46 genes, and evidence for abnormal growth (pre- and post-natal) was obtained for 279 genes. Most notably, 161 genes are known to be associated with heart defects when mutated, and 112 genes are known to play causal roles in neural tube defects. This is only a fraction (35%) of the more than 300 NTD genes contained in the MGI database (as of October 1, 2008). Similarly, from the published collection of 170 mouse mutants with neural tube defects [37] for which the underlying molecular defect is known, 55 genes (32% of 170) were identified in Experiment II. Taken together, these results indicate that maternal diabetes affects specific pathogenic pathways leading to NTDs. Except for two genes, all NTD genes exhibited decreased expression on the arrays. In summary, the main findings of the initial microarray experiment were confirmed.

Indeed, of the 126 genes whose expression was altered by more than 2-fold in Experiment I, 67 were also recovered above the 2-fold change cut-off in Experiment II. Of the 378 probe sets with expression level altered between 1.5-fold and 2-fold, 187 were shared. Thus, of the 504 probe sets with altered expression in Experiment I, more than half (254) were confirmed in Experiment II, providing independent validation for the major results of the first experiment. This 50% confirmation rate for independent microarray experiments in the same biological paradigm agrees well with similar findings for independent yeast microarray results [38]. Thus, employing individual embryo samples as well as a pooling strategy, we have identified molecular targets in the embryo that respond to maternal diabetes. Also noteworthy is that Experiment II confirmed our earlier candidate gene studies that showed components of the Wnt pathway altered in diabetes-exposed embryos [25]. In fact, 43 genes with roles in Wnt signaling are affected by maternal diabetes (Table 4); with exception of Cank2d, Rbx1 and Tcf7l2 (which are upregulated), the expression levels of all of these genes are decreased in diabetes-exposed embryos compared to controls. This finding provides further support for our hypothesis that maternal diabetes affects specific developmental programs.

Using mRNA from whole individual embryos allowed us to survey a broad range of embryonic tissues that are potentially affected by maternal diabetes. This approach might have "missed" effects on genes that are expressed only in small cell populations of the embryo. However, in Experiment II, we identify 19 of the 47 published genes that were found altered more than 1.5-fold in microarray analysis of cranial neural tube tissue from diabetes-exposed embryos with neural tube defects at E 11.5 [39], and four of those genes are shared with Experiment I. Concordance was found for increased expression of Bnip3, and for decreased levels of En2, Hes6, Ina, Map3k7, Med1, Msx1, Mtap1B, Ngn2, Notch1, TgfβII, Doublecortin, Protocadherin18, Tgfβρεχεπτορ1, TopoIIβ, with the latter four genes confirmed also in Experiment I. Notch3, Nr2f2, Shh, and Tial1, were increased in dissected neural tube [39] but decreased in whole embryos, indicating that they may be deregulated in multiple tissues. Nonetheless, the overlap between results from different laboratories, despite differences in experimental design, provides additional validation to our findings.

We cannot currently distinguish which of the changes in gene expression are in direct response to the diabetic milieu, and which are indirect changes downstream of altered transcription factor expression, potentially increased hypoxia [14] or alterations in yolk sac [40] or placenta [Salbaum, Kruger, Pavlinkova, Zhang, and Kappen, manuscript submitted]. In this regard, it is interesting to note that we find no congruence to genes reported as affected by maternal diabetes in yolk sac of E12 rat embryos [41]. This indicates not only that both yolk sac and embryo gene expression are affected by maternal diabetes, but that extra-embryonic tissues respond differently than the embryo proper. It is noteworthy that among the deregulated genes with known phenotypes in mouse mutants, over 100 have been reported to be associated with placental alterations. Even though we have only surveyed the embryo proper, this is suggestive evidence that placental gene expression may also be altered in diabetic pregnancies. Our findings are consistent with the idea that altered gene expression in the embryo, as de-regulated by maternal diabetes, plays an important role in the pathogenesis of diabetes-induced birth defects [2,42].

Implications for Prevention of Adverse Outcomes From Diabetic Pregnancies

High glucose levels during critical periods of morphogenesis appear to be the major teratogen in diabetic pregnancy. In experimental animals, excess glucose is sufficient to cause dysmorphogenesis of embryos in glucose-injected dams or in whole embryo culture [11,43-45]. The precise mechanism(s) by which hypergly-cemia induces diabetic embryopathy is (are) not clear, although involvement of the Glut2 (Slc2a2) transporter has been demonstrated [45]. Several studies report increased oxidative stress in embryos in a diabetic environment, and the admin-istration of antioxidants, such as vitamins C or E, can reduce the occurrence of developmental defects [13,17,46,47]. Which genes are functionally involved in these responses in diabetes-exposed embryos, and which mechanisms provide for the protective effect of anti-oxidant treatment in diabetic embryopathy remains to be investigated, but it is likely that one or more of the genes we have identi-fied constitute targets in the antioxidant response. Similarly, folate supplemen-tation has been shown to be protective against NTDs in diabetic pregnancies [46,48]. Interestingly, the gene encoding platelet derived growth factor receptor α (Pdgfrα), mutants of which exhibit neural tube defects [49], is folate-responsive in mice [50]. Genes whose expression is altered in diabetes-exposed embryos thus represent excellent candidates for folate-responsive genes, and may mediate the beneficial effect of folate in the prevention of neural tube and other developmen-tal defects.

Methods

Animals

Diabetes was induced in 7–9 week old female FVB mice by two intraperitoneal injections of 100 mg/kg body weight Streptozotocin in 50 mM sodium citrate buffer at pH4.5 (STZ; Sigma, St. Louis, MO) within a one-week interval. The dams were set up for mating no earlier than 7 days after the last injection, and the day of detection of a vaginal plug was counted as day 0.5 of gestation. We used embryos only from dams (n = 11) whose blood glucose levels exceeded 250 mg/dl; average glucose levels were 148 mg/dl(± 18) before STZ treatment, 337 mg/dl(± 79) on the day of mating, and 528 mg/dl(± 70) on the day of embryo harvest.

Microarray Analysis

Total RNA was isolated from embryos at embryonic day 10.5 (E10.5) using Tri-zol® (Invitrogen, Carlsbad, CA). We processed 2 controls and five diabetes-exposed embryos; the latter group included two specimen with neural tube defects (NTD) so as to capture the full phenotype spectrum of diabetes-exposure in pregnancy. Individual RNA samples (5 µg) from whole embryos were reverse transcribed (Invitrogen) and labeled (Affymetrix, Santa Clara, CA). In Experiment I, samples were individually hybridized to 7 Affymetrix430A2.0 chips, which were scanned using a GeneChip3000 scanner; Affymetrix GCOS imaging software was used for quality control. In Experiment II, equal amounts of RNA prepared from 3 individual embryos were pooled into one sample; each embryo was from a different pregnancy and three such pools were constructed for a total of 9 control embryos, and independently, for 9 diabetes-exposed embryos; all embryos were morphologically normal. Expression levels and "Present", "Marginal", "Absent" flags were determined with default parameters through comparison of matched and mismatched oligonucleotides for the respective gene sequence.

Statistical analyses were performed using GeneSpring7 (Silicon Genetics, Redwood City, CA) and CyberT [51], http://cybert.microarray.ics.uci.edu. We grouped the data for control embryos and those for diabetes-exposed embryosm respectively, and filtered results in three steps: (i) "expression", i.e. "present" or "marginal" in at least one of seven samples; (ii) "statistical significance" between control and experimental samples of $P < 0.05$ in both CyberT and the t-test in GeneSpring; and (iii) "fold change", i.e. difference between control and diabetes-exposed samples of beyond either two-fold or 1.5-fold. The rationale for employing complementary data analysis packages and details of data transformation have been described elsewhere [52].

Of 22690 probe sets present on the arrays, http://www.affymetrix.com/products_ services/arrays/specific/mouse430a_2.affx, 15364 probes exhibited signals in at least one of the 7 samples, and 302 probe sets differed by more than two fold between these samples. Of these, 180 probe sets passed the t-test in GeneSpring ($P < 0.05$; Welch's test assuming unequal variances; false discovery rate set at 0.05), and 174 probes yielded P-values below $P < 0.05$ in Cyber-T. Permutation of the order of tests (significance first, fold-change second) identified differential signals from 2262 probe sets (Cyber-T), with 575 probes displaying differences between 1.5 and 2-fold, and 174 probes with differences greater than 2-fold between controls and exposed embryos. Regardless of order of filtering criteria, identical sets of probes were recovered, thus validating the analysis process. Between Cyber-T and GeneSpring, 145 sets passed both statistical filters, and after removal of duplicates, 126 genes were found to be differentially expressed above the 2-fold cut-off criterion.

For the second experiment, Affymetrix Mouse 430 2.0 arrays were used, which contain 45101 probe sets http://www.affymetrix.com/products_services/arrays / specific/mouse430_2.affx. 29687 probe sets exhibited signals in at least one of 6 samples; differences reached statistical significance at $p < 0.05$ for 9835 probes in the t-test ($P < 0.05$; Welch's test assuming unequal variances; false discovery rate set at 0.05) implemented in GeneSpring (GX version 9). 5915 probe sets exhibited differences greater than 1.5-fold, with 2796 differentially expressed greater than 2-fold. Cyber-T identified 13770 probe sets with statistical significance, of which 3992 exceeded the 1.5-fold change level, and an additional 4601 exceeded the greater than 2-fold criterion. After removal of internal controls, 5688 probe sets passed the filtering criteria for statistical significance in both Cyber-T and GeneSpring and exhibited >1.5-fold change between experimental and control samples, of which 2634 probe sets were identified with greater than 2-fold differential expression. Reduction of duplicates for a given gene was done by judgement call factoring in signal intensity, P-value, distribution of calls ("Present" was judged as more reliable than "Marginal") and fold-change; only one entry per gene was retained for a total of 2231 transcripts with differential expression greater than 2-fold.

The Primary Data Files are Available at the NCBI Gene Expression Omnibus Repository (Accession Number GSE9675).

Quantitative Real-time PCR

Quantitative Real-Time PCR (Q-RT-PCR) using an ABI Prism7000 instrument was performed as described [53] on cDNA samples from individual diabetes-exposed embryos (5 litters) and controls (4 litters), or pools of 4–5 control embryos from the same litter (4 litters) isolated at E10.5 (for details, see legend to Table 2). At E9.5, 6 control and 9 diabetes-exposed embryos were selected from 3 litters each, respectively, and E8.5 embryos were from 4 litters (10 controls) and 5 litters (9 diabetes-exposed embryos). All embryos used for Q-RT-PCR were morphologically normal. Normalization was done to Polymerase epsilon 4 (Pole4) cDNA in the same sample; Pole4 levels were unaffected by maternal diabetes on Experiment I and Experiment II arrays. Differences between samples (n = individual embryos except where noted otherwise) were evaluated for statistical significance using an unpaired two-tailed t-test. Primers were positioned across exon-exon junctions to exclude amplification of potentially contaminating DNA. The amplification products were designed to originate from a different region of the mRNA than that detected by probes on the microarray, in order to provide independent confirmation of expression measurements.

Annotation for Tissue Expression, Function and Mutant Mouse Phenotypes

Information on gene expression in embryos, where available, was collected from MGI, http://www.informatics.jax.org. Molecular function attributes were based on GO-annotation (NetAffx™, https://www.affymetrix.com/analysis/netaffx/index.affx, updated as of July 21, 2008), supplemented with information from ENSEMBL and UCSC genome browsers and PubMed. Information on mutant phenotypes was obtained from MGI (as of October 21, 2008) for null and conditional alleles.

Transcription Factor Binding Site Prediction

Whole Genome rVISTA, http://genome.lbl.gov/vista/index.shtml was used to identify transcription factor binding sites that are conserved between mouse and human and are over-represented in the 5 Kb upstream regions of our maternal diabetes affected genes relative to all 5 Kb upstream regions in the human genome (P-value < 0.006).

Authors' Contributions

GP performed microarray analyses and PCR assays, collected information for annotations and drafted a first version of the manuscript, JMS oversaw the statistical analysis by Cyber-T, designed experiment II and led the annotation effort, CK conceived of the study, performed the annotation for experiment II and wrote the manuscript.

Acknowledgements

We are grateful for technical assistance by Diane Costanzo, Dana S'aulis, and the UNMC microarray core facility, which received support from the NCRR through P20RR016469 and P20RR018788. We are also grateful to Drs. Claudia Kruger and Daniel Geschwind (UCLA) for advice on Q-RT-PCR and microarray interpretation, respectively. G.P. was funded through a supplement to RO1-HD34706 to C.K., and J.M.S. was funded through RO1-HD055528. All authors have read and approved the final version of the manuscript.

References

1. Kucera J: Rate and type of congenital anomalies among offspring of diabetic women. J Reprod Med 1971, 7(2):73–82.

2. Martinez-Frias ML: Epidemiological analysis of outcomes of pregnancy in diabetic mothers: identification of the most characteristic and most frequent congenital anomalies. Am J Med Genet 1994, 51:108–113.

3. Reece EA, Ma XD, Zhao Z, Wu YK, Dhanasekaran D: Aberrant patterns of cellular communication in diabetes-induced embryopathy in rats: II, apoptotic pathways. Am J Obstet Gynecol 2005, 192(3):967–972.

4. Phelan SA, Ito M, Loeken MR: Neural tube defects in embryos of diabetic mice: role of the Pax-3 gene and apoptosis. Diabetes 1997, 46(7):1189–1197.

5. Wentzel P, Eriksson UJ: A diabetes-like environment increases malformation rate and diminishes prostaglandin E(2) in rat embryos: reversal by administration of vitamin E and folic acid. Birth Defects Res A Clin Mol Teratol 2005, 73(7):506–511.

6. Wentzel P, Welsh N, Eriksson UJ: Developmental damage, increased lipid peroxidation, diminished cyclooxygenase-2 gene expression, and lowered prostaglandin E2 levels in rat embryos exposed to a diabetic environment. Diabetes 1999, 48(4):813–820.

7. Piddington R, Joyce J, Dhanasekaran P, Baker L: Diabetes mellitus affects prostaglandin E2 levels in mouse embryos during neurulation. Diabetologia 1996, 39(8):915–920.

8. Goldman AS, Baker L, Piddington R, Marx B, Herold R, Egler J: Hyperglycemia-induced teratogenesis is mediated by a functional deficiency of arachidonic acid. Proc Natl Acad Sci USA 1985, 82(23):8227–8231.

9. Sussman I, Matschinsky FM: Diabetes affects sorbitol and myo-inositol levels of neuroectodermal tissue during embryogenesis in rat. Diabetes 1988, 37(7):974–981.

10. Khandelwal M, Reece EA, Wu YK, Borenstein M: Dietary myo-inositol therapy in hyperglycemia-induced embryopathy. Teratology 1998, 57(2):79–84.

11. Wentzel P, Wentzel CR, Gareskog MB, Eriksson UJ: Induction of embryonic dysmorphogenesis by high glucose concentration, disturbed inositol metabolism, and inhibited protein kinase C activity. Teratology 2001, 63(5):193–201.

12. Yang X, Borg LA, Eriksson UJ: Altered metabolism and superoxide generation in neural tissue of rat embryos exposed to high glucose. Am J Physiol 1997, 272(1 Pt 1):E173–180.

13. Reece EA, Wu YK, Zhao Z, Dhanasekaran D: Dietary vitamin and lipid therapy rescues aberrant signaling and apoptosis and prevents hyperglycemia-induced diabetic embryopathy in rats. Am J Obstet Gynecol 2006, 194(2):580–585.

14. Li R, Chase M, Jung SK, Smith PJ, Loeken MR: Hypoxic stress in diabetic pregnancy contributes to impaired embryo gene expression and defective development by inducing oxidative stress. Am J Physiol Endocrinol Metab 2005, 289(4):E591–599.

15. Sakamaki H, Akazawa S, Ishibashi M, Izumino K, Takino H, Yamasaki H, Yamaguchi Y, Goto S, Urata Y, Kondo T, et al.: Significance of glutathione-dependent antioxidant system in diabetes-induced embryonic malformations. Diabetes 1999, 48(5):1138–1144.

16. Sivan E, Lee YC, Wu YK, Reece EA: Free radical scavenging enzymes in fetal dysmorphogenesis among offspring of diabetic rats. Teratology 1997, 56(6):343–349.

17. Cederberg J, Siman CM, Eriksson UJ: Combined treatment with vitamin E and vitamin C decreases oxidative stress and improves fetal outcome in experimental diabetic pregnancy. Pediatr Res 2001, 49(6):755–762.

18. Chang TI, Horal M, Jain SK, Wang F, Patel R, Loeken MR: Oxidant regulation of gene expression and neural tube development: Insights gained from diabetic pregnancy on molecular causes of neural tube defects. Diabetologia 2003, 46(4):538–545.

19. Zangen SW, Ryu S, Ornoy A: Alterations in the expression of antioxidant genes and the levels of transcription factor NF-Kappa B in relation to diabetic embryopathy in the Cohen Diabetic rat model. Birth Defects Res A Clin Mol Teratol 2006, 76(2):107–114.

20. Greene MF: Diabetic embryopathy 2001: moving beyond the "diabetic milieu". Teratology 2001, 63:116–118.

21. Epstein DJ, Vekemans M, Gros P: Splotch (Sp2H), a mutation affecting development of the mouse neural tube, shows a deletion within the paired homeodomain of Pax-3. Cell 1991, 67(4):767–774.

22. Epstein DJ, Vogan KJ, Trasler DG, Gros P: A mutation within intron 3 of the Pax-3 gene produces aberrantly spliced mRNA transcripts in the splotch (Sp) mouse mutant. Proc Natl Acad Sci USA 1993, 90(2):532–536.

23. Morgan SC, Relaix F, Sandell LL, Loeken MR: Oxidative stress during diabetic pregnancy disrupts cardiac neural crest migration and causes outflow tract defects. Birth Defects Res A Clin Mol Teratol 2008, 82(6):453–463.

24. Morgan SC, Lee HY, Relaix F, Sandell LL, Levorse JM, Loeken MR: Cardiac outflow tract septation failure in Pax3-deficient embryos is due to p53-dependent regulation of migrating cardiac neural crest. Mech Dev 2008, 125(9–10):757–767.

25. Pavlinkova G, Salbaum JM, Kappen C: Wnt signaling in caudal dysgenesis and diabetic embryopathy. Birth Defects Res A Clin Mol Teratol 2008, 82:710–719.

26. Stearne PA, Pietersz GA, Goding JW: cDNA cloning of the murine transferrin receptor: sequence of trans-membrane and adjacent regions. J Immunol 1985, 134(5):3474–3479.

27. Yang L, Lanier ER, Kraig E: Identification of a novel, spliced variant of CREB that is preferentially expressed in the thymus. J Immunol 1997, 158(6):2522–2525.

28. Gray PA, Fu H, Luo P, Zhao Q, Yu J, Ferrari A, Tenzen T, Yuk DI, Tsung EF, Cai Z, et al.: Mouse brain organization revealed through direct genome-scale TF expression analysis. Science 2004, 306(5705):2255–2257.

29. Schug J, Schuller W-P, Kappen C, Salbaum JM, Bucan M, Stoeckert CJ: Promoter Features Related to Tissue Specificity as Measured by Shannon Entropy. Genome Biology 2005, 6(4):R33.

30. Reece EA, Homko CJ, Wu YK, Wiznitzer A: The role of free radicals and membrane lipids in diabetes-induced congenital malformations. J Soc Gynecol Investig 1998, 5(4):178–187.

31. Sakamaki H, Akazawa S, Ishibashi M, Izumino K, Takino H, Yamasaki H, Yamaguchi Y, Goto S, Urata Y, Kondo T, et al.: Significance of glutathione-dependent antioxidant system in diabetes-induced embryonic malformations. Diabetes 1999, 48(5):1138–1144.

32. Iyer NV, Kotch LE, Agani F, Leung SW, Laughner E, Wenger RH, Gassmann M, Gearhart JD, Lawler AM, Yu AY, et al.: Cellular and developmental control of O2 homeostasis by hypoxia-inducible factor 1 alpha. Genes Dev 1998, 12(2):149–162.

33. Semenza GL: Targeting HIF-1 for cancer therapy. Nat Rev Cancer 2003, 3(10):721–732.

34. Loffredo CA, Wilson PD, Ferencz C: Maternal diabetes: an independent risk factor for major cardiovascular malformations with increased mortality of affected infants. Teratology 2001, 64(2):98–106.

35. Becerra JE, Khoury MJ, Cordero JF, Erickson JD: Diabetes mellitus during pregnancy and the risks for specific birth defects: a population-based case-control study. Pediatrics 1990, 85(1):1–9.

36. Jovanovic L, Knopp RH, Kim H, Cefalu WT, Zhu XD, Lee YJ, Simpson JL, Mills JL: Elevated pregnancy losses at high and low extremes of maternal

glucose in early normal and diabetic pregnancy: evidence for a protective adaptation in diabetes. Diabetes Care 2005, 28(5):1113–1117.

37. Harris MJ, Juriloff DM: Mouse mutants with neural tube closure defects and their role in understanding human neural tube defects. Birth Defects Res A Clin Mol Teratol 2007, 79(3):187–210.

38. Marguerat S, Jensen TS, de Lichtenberg U, Wilhelm BT, Jensen LJ, Bahler J: The more the merrier: comparative analysis of microarray studies on cell cycle-regulated genes in fission yeast. Yeast 2006, 23(4):261–277.

39. Jiang B, Kumar SD, Loh WT, Manikandan J, Ling EA, Tay SS, Dheen ST: Global gene expression analysis of cranial neural tubes in embryos of diabetic mice. J Neurosci Res 2008, 86(16):3481–3493.

40. Reece EA, Pinter E, Homko C, Wu Y-K, Naftolin F: The Yolk Sac Theory: Closing the Circle on Why Diabetes-Associated Malformations Occur. J Soc Gynecol Investig 1994, 1(1):3–13.

41. Reece EA, Ji I, Wu YK, Zhao Z: Characterization of differential gene expression profiles in diabetic embryopathy using DNA microarray analysis. Am J Obstet Gynecol 2006, 195(4):1075–1080.

42. Goto MP, Goldman AS: Diabetic embryopathy. Curr Opin Pediatr 1994, 6(4):486–491.

43. Fine EL, Horal M, Chang TI, Fortin G, Loeken MR: Evidence that elevated glucose causes altered gene expression, apoptosis, and neural tube defects in a mouse model of diabetic pregnancy. Diabetes 1999, 48(12):2454–2462.

44. Kumar SD, Dheen ST, Tay SS: Maternal diabetes induces congenital heart defects in mice by altering the expression of genes involved in cardiovascular development. Cardiovasc Diabetol 2007, 6(1):34.

45. Li R, Thorens B, Loeken MR: Expression of the gene encoding the high-Km glucose transporter 2 by the early postimplantation mouse embryo is essential for neural tube defects associated with diabetic embryopathy. Diabetologia 2007, 50(3):682–689.

46. Gareskog M, Eriksson UJ, Wentzel P: Combined supplementation of folic acid and vitamin E diminishes diabetes-induced embryotoxicity in rats. Birth Defects Res A Clin Mol Teratol 2006, 76(6):483–490.

47. Reece EA, Wu YK: Prevention of diabetic embryopathy in offspring of diabetic rats with use of a cocktail of deficient substrates and an antioxidant. Am J Obstet Gynecol 1997, 176(4):790–797.

48. Wentzel P, Gareskog M, Eriksson UJ: Folic acid supplementation diminishes diabetes- and glucose-induced dysmorphogenesis in rat embryos in vivo and in vitro. Diabetes 2005, 54(2):546–553.

49. Pickett EA, Olsen GS, Tallquist MD: Disruption of PDGFRalpha-initiated PI3K activation and migration of somite derivatives leads to spina bifida. Development 2008, 135(3):589–598.

50. Spiegelstein O, Cabrera RM, Bozinov D, Wlodarczyk B, Finnell RH: Folate-regulated changes in gene expression in the anterior neural tube of folate binding protein-1 (Folbp1)-deficient murine embryos. Neurochem Res 2004, 29(6):1105–1112.

51. Baldi P, Long AD: A Bayesian framework for the analysis of microarray expression data: regularized t-test and statistical inferences of gene changes. Bioinformatics 2001, 17(6):509–519.

52. Kappen C, Pavlinkova G, Kruger C, Salbaum JM: Analysis of altered gene expression in diabetic embryopathy. In Comprehensive Toxicology. 2nd edition. Edited by: McQueen CA. Oxford, United Kingdom: Elsevier; 2008.

53. Kruger C, Talmadge C, Kappen C: Expression of folate pathway genes in the cartilage of Hoxd4 and Hoxc8 transgenic mice. Birth Defects Res A Clin Mol Teratol 2006, 76(4):216–229.

Increased Expression of Heat Shock Protein 105 in Rat Uterus of Early Pregnancy and Its Significance in Embryo Implantation

Jin-Xiang Yuan, Li-Juan Xiao, Cui-Ling Lu, Xue-Sen Zhang, Tao Liu, Min Chen, Zhao-Yuan Hu, Fei Gao and Yi-Xun Liu

ABSTRACT

Background

Heat shock proteins (Hsps) are a set of highly conserved proteins, Hsp105, has been suggested to play a role in reproduction.

Methods

Spatio-temporal expression of Hsp105 in rat uterus during peri-implantation period was examined by immunohistochemistry and Western blot, pseudopregnant uterus was used as control. Injection of antisense oligodeoxy-nucleotides to Hsp105 into pregnant rat uteri was carried out to look at effect of Hsp105 on embryo implantation.

Results

Expression of Hsp105 was mainly in the luminal epithelium on day 1 of pregnancy, and reached a peak level on day 5, whereas in stroma cells, adjacent to the implanting embryo, the strongest expression of Hsp105 was observed on day 6. The immunostaining profile in the uterus was consistent with that obtained by Western blot in the early pregnancy. In contrast, no obvious peak level of Hsp105 was observed in the uterus of pseudopregnant rat on day 5 or day 6. Furthermore, injection of antisense oligodeoxynucle-otides to Hsp105 into the rat uterine horn on day 3 of pregnancy obviously suppressed the protein expression as expected and reduced number of the implanted embryos as compared with the control.

Conclusion

Temporal and spatial changes in Hsp105 expression in pregnant rat uterus may play a physiological role in regulating embryo implantation.

Background

Heat shock proteins (Hsps) have been identified in all eukaryotic and prokaryotic organisms[1]. They may act as molecular chaperones by preventing aggregation and assisting refolding of misfolded proteins [2-4]. Hsps could be induced in response to a physiological effect or environmental effect of stress, such as elevation in temperature, oxidative stress, viral infection, nutritional deficiency, or toxic chemical exposure [5,6]. On the basis of molecular weight, mammalian Hsps have been classified into various families, including Hsp105, 90, 70, 60, and other small Hsps [4,7]. The 105 kDa protein is one of the major mammalian Hsp which belongs to the family of higher molecular mass, and is composed of 858 amino acid residues [8]. Hatayama et al. [9] demonstrated a role of this protein in protecting neuronal cells against stress-induced apoptosis in rat neuronal PC12 cells, suggesting that this protein may be a novel anti-apoptotic neuroprotective factor in the mammalian brain [10,11]. Increasing evidences indicate that Hsps could regulate cell apoptosis either by directly promoting cell apoptosis or by inhibiting apoptotic response as a chaperone of a key signaling protein [12,13]. We have demonstrated that Hsp105 was expressed in monkey testis and may play an important role in regulation of germ cell apoptosis induced by heat stress [14]. Hsp105 may function as a pro-apoptotic factor [15] or as an anti-apoptotic factor depending on cell type in mammals [16].

The evidences from our previous studies both on rhesus monkey and human being demonstrated that a relatively high frequency of apoptosis occurs in the secretory endometrium, correlated to the period of formation of implantation window [17] which was a limited period of endometrial receptivity to blastocyst stimulus[18,19]. The time surrounding the window of receptivity in the rat is referred to as the peri-implantation period and involves days 4, 5, and 6 of pregnancy. In response to implanting embryos the underlying endometrial stromal cells undergo decidualization that involves proliferation and differentiation through cell division and apoptosis [20,21]. Apoptosis is a physiological process which remodels tissue by removing expendable cells without allowing the entry of proteolytic enzymes and other harmful or corrosive substances into the surrounding tissue, and thus reducing the likelihood of an inflammatory response[22,23].

Localization of apoptotic cells in relation to the expression of apoptosis-related molecules, such as Fas/FasL, Bcl-2/Bax, and P53 have been demonstrated in the materno-fetal boundary of rhesus monkeys in pregnancy [24,25]. Apoptotic nuclei were observed mainly in the glandular cells and the blood vessel endothelial cells in decidua [26]. A transient increase in Hsp105 expression during mouse embryogenesis was observed in the embryonic tissues [9]. Human endometrium [27], deciduas [28,29] and trophoblast tissues have been also reported to be capable of expressing Hsps during the first trimester of pregnancy[30,31], however, to the best of our knowledge, no studies about an action of Hsp105 in mammalian uterus during implantation have been reported. In the present study, we have analyzed Hsp105 protein expression in rat uterus of early pregnancy, and examined the effect of injection of antisense Hsp105 oligodeoxynucleotides into the pregnant uterine horn on embryo implantation.

Methods

Animals

Spague Dawley rats were obtained from the Animal Facility of Institute of Zoology, Chinese Academy of Sciences. The Guidelines for the Care and Use of Animals in Research enforced by Beijing Municipal Science and Technology Commission were followed. All protocols have been approved by the Animal Care and Use Committee of Institute of Zoology, Chinese Academy of Sciences. The rats were caged in a controlled environment with a 14 hr light:10 hr dark cycle. The adult females were mated with fertile males of the same strain to induce pregnancy (day 1, D1 = day of vaginal plug positive). Pregnancy on D1–5 was confirmed by flushing embryos from the reproductive tracts. The implantation sites on D6–7 were identified by intravenous injection of 1% (w/v) trypan blue (Sigma Chemical

Company, St. Louis, MO) in 0.85% (w/v) sodium chloride, according to the procedures described by Chun et al. and Xiao et al. [32,33]. In several experiments, some male rats were vasectomized, and after 14 days they were used to mate with females to induce pseudo-pregnancy (PD, PD1 = day of vaginal plug positive).

Immunohistochemistry

In the designed time points the animals were killed by cervical dislocation under anaesthetic and the uteri were collected. In some experiments the implantation sites on day 6 and 7 were separated from the inter-implantation segments, the corrected uterine materials were fixed immediately in 10% neutral buffered formalin solution (Beijing Chemical Reagents Co. Beijing, China) overnight, and then embedded in paraffin. Serial 5 μm sections of the uterine tissues were deparaffinized and rehydrated through degraded ethanol. Antigen retrieval was performed by incubating the sections in 0.01 M citrate buffer (pH 6.0) at 98°C for 20 min, followed by cooling at room temperature for 20 min. Non-specific binding was blocked with 5% (v/v) normal goat serum (Santa Cruz Biotechnology, Inc) in PBS for 1 h. The sections were incubated with the primary antibodies against Hsp105 (sc-6241, Santa Cruz Biotechnology, Santa Cruz, CA 1: 200) in 10% goat serum overnight at 4°C. The sections were then washed three times with PBS (10 min each) and incubated with biotin labeled secondary antibody (goat anti-rabbit IgG, RT, 30 min, 1:200), After three times washes with PBS, the sections were incubated with avidin-AP complex (1:200, RT, 20 min). After three more washes, the sections were developed with Vector Red AP substrates according to the manufacturer's protocol (Vectastain ABCAP kit, Vector Laboratories, Burlingame, CA). Endogenous AP activity was inhibited by supplement of 1 mM levamisole (Sigma Corp., St. Louis, MO) into the substrate. The sections stained with Vector Red substrates were counter-stained with haematoxylin. The sections incubated with normal IgG instead of the primary antibody served as the negative controls.

Western Blot Analysis

The uteri from various groups were homogenized respectively in the lysis buffer (5 mmol/L phosphate buffer, pH 7.2, containing 0.1% Triton X-100, 1 mM phenylmethylsulfonylfluoride, and 1 mg/L chymostatin), and the concentration of protein in the supernatant after centrifugation was determined by UV spectrophotometer. The sample lysates in each group were mixed with the loading buffer (62.5 mM 1,4- dithiothreitol, 5% sodium dodecyl sulfate (SDS), and 10% glycerol), boiled for 8 min, and then separated by SDS-polyacrylamide gel electrophoresis (PAGE) (50 µg total protein/lane). The separated proteins were transferred electrophoretically onto a pure nitrocellulose blotting membrane (Pall Corporation, Pensacola, FL), and then incubated with blocking buffer (3% BSA (v/v) in TBST for 1 h at room temperature. The membrane was subsequently incubated with the anti-Hsp105 antibodies overnight at 4°C, washed for three times with TBST, 15 min each time, and further incubated for 1 h at room temperature with TBST containing alkaline phosphatase-conjugated secondary antibodies, and then washed three times with TBST. After one more time of wash with TBS, then the membrane was subjected to an alkaline phosphatase color reaction by a standard method. Actin protein was used as the internal control for cytosolic protein. Band intensity was determined by Quantity One Software (Bio-Rad, Hercules, CA).

Design of Oligodeoxynucleotides Specific Sequence for Hsp105

The sequence(s) of oligodeoxynucleotides (ODNs) (16 nucleotides in length) were: sense oligodeoxynucleotides(S-ODNs):5'-AGCCATGTCGGTGGTT-3, antisense oligodeoxy- nucleotides (A-ODNs): 5'-AACCACCGACATGGCT-3'. The A-ODNs were complementary to bases 191–206 bp within exon I of the rat Hsp105 (GenBank Accession Number: NM_001011901). All the sequences were thiophosphate-modified for their long half-lives in cells. FITC/A-ODNs and FITC/S-ODNs are FITC-conjugated ODNs at the 3' end. All the ODNs were synthesized by SBS Genetech Co., Ltd. (Beijing, China). Analysis of homology between the synthesized oligomer and the rodent sequences present in the GenBank data bases (release 73.0) by the Genetics Computer Group sequence analysis software package revealed that the synthetic oligomers were fully complementary only to their own specific mRNA.

Tracking of FITC- Labeled ODNs in the Tissue of Rat Uterus

Pregnant rats were injected with Hsp105 S-ODNs or A-ODNs according to the procedures described by Zhu et al. [34]. Tracking of FITC-labeled ODNs was performed according to the procedures described by Luu et al. [35]. Uterine penetration of the ODNs and cross-contamination between the two horns were assessed by injecting 10 µg (in 100 µL distilled water) of FITC-A-ODNs or FITC-S-ODNs into one horn, with either unlabeled standard control A-ODNs or S-ODNs alone into the contralateral horn of the uterus. The uteri were excised and frozen in OCT compound (QIAGEN N.V.) at 2.5 h, 24 h and 48 h after injection of the ODNs. 6 μm thick frozen sections were then analyzed under a fluorescence microscope at 488 nm.

The ODNs experiments were carried out in the afternoon on day 3 of pregnancy. The animals were divided into two groups, each group (n = 12/ ODNs pair, 6 for S-ODNs and the other 6 for A-ODNs) was subject to a surgical operation and each uterine horn was injected with 10 μg of A-ODNs targeted against exon I of the Hsp105 or the corresponding S-ODNs or double distilled water (DD water). The animals were killed at 24 h and 48 h, respectively, after the operation, the uteri were fixed immediately for overnight in 10% neutral buffered formalin solution (Beijing Chemical Reagents Co. Beijing, China) and embedded in paraffin. Serial 5 μm sections of the uterine tissues were deparaffinized and rehydrated through graded ethanol for immunohistochemical analysis.

Microscopic Assessment and Statistical Analysis

The uterine samples from 3 rats in each group were analyzed. Experiments were repeated at least three times, from which one taken from at least three similar results was presented as a representative of the immunocytochemical data in the group. Signal intensities of Hsp105 detected by immunohistochemistry were quantified by computer-aided laser-scanning densitometry (Personal Densitometer SI; Molecular Dynamics, Inc., Sunnyvale, CA). In order to make the statistical significance of quantitative difference credible, three slides from each of six animals of each group were examined (n = 6), and 40 spots were randomly selected in every specific location of the specific cell types. The gray level of intercellular substance was considered as background. Statistical analysis was carried out with SPSS (version 10.0, SPSS Inc., Chicago, IL), and one-way ANOVA was used followed by Post-Hoc comparisons for analyzing the data in different groups. P values lower than 0.05 were considered statistically significant. To estimate specific staining in various cells of the uteri, a semi-quantitative subjective scoring was also performed by three blinded investigators using a 4-scale system with – (nil), +/- (weak), + (moderate), and ++ (strong), as described by Yue et al. [36]. The statistical data of Western blot from three individual experiments were analyzed by using Statistical Package for Social Science (SPSS for Windows package release 10.0, SPSS Inc., Chicago, IL). Statistical significance was determined by one-way ANOVA. Post-Hoc comparisons between groups were made using Fisher's protected least-significance-difference test. Values were means ± SEM. P values lower than 0.05 were considered statistically significant.

Results

Hsp105 Expression in Rat Uterus During Early Pregnancy

In order to examine developmental expression of Hsp105 in rat uterus of normal pregnancy, we performed immunohistochemistry using an antibody against rat Hsp105 protein. The results showed that Hsp105 expression was mainly localized in the luminal epithelium on day 1 of pregnancy (Fig. 1, D1), and increased in the glandular epithelium on days 2 and 3 (Fig. 1, D2). On days 4 and 5, additional staining was observed in the stromal cells immediately underneath the luminal epithelium, reaching a peak level on day 5 (Fig. 1, D4, D5). The strongest expression of this protein was detected in the decidual cells adjacent to the implanting embryo on day 6 (Fig. 1, D6). Localization and average score of Hsp105 protein at the various uterine locations are summarized in Table 1.

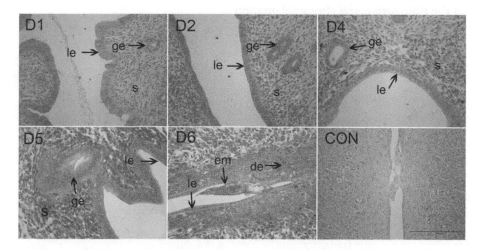

Figure 1. Immunohistochemistry of Hsp105 in rat uterus during early pregnancy. Hsp105 protein was observed mainly in the luminal epithelium on day 1 of pregnancy (D1) and moderately expressed in the luminal epithelium and the glandular epithelium from day 2 to day 3 (D2). Hsp105 staining was also detected in the stromal cells, immediately underneath the luminal epithelium on day 4 and day 5 of pregnancy (D4, D5), the staining was increased markedly on day 5 just before implantation (D5). On day 6, the protein staining was mainly observed in the implanted blastocyst and the stromal cells around the implantation site (D6), while its expression in the luminal epithelium reduced to an undetected level. CON, negtive control; le, luminal epithelium; ge, glandular epithelium; s, stroma; de, decidua; em, embryo. Bar = 200 μm.

Table 1. Semi-quantitative estimation of Hsp105 expression in the various uterine cells during early pregnancy

Cell Types	Days of early pregnancy					
	1	2	3	4	5	6
Luminal epithelium	+	+	+	+	++	-
Glandular epithelium	+/-	+	+	+	+++	-
Stromal cells				+	+++	-
Primary Deciduas						+++
Embryo						+++
Secondary Deciduas						+++

Western Blot Analysis of Hsp105 Expression in Uterus During Early Pregnancy

The quantitative change in uterine Hsp105 expression was estimated by Western blot, as shown in Fig. 2. The protein level in the uterus was increased in a time-dependent manner, the highest expression was observed on day5 and day 6, just around the time before and after implantation.

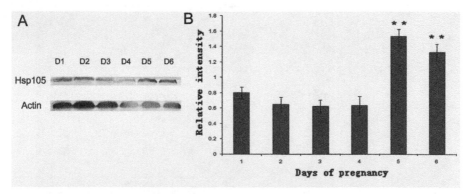

Figure 2. Western blot analysis of Hsp105 protein in uterus during early pregnancy. A: Representative Western Blot analysis of Hsp 105 protein. Actin protein was used as an internal control. B: The bar graph represents the densitometric analysis of the Hsp 105. The relative intensity was determined by the ratio of Hsp105 protein to its corresponding internal control as measured by densitometry. Data are presented as mean ± SEM (n = 3). Statistical analysis was performed using one-way ANOVA followed by the Fisher's protected least-significance-difference test. Bar with ** is significantly different from D1 of pregnancy (P < 0.01).

Hsp105 Expression in Rat Uterus During Pseudo-Pregnancy

To further confirm specific expression of Hsp105 in relation to implantation, we performed an experiment with pseudopregnant rats. The protein was mainly localized in the luminal epithelium on day 1 (Fig. 3, PD1), with the staining increased in both the luminal and the glandular epithelium on day 2 and 3

(Fig. 3, PD2, PD3), sharply decreased on day 4, and remaining at a low level on day 5 to 7 (Fig. 3, PD4, PD5, PD6, PD7). No peak level expression of this protein was observed in the pseudopregnant uterus. The score of the specific cell staining for Hsp105 in the uterus during pseudopregnancy is summarized in Table 2.

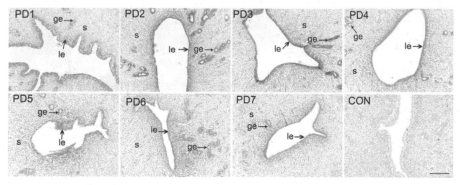

Figure 3. Immunohistochemistry of Hsp105 during pseudopregnancy. The adult male animals were vasectomized and 14 days later were used to mate with the adult females of the same strain to induce pseudopregnancy (pseudo-pregnancy day 1 (PD1) = day of vaginal plug positive). Hsp150 protein was mainly detected in the luminal epithelium on day 1 (PD1). The protein staining was increased in the luminal epithelium and glandular epithelium from day 2 to 3 (PD2, PD3). The protein expression was decreased from day 4 to 7 (PD4, PD5, PD6, PD7). CON, negtive control from day 2 of pseudopregnancy; le, luminal epithelium; ge, glandular epithelium; s, stroma. Bar = 200 μm.

Table 2. Semi-quantitative estimation of Hsp105 expression at the various uterine cells during pseudopregnancy

Cell Types	Days of pseudopregnancy						
	1	2	3	4	5	6	7
Luminal epithelium	+	+	+	+/-	+/-	+/-	+/-
Glandular epithelium	+/-	+	+	+/-	+/-	+/-	+/-
Stromal cells	-	-	-	-	-	-	-

Comparison of Hsp105 Protein Expression in Uterus Between Implantation Site and Inter-Implantation Segment

In order to know whether Hsp105 expression is related to implantation, we analyzed its expression in both implantation site and the inter-implantation segment on day 6 by immunohistochemistry. The results showed that the expression of this protein at the implantation site (Fig. 4, D6m) was much stronger than that in the interimplantation segment (Fig. 4, D6n), as summarized in Table 3.

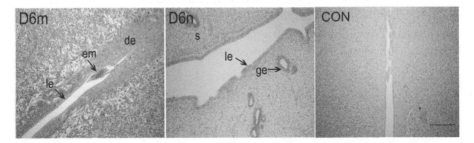

Figure 4. Immunohistochemistry of Hsp105 in implantation site and inter-implantation segment. The pregnant rats were killed on Day 6, the implantation sites (D6m) and the inter-implantation segments (D6n) were isolated for preparation of immunohistochemistry. Hsp105 expression in endometrium at the implantation site was much stronger than that at the inter-implantation segment. CON, negtive control; le, luminal epithelium; ge, glandular epithelium; s, stroma; de, decidua; em, embryo. Bar = 200 μm.

Table 3. Semi-quantitative estimation of Hsp110 expression at the various uterine cells in implantation site and interimplantation segment on day 6 of pregnancy

Cell Types	Implantation site	Interimplantation segment
Luminal epithelium	-	+
Glandular epithelium	-	+
Primary decidua	+++	
Embryo	+++	
Secondary decidua	+++	

Suppression of Hsp105 Expression in Pregnant Rat Uterus by Antisense ODNs

Using an A-ODNs as a blocker we examined effect of blockage of Hsp105 gene expression on rat implantation. To assess A-ODNs penetrating capacity, the Hsp105 FITC-ODNs was first injected into the uterine lumen, and then the uteri were taken for preparing sections at the indicated time points for FITC-ODNs examination by fluorescence microscopy. Strong green fluorescence representing cellular uptake of FITC-ODNs was observed in the luminal epithelium at 2.5 hours after injection (Fig. 5A). A detectable fluorescence in the underlying stroma was detected 48 hours later (Fig. 5B), indicating the penetration of Hsp105 ODNs into these cells in vivo. No fluorescence was observed in the contralateral horn treated with the unlabeled ODNs as the control (Fig. 5C).

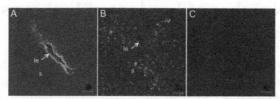

Figure 5. Fluorescence micrographs of rat uteri after intrauterine administration of FITC-ODNs. (A) The uterine horn treated with FITC- ODNs (either sense or antisense) displaying high level of fluorescence only in the luminal epithelium 2.5 hours after administration. (B) 48 hours later, a moderate fluorescence in the stromal compartment was observed after treatment with FITC- ODNs (either sense or antisense). (C) Uterine horn treated with unlabeled control A-ODNs displaying no fluorescence. le, luminal epithelium; s, stroma. Bar = 100 μm.

Based on Hsp105 expression profile in the uterus, the time window of Hsp105 ODNs administration should be between days 3 and 5 of gestation for allowing blockage of its protein expression. The pregnant rat uteri were injected with either DD water, or Hsp105 S-ODNs or Hsp105 A-ODNs on day 3 of pregnancy, the uteri were collected 24 h and 48 h later, and then subjected to immunostaining analysis. As shown in Fig. 6A (a, b), an intensive staining was observed mainly in the luminal epithelium and glandular epithelial cells in the uterus treated with water and S-ODNs respectively. In contrast, the contralateral horn treated with A-ODNs showed only low level of Hsp105 staining on day 4, 24h after injection of ODNs (Fig. 6A (c)). A marked decrease in Hsp105 immunostaining was noted on day 5 after treatment with A-ODNs (data not shown). Statistical analysis by the computer-aided laser-scanning densitometry showed that the Hsp105 levels between the uteri treated with DD water, S-ODNs and A-ODNs were significant different in the luminal epithelium and the glandular epithelium (Fig. 6B).

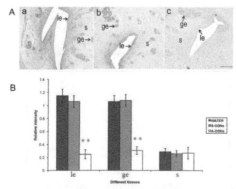

Figure 6. Hsp 105 immunohistochemistry staining analysis of the pregnant rat uterus after antisense or sense ODNs treatment. A: The figure was one representative from at least three similar independent experiment results. The immunohistochemistry staining was greatly decreased when the uteri were treated with A-ODNs (c) as compared with S-ODNs (b) and DD water (a). B: Statistical analysis of Hsp105 protein level in the uterus treated with DD water, S-ODNs or A-ODNs. Data are presented as mean ± SEM (n = 6). Statistical analysis was performed using one-way ANOVA followed by Post-Hoc comparisons. Bar with ** is significantly different from S-ODNs treated and DD water treated control (P < 0.01). le, luminal epithelium; ge, glandular epithelium; s, stroma. Bar = 200 μm.

Decreasing Number of Implanted Embryos by Antisense Hsp105 ODNs Treatment

We further examined whether inhibition of Hsp105 expression could influence embryo implantation. After administration of either the antisense or the corresponding sense Hsp105 ODNs or distilled water into the respective unilateral uterine horns of pregnant rats on day 3, the animals were killed on day 9, and the uteri were examined for the number of implanted embryos as well as their morphological status. One representative picture of the A-ODNs- and the S-ODNs-treated uteri was shown (Fig. 7A). Ten and 9 embryos were observed in the S-ODNs-treated horns (n = 8) (a: left horn, b: right horn), while only 3 and 4 embryos (a: right horn, b: left horn) were observed in the contralateral A-ODNs-treated horns. However, all the embryos in both treated horns were normal by appearance and size. The water-injected rats contained eight to ten normal implanted embryos in each uterine horn in average (Fig. 7A(c)). No significant changes in the number of implanted embryos or the embryo normality were observed in the S-ODNs-treated horns as compared with that in the water-treated control group, indicating that the dose of ODNs used in this study was non-toxic to the embryo implantation. In contrast, as shown in Fig. 7B, a significant reduction in the number of implanted embryos in the A-ODNs-treated group was observed (60%, P < 0.01) as compared with that of the S-ODNs-treated group, but no embryo abnormality in the A-ODNs treated animals was observed.

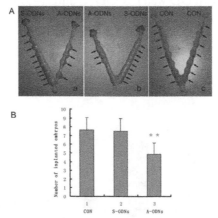

Figure 7. Effect of antisense ODNs on number of implanted embryo. A: Three representative uteri: (a) Pregnant rat was injected with sense Hsp105 ODNs in the left horn, with antisense Hsp105 ODNs in the right horn (10 ug in100 μl DD water) on day 3 of pregnancy; (b): Pregnant rat was injected in the left horn with antisense, the right horn with sense Hsp105 ODNs (10 ug in 100 μl DD water) on day 3 of pregnancy; (c): Pregnant rat was injected in both horns with DD water (100 μl). B: Statistical analysis of implanted embryo numbers in the uteri with the various treatments. Vertical axis represents the number of implanted embryos in the unilateral uterine horn. Data are presented as mean ± SEM (n = 8). Statistical analysis was performed using one-way ANOVA followed by Post-Hoc comparisons Bar with ** is significantly different from S-ODNs and DD water treated control (P < 0.01).

Discussion

The results of the present study indicate that time-dependent expression of Hsp105 in the uterine luminal, glandular epithelium and stromal cells during periimplantation period might be essential for regulation of embryo implantation. Our data also show that the presence of embryo in uterus as a stimulus may be important for increasing Hsp105 expression. If Hsp105 is involved in regulation of endometrial differentiation for embryo implantation, one may expect that a reduction of its expression could prevent acquisition of receptive state of the endometrium leading to a failure of implantation. Therefore, we designed an experiment with Hsp105 antisense oligodeoxynucleotides directly injecting into pregnant rat uterus at early pregnancy, which allowed us to investigate a function of this protein in the process of implantation. Technically, it would be important to do such an experiment to know the transient nature of Hsp105 gene expression in the uterus. In order to select an appropriate time window of ODNs administration for blockage of Hsp105 expression, we reasoned that the time window should be immediately preceding that of Hsp105 induction, i.e., between days 3 and 6 of gestation. The precise half-life of the Hsp105 mRNA or its protein in uterus has not yet been determined, nevertheless, the modified Hsp105 ODNs are known to have a half life of 24–48 hours in certain tissues [37]. Therefore, ODNs were designed for injection in the afternoon of day 3 of pregnancy, one may expect the tissue on observation to survive for the subsequent 3–4 days of gestation, for an effective suppression of the surge of Hsp105 expression. Because rat embryos were observed to be also capable of expressing Hsp105 (unpublished data), we examined a potential effect of ODNs on embryonic development by observing its normality. Therefore we selected a much later time point (Day 9) to count and examine the embryos. The statistical analysis of the difference of the numbers of implanted embryo between the antisense- and the sense ODNs-treated groups indicated that embryo implantation was indeed prohibited by the antisense ODNs ($P < 0.01$). These results together with the other observations suggest that treatment with antisense Hsp105 ODNs, but not with complementary sense ODNs, could severely impair the process of embryo implantation, but no effect on the normality of implanted embryos was observed on day 9 of pregnancy. However, Nakamura et al. just recently generated the Hsp105 knockout mice which did not appear a problem with reproduction [38], implying that Hsp105 may be not the necessary gene required for implantation in mouse. However, the authors did not specifically pay attention to examine if the animals had any implantation defect present. Hsp105 family has another two members, APG1 and APG which have shown a similar function with Hsp105, and may rescue its function in the absence of Hsp105. We have demonstrated previously that plasminogen activator is important in ovulation of rat and monkeys both in vivo and in vitro[39], however, double knockout of tPA and uPA in mice showed only 26% inhibition of ovulation could be observed[40]. Our further studies showed that mouse ovary produces not tPA and uPA, but also MMPs which have also shown to play a role in ovulation. It is possible that MMPs could rescue the function in absence of tPA or uPA. Implantation is a very complex event, which involves various processes, such as blastocyst adhesion, trophoblast invasion, decidualization and cell-to-cell interaction, controlled by a variety of molecules produced by endometrium, embryo and ovary [41]. During mammalian implantation stroma of the endometrium undergoes severe remodeling, involving apoptosis, proteolysis and angiogenesis [41,42]. Endometrial cells rapidly proliferate and differentiate to form the decidua tissue which accommodates and protects implanted embryos [43]. In our previous reports, analysis of the endometrium of both rhesus monkey and human during peri-implantation period has demonstrated that a relatively high frequency of apoptosis occurs in the secretory endometrium and is correlated to increased expression of apoptosis related molecules [25,26], while only limited numbers of the apoptotic cells were observed in the other phases of the cycle. It appears that endometrial apoptosis and the cyclic changes in endometrial growth and regression during establishment of implantation window might be regulated precisely and coordinately, not only by Fas, FasL, BcL-2 and Bax [17], but also by Hsp105, because the profile of these molecules is well correlated with that of the Hsp105 expression in rat uterus as demonstrated in the present study. Evidence has shown that Hsp105 is capable of enhancing cell apoptosis in mouse embryonal F9 cells[15,44] and murine embryos during embryogenesis[9,45]. On the contrary, the Hsp protein was also observed to inhibit cell apoptosis in rat testis and some experimental cell models [16,46-48]. These observations suggest that Hsp105 may be involved in regulation of murine uterine cell apoptosis. Since the cell types, species used in the individual studies were different, some unknown factors as well as cellular environment present in the various studies might determine an inhibitory or a promotional effect of the Hsp protein on cell apoptosis. However, the molecular mechanism of Hsp105 in regulating uterine cell apoptosis during rat periimplantation period remains to be further investigated. In summary, our data have demonstrated a significant increase in Hsp105 expression on day 5. It seems that the protein might be able to induce luminal cell apoptosis which in turn destabilizes epithelial barrier at implantation site and facilitates trophoblast invasion and implantation. On D6 of pregnancy Hsp105 expression was down-regulated in luminal epithelium, and upregulated in stroma cells adjacent to the embryo attachment. The dynamic changes in the level of expression of this protein during early implantation period suggest involvement of this protein in certain cellular events in luminal epithelial and stromal cells, which are essential for implantation.

Competing Interests

The authors declare that they have no competing interests.

Authors' Contributions

JXY participated together with LJX and YXL in the design of the study. The experiments were carried out by JXY, LJX, CLL, XSZ, TL, MC and FG. Data analysis was performed by JXY. The manuscript was written by JXY, FG and YXL.

Acknowledgements

This project was supported by the Major Research Plan (2006CB944000), the "973" Project (2006CB504000), the CAS Innovation Project (KSCA2-YW-R-55), the National Nature Science Foundation of China (No: 30370196, 30770284, 30570921, 90208025), and the WHO/Rockefeller Foundation projects. The authors are thankful to Dr. Petter Leung for the proofreading and corrections and thankful to Yungang Liu for his help by improving the English language of this manuscript.

References

1. Minohara M: [Heat shock protein 105 in multiple sclerosis]. Nippon Rinsho 2003, 61(8):1317–1322.

2. Wu C: Heat shock transcription factors: structure and regulation. Annu Rev Cell Dev Biol 1995, 11:441–469.

3. Georgopoulos C, Welch WJ: Role of the major heat shock proteins as molecular chaperones. Annu Rev Cell Biol 1993, 9:601–634.

4. Bukau B, Horwich AL: The Hsp70 and Hsp60 chaperone machines. Cell 1998, 92(3):351–366.

5. Pockley AG: Heat shock proteins, inflammation, and cardiovascular disease. Circulation 2002, 105(8):1012–1017.

6. Hendrick JP, Hartl FU: Molecular chaperone functions of heat-shock proteins. Annu Rev Biochem 1993, 62:349–384.

7. Tang D, Khaleque MA, Jones EL, Theriault JR, Li C, Wong WH, Stevenson MA, Calderwood SK: Expression of heat shock proteins and heat shock protein

messenger ribonucleic acid in human prostate carcinoma in vitro and in tumors in vivo. Cell Stress Chaperones 2005, 10(1):46–58.

8. Yasuda K, Nakai A, Hatayama T, Nagata K: Cloning and expression of murine high molecular mass heat shock proteins, HSP105. J Biol Chem 1995, 270(50):29718–29723.

9. Hatayama T, Takigawa T, Takeuchi S, Shiota K: Characteristic expression of high molecular mass heat shock protein HSP105 during mouse embryo development. Cell Struct Funct 1997, 22(5):517–525.

10. Yasuda K, Ishihara K, Nakashima K, Hatayama T: Genomic cloning and promoter analysis of the mouse 105-kDa heat shock protein (HSP105) gene. Biochem Biophys Res Commun 1999, 256(1):75–80.

11. Wakatsuki T, Hatayama T: Characteristic expression of 105-kDa heat shock protein (HSP105) in various tissues of nonstressed and heat-stressed rats. Biol Pharm Bull 1998, 21(9):905–910.

12. Neuer A, Spandorfer SD, Giraldo P, Dieterle S, Rosenwaks Z, Witkin SS: The role of heat shock proteins in reproduction. Hum Reprod Update 2000, 6(2):149–159.

13. Sreedhar AS, Csermely P: Heat shock proteins in the regulation of apoptosis: new strategies in tumor therapy: a comprehensive review. Pharmacol Ther 2004, 101(3):227–257.

14. Zhang XS, Lue YH, Guo SH, Yuan JX, Hu ZY, Han CS, Hikim AP, Swerdloff RS, Wang C, Liu YX: Expression of HSP105 and HSP60 during germ cell apoptosis in the heat-treated testes of adult cynomolgus monkeys (macaca fascicularis). Front Biosci 2005, 10:3110–3121.

15. Yamagishi N, Ishihara K, Saito Y, Hatayama T: Hsp105alpha enhances stress-induced apoptosis but not necrosis in mouse embryonal f9 cells. J Biochem 2002, 132(2):271–278.

16. Hatayama T, Yamagishi N, Minobe E, Sakai K: Role of hsp105 in protection against stress-induced apoptosis in neuronal PC12 cells. Biochem Biophys Res Commun 2001, 288(3):528–534.

17. Fei G, Peng W, Xin-Lei C, Zhao-Yuan H, Yi-Xun L: Apoptosis occurs in implantation site of the rhesus monkey during early stage of pregnancy. Contraception 2001, 64(3):193–200.

18. Tabibzadeh S, Babaknia A: The signals and molecular pathways involved in implantation, a symbiotic interaction between blastocyst and endometrium involving adhesion and tissue invasion. Hum Reprod 1995, 10(6):1579–1602.

19. Bulletti C, Flamigni C, de Ziegler D: Implantation markers and endometriosis. 2005, 11(4):464–468.

20. Quinn CE, Simmons DG, Kennedy TG: Expression of Cystatin C in the rat endometrium during the peri-implantation period. Biochem Biophys Res Commun 2006, 349(1):236–244.

21. Psychoyos A: Hormonal control of ovoimplantation. Vitam Horm 1973, 31:201–256.

22. Wyllie AH, Kerr JF, Currie AR: Cell death: the significance of apoptosis. Int Rev Cytol 1980, 68:251–306.

23. Dmowski WP, Ding J, Shen J, Rana N, Fernandez BB, Braun DP: Apoptosis in endometrial glandular and stromal cells in women with and without endometriosis. Hum Reprod 2001, 16(9):1802–1808.

24. Wei P, Yuan JX, Jin X, Hu ZY, Liu YX: Molsidomine and N-omega-nitro-L-arginine methyl ester inhibit implantation and apoptosis in mouse endometrium. Acta Pharmacol Sin 2003, 24(12):1177–1184.

25. Wei P, Jin X, Zhang XS, Hu ZY, Han CS, Liu YX: Expression of Bcl-2 and p53 at the fetal-maternal interface of rhesus monkey. Reprod Biol Endocrinol 2005, 3:4.

26. Wei P, Tao SX, Zhang XS, Hu ZY, Yi-Xun L: [Effect of RU486 on apoptosis and p53 expression at the boundary of fetal-maternal interface of rhesus monkey (Macaca mulatta)]. Sheng Li Xue Bao 2004, 56(1):60–65.

27. Tabibzadeh S, Kong QF, Satyaswaroop PG, Babaknia A: Heat shock proteins in human endometrium throughout the menstrual cycle. Hum Reprod 1996, 11(3):633–640.

28. Ron A, Birkenfeld A: Stress proteins in the human endometrium and decidua. Hum Reprod 1987, 2(4):277–280.

29. Shah M, Stanek J, Handwerger S: Differential localization of heat shock proteins 90, 70, 60 and 27 in human decidua and placenta during pregnancy. Histochem J 1998, 30(7):509–518.

30. Divers MJ, Bulmer JN, Miller D, Lilford RJ: Placental heat shock proteins: no immunohistochemical evidence for a differential stress response in preterm labour. Gynecol Obstet Invest 1995, 40(4):236–243.

31. Ziegert M, Witkin SS, Sziller I, Alexander H, Brylla E, Hartig W: Heat shock proteins and heat shock protein-antibody complexes in placental tissues. Infect Dis Obstet Gynecol 1999, 7(4):180–185.

32. Teng CB, Diao HL, Ma H, Cong J, Yu H, Ma XH, Xu LB, Yang ZM: Signal transducer and activator of transcription 3 (Stat3) expression and activation in rat uterus during early pregnancy. Reproduction 2004, 128(2):197–205.

33. Xiao LJ, Yuan JX, Song XX, Li YC, Hu ZY, Liu YX: Expression and regulation of stanniocalcin 1 and 2 in rat uterus during embryo implantation and decidualization. Reproduction 2006, 131(6):1137–1149.

34. Zhu LJ, Bagchi MK, Bagchi IC: Attenuation of calcitonin gene expression in pregnant rat uterus leads to a block in embryonic implantation. Endocrinology 1998, 139(1):330–339.

35. Luu KC, Nie GY, Salamonsen LA: Endometrial calbindins are critical for embryo implantation: evidence from in vivo use of morpholino antisense oligonucleotides. Proc Natl Acad Sci USA 2004, 101(21):8028–8033.

36. Yue ZP, Yang ZM, Wei P, Li SJ, Wang HB, Tan JH, Harper MJ: Leukemia inhibitory factor, leukemia inhibitory factor receptor, and glycoprotein 130 in rhesus monkey uterus during menstrual cycle and early pregnancy. Biol Reprod 2000, 63(2):508–512.

37. Wagner RW: Gene inhibition using antisense oligodeoxynucleotides. Nature 1994, 372(6504):333–335.

38. Nakamura J, Fujimoto M, Yasuda K, Takeda K, Akira S, Hatayama T, Takagi Y, Nozaki K, Hosokawa N, Nagata K: Targeted disruption of Hsp110/105 gene protects against ischemic stress. Stroke 2008, 39(10):2853–2859.

39. Liu YX: Plasminogen activator/plasminogen activator inhibitors in ovarian physiology. Front Biosci 2004, 9:3356–3373.

40. Leonardsson G, Peng XR, Liu K, Nordstrom L, Carmeliet P, Mulligan R, Collen D, Ny T: Ovulation efficiency is reduced in mice that lack plasminogen activator gene function: functional redundancy among physiological plasminogen activators. Proc Natl Acad Sci USA 1995, 92(26):12446–12450.

41. Liu YX, Gao F, Wei P, Chen XL, Gao HJ, Zou RJ, Siao LJ, Xu FH, Feng Q, Liu K, et al.: Involvement of molecules related to angiogenesis, proteolysis and apoptosis in implantation in rhesus monkey and mouse. Contraception 2005, 71(4):249–262.

42. Liu Y: Endometrium implantation and ectopic pregnancy. Sci China C Life Sci 2004, 47(4):293–302.

43. Kim JJ, Jaffe RC, Fazleabas AT: Blastocyst invasion and the stromal response in primates. Hum Reprod 1999, 14(Suppl 2):45–55.

44. Yamagishi N, Saito Y, Ishihara K, Hatayama T: Enhancement of oxidative stress-induced apoptosis by Hsp105alpha in mouse embryonal F9 cells. Eur J Biochem 2002, 269(16):4143–4151.

45. Evrard L, Vanmuylder N, Dourov N, Glineur R, Louryan S: Cytochemical identification of HSP110 during early mouse facial development. J Craniofac Genet Dev Biol 1999, 19(1):24–32.

46. Yamagishi N, Saito Y, Hatayama T: Mammalian 105 kDa heat shock family proteins suppress hydrogen peroxide-induced apoptosis through a p38 MAPK-dependent mitochondrial pathway in HeLa cells. Febs J 2008, 275(18):4558–4570.

47. Yamagishi N, Ishihara K, Saito Y, Hatayama T: Hsp105 family proteins suppress staurosporine-induced apoptosis by inhibiting the translocation of Bax to mitochondria in HeLa cells. Exp Cell Res 2006, 312(17):3215–3223.

48. Ishihara K, Yamagishi N, Saito Y, Adachi H, Kobayashi Y, Sobue G, Ohtsuka K, Hatayama T: Hsp105alpha suppresses the aggregation of truncated androgen receptor with expanded CAG repeats and cell toxicity. J Biol Chem 2003, 278(27):25143–25150.

Ovarian Hyperstimulation Syndrome and Prophylactic Human Embryo Cryopreservation: Analysis of Reproductive Outcome Following Thawed Embryo Transfer

Eric Scott Sills, Laura J. McLoughlin, Marc G. Genton,
David J. Walsh, Graham D. Coull and Anthony P. H. Walsh

ABSTRACT

Objective

To review utilisation of elective embryo cryopreservation in the expectant management of patients at risk for developing ovarian hyperstimulation syndrome (OHSS), and report on reproductive outcome following transfer of thawed embryos.

Materials and Methods

Medical records were reviewed for patients undergoing IVF from 2000–2008 to identify cases at risk for OHSS where cryopreservation was electively performed on all embryos at the 2 pn stage. Patient age, total number of oocytes retrieved, number of 2 pn embryos cryopreserved, interval between retrieval and thaw/transfer, number (and developmental stage) of embryos transferred (ET), and delivery rate after IVF were recorded for all patients.

Results

From a total of 2892 IVF cycles undertaken during the study period, 51 IVF cases (1.8%) were noted where follicle number exceeded 20 and pelvic fluid collection was present. Elective embryo freeze was performed as OHSS prophylaxis in each instance. Mean (± SD) age of these patients was 32 ± 3.8 yrs. Average number of oocytes retrieved in this group was 23 ± 8.7, which after fertilisation yielded an average of 14 ± 5.7 embryos cryopreserved per patient. Thaw and ET was performed an average of 115 ± 65 d (range 30–377 d) after oocyte retrieval with a mean of 2 ± 0.6 embryos transferred. Grow-out to blastocyst stage was achieved in 88.2% of cases. Delivery/livebirth rate was 33.3% per initiated cycle and 43.6% per transfer. Non-transferred blastocysts remained in cryostorage for 24 of 51 patients (46.1%) after ET, with an average of 3 ± 3 blastocysts refrozen per patient.

Conclusion

OHSS prophylaxis was used in 1.8% of IVF cycles at this institution; no serious OHSS complications were encountered during the study period. Management based on elective 2 pn embryo cryopreservation with subsequent thaw and grow-out to blastocyst stage for transfer did not appear to compromise embryo viability or overall reproductive outcome. For these patients, immediate elective embryo cryopreservation and delay of ET by as little as 30 d allowed for satisfactory conclusion of the IVF sequence, yielding a livebirth-delivery rate (per ET) >40%.

Introduction

Ovarian hyperstimulation syndrome (OHSS) is the most serious consequence of ovulation induction and in vitro fertilisation (IVF), potentially resulting in death in its extreme manifestation [1]. How best to manage this condition has been the subject of considerable study, with primary emphasis on risk recognition before commencing the IVF stimulation sequence [2,3]. The exact etiology of OHSS remains unknown. Since pregnancy can worsen OHSS, embryo transfer is sometimes intentionally postponed by electively freezing embryos until symptoms have resolved and the clinical picture improves [1]. In this study, data collected at one IVF referral centre during a nine-year period were used to assess a conservative strategy for OHSS prophylaxis and to investigate how empiric embryo cryopreservation and delayed transfer might impact reproductive outcome.

Materials and Methods

Patient Selection and Study Design

Patient records for all ovulation induction performed at the Sims International Fertility Clinic were retrospectively reviewed for the period 2000–2008, including all IVF patients (n = 2892). No case of OHSS was diagnosed in patients undergoing gonadotropin treatment for IVF. In this population, OHSS prophylaxis using elective embryo cryopreservation was instituted when the number of follicles with mean diameter ≥ 15 mm exceeded 20 [1] and when an intraperitoneal pelvic fluid collection was present measuring >5 cm (in any diameter) on transvaginal ultrasound. Serum oestradiol measurements were not obtained for every patient for the entire duration of the study period, so this parameter was not included for analysis.

All IVF patients included for study received a focused physical examination, and saline infusion sonogram, which were normal before initiating gonadotropin therapy. Controlled ovarian hyperstimulation regimens were developed from factors including historical response to medications, patient age, BMI and ovarian reserve assessment. Pituitary downregulation was achieved with oral contraceptives and GnRH agonist, followed by daily administration of gonadotropins (daily dose ≤ 150 IU/d) with periodic monitoring as previously described [4]. Treatment continued until adequate ovarian response was attained, defined as at least three follicles with mean diameter ≥ 17 mm. Transvaginal sonogram-guided oocyte retrieval was accomplished 36 h after subcutaneous administration of 10,000 IU hCG. Immediately after retrieval oocyte-cumulus complexes were placed into Universal IVF medium (MediCult; Jyllinge, Denmark), with insemination (including ICSI) also carried out using this reagent under washed liquid paraffin oil (MediCult, Denmark). Fertilisation was assessed after 16–18 h and was considered normal when two distinct pronuclei were noted.

For patients considered at risk for OHSS (based on criteria outlined above), extensive and immediate counselling was provided during the IVF cycle to review the potentially grave risks associated with fresh embryo transfer (as originally planned at cycle initiation). Since fresh transfer was not regarded as safe when OHSS might develop, alternate options of cycle cancellation and elective embryo cryopreservation were carefully outlined. While complete cycle cancellation was uniformly offered, no patients at risk for OHSS elected to do this during the study interval. Consequently, these cases were managed via elective embryo freeze (see Figure 1). Once this "freeze all" decision was made, this was documented and communicated to embryology staff. There were no additional OHSS cases that developed after embryo transfer who were not previously recognised during follicular recruitment and ovulation induction with gonadotropins.

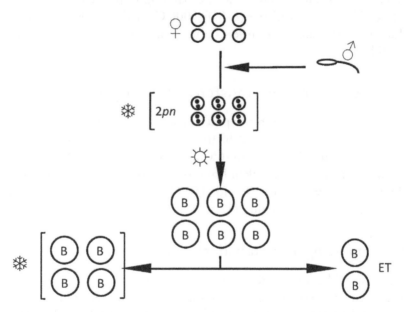

Figure 1. Schematic for prophylactic embryo cryopreservation at the 2pn stage, followed by extended culture to blastocyst stage (B) and subsequent transfer (ET). Non-transferred blastocysts are re-frozen for subsequent use (lower left).

Embryo Cryopreservation Sequence

Following confirmation of normal fertilisation by the presence of two distinct pronuclei, embryos were placed in cryoprotectant (Embryo Freezing Pack, Medi-Cult, Denmark) at room temperature and cooled to -7°C at a rate of 2°C/min. Manual seeding followed after 5 min, then the embryos were cooled from -7°C to -30°C at a rate of 0.3°C/min. The final rapid cooling step brought the embryos from -30°C to -190°C at 50 C/min; they were next transferred to liquid N2 for long-term storage and maintained at -196°C.

Thaw, Culture & Transfer Protocols

2 pn embryos were removed from liquid N2 storage and kept at room temperature ×30 sec before being placed in H2O bath at 30°C for 1 min. Embryos were placed in 1,2 propanediol/sucrose-based thaw media (Embryo Thawing Pack, MediCult, Denmark) at room temperature for a total of 20 min. Culture was maintained to day five in microdrops of BlastAssist media I and II (MediCult, Denmark) under washed paraffin oil in a 5%CO2 + 5%O2 atmosphere at 95% humidity. Embryos were assessed daily for cell number, degree of fragmentation, and compaction. Day five blastocysts selected for in utero transfer generally demonstrated a well-defined inner cell mass and highly cellular, expanding trophoectoderm. Blastocysts were loaded into an ET catheter (K-Soft-5000 Catheter; Cook Medical Inc., Spencer, Indiana USA), and all transfers occurred under direct transabdominal sonogram guidance.

Secondary Freeze for Non-Transferred Blastocysts

Supernumary blastocysts selected for (repeat) cryopreservation were incubated in 5–6% CO2 atmosphere at 37°C × 2 h, then placed in cryoprotectant (Blast-Freeze, MediCult, Denmark) cooled to -6°C at a rate of 2°C/min. After manual seeding, embryo temperature was taken from -6°C to -40°C at 0.3°C/min. Final rapid cooling of blastocysts from -40°C to -150°C at 35°C/min was followed by transfer to long-term storage in liquid N2 at -196°C.

Outcomes Reporting and Statistical Analysis

Primary endpoints of the study were patient age, total number of oocytes retrieved, number of 2 pn embryos cryopreserved, interval between retrieval and thaw/transfer, number (and developmental stage) of embryos transferred (ET), and livebirth-delivery rate. All data were tabulated as mean ± SD. Patients were periodically followed during pregnancy, or contact was established with their delivering obstetrician to determine delivery status. In the event that contact could not be made and delivery status remained unknown, this was also noted in the record.

Results

A total of 2892 IVF cycles proceeded to oocyte retrieval during the study period. Of these, 51 patients (1.8%) were judged to be at risk for developing OHSS and prophylactic embryo freezing was performed. While none of these patients qualified for fresh embryo transfer according to medical centre policy, there were some patients requesting elective embryo cryopreservation who were not at risk for OHSS. Reasons for empiric embryo cryopreservation in these cases included incidental surgery unrelated to fertility, diagnosis of malignancy, and divorce. Reproductive outcomes for these patients not considered at-risk for OHSS were excluded from the calculation of delivery rates in this study.

The mean (± SD) age of patients at risk for OHSS during the study period was 32 ± 3.8 yrs. Both ovaries were present and morphologically normal at baseline for all patients at risk for OHSS. All patients underwent ultrasound-guided transvaginal oocyte retrieval at our facility without incident; the average number of oocytes retrieved per patient was 23 ± 8.7. After fertilisation either by conventional insemination or ICSI, an average of 14 ± 5.7 2 pn embryos were cryopreserved per patient. Embryo cryopreservation was successfully carried out for approximately 61% of the total number of retrieved oocytes in this population.

Over the next 30 d, patients at risk for OHSS were periodically re-evaluated after elective cryopreservation of their embryos to document clinical improvement and resolution of symptoms. When there was no laboratory evidence of haemoconcentration, pelvic fluid collections had cleared, and ovarian quiescence was noted via ultrasound, it was considered safe to resume the IVF treatment sequence: thaw and grow-out to blastocyst stage was carried out. Development to blastocyst stage was achieved for 88.2% of cases, but when the cohort of thawed embryos did not advance to blastocyst stage the most developed embryos were transferred. This affected six cases after thaw, five of whom had day three embryos and one case where a 'mixed transfer' consisting of one morula + one blastocyst was performed. Embryo transfer was performed an average of 115 ± 65 d (range 30–377 d) after oocyte retrieval. In these patients at risk for OHSS, the mean number of embryos transferred was 2 ± 0.6. For 24 of these (46.1%), supernumary blastocysts remained after transfer and were returned to cryostorage.

For patients considered at risk for OHSS where elective 2 pn embryo cryopreservation was performed, the live birth delivery rate was 33.3% (17/51) per initiated cycle and 43.6% (17/39) per transfer. No twin or triplet deliveries occurred in this series. Follow-up with patients after delivery identified no long-term OHSS sequela, and there were no malformations or developmental anomalies reported among offspring.

Discussion

Ovarian hyperstimulation syndrome (OHSS) is a potentially fatal iatrogenic condition resulting from excessive stimulation of the ovaries [5]. The vast majority of OHSS develops in the setting of injectable gonadotrophins used in IVF, although in the absence of proper monitoring oral clomiphene treatment can also result in massive ovarian hyperstimulation necessitating surgical removal of the ovary [6].

According to the World Health Organization (WHO), severe OHSS develops in 0.2–1% of all stimulated ART cycles [7]. Several methods to prevent OHSS have been advocated including elective embryo cryopreservation, using low-dose hCG or GnRH-agonist for triggering oocyte maturation, "coasting" gonadotropin use, and cycle cancellation. To date, no single investigation has compared patient outcome and pregnancy rates among these varied interventions. In the present study, we implemented elective embryo freezing as an OHSS prophylactic measure in 1.8% of our IVF patients, a figure parallel to previous reports of actual OHSS incidence [8,9]. The pathophysiology of OHSS is complex; it likely involves a disruption of inflammatory processes normally evoking ovulation. The characteristic capillary extravasation of OHSS seems to be mediated by interactions of prolactin, prostaglandins, the ovarian prorenin-renin-angiotensin system, vascular endothelial growth factor (VEGF), angiogenin, the kinin-kallikrein system, selectins, von Willebrand factor, and/or endothelin [2]. The altered vascular permeability of OHSS may also be modulated by VE-cadherin [10], an inter-endothelial adhesion molecule or serum soluble ICAM-1 [11]. Of note, significantly higher IL-18 levels have been detected in the serum and extravascular fluids of patients with severe OHSS compared to non-OHSS controls [12].

When a patent is considered at risk for OHSS at our centre, we do not typically proceed with fresh embryo transfer (ET). During patient counselling, we explain that the strategy of empiric embryo freezing to minimise OHSS risk is not new [1,3,13,14], but has considerably lower risk than proceeding with fresh ET as originally planned. The rationale for delaying ET derives from the intent to delay pregnancy, since hCG increases VEGF which in turn facilitates the endothelial permeability associated with OHSS [15]. One of the first prospective studies to demonstrate the therapeutic benefit of elective early embryo cryopreservation in OHSS patients randomised subjects to undergo either fresh ET or receive delayed ET after cryopreservation (and thaw) of all embryos. No cases of OHSS developed in the setting of elective embryo freeze; pregnancy rates were comparable between the two groups [16].

OHSS risk is not always eliminated by elective freezing of embryos. An earlier review of precautionary cryopreservation of all embryos at the 2 pn stage noted that OHSS developed anyway in 27% of cases [13]. Some have speculated that elective embryo freezing may reduce the severity – but not lower the incidence of – symptomatic OHSS [14].

One unexpected finding from the present study was the large variation in time interval between oocyte retrieval and thaw/transfer among patients at risk for OHSS, which ranged from 30 to 377 days. While neither duration of cryostorage nor type of ovulation induction protocol has been found to affect reproductive outcome in IVF [17,18], we nevertheless remain curious why any IVF patient would wait for more than a year to initiate a thaw-transfer sequence – risk of OHSS notwithstanding. It is our belief that such extended (>90 d) delays are a function of patient scheduling preferences rather than medical factors, but the question forms the basis of ongoing research here. As our data show, elective embryo cryopreservation may lead to considerable delay in treatment (and consequently postpones the potential for pregnancy) and many patients at risk for OHSS are understandably discouraged by the prospect of elective embryo freeze. However, patients should be made aware that while live birth delivery rates among IVF patients at risk for OHSS can be impressive [19], the dangers accompanying OHSS are not be underestimated. For example, one series in Ireland recently demonstrated a 4% case fatality rate for the condition [9].

Our study has several limitations which should be acknowledged. We focused more on OHSS prevention rather than development of the condition itself. Serum oestradiol has been used as a marker for OHSS risk for many years [1,2,5,7,9], but this method of screening was not regularly available at our institution throughout the nine-year study period and thus was not part of this analysis. Additionally, the role of paracentesis or albumin/hespan infusion could not be specifically studied in this report because medical records were not electronically searchable for these terms throughout the study period. While 2 pn embryos for cryopreservation were produced from approximately 61% of retrieved oocytes in this series, our OHSS cases were not stratified according to ICSI vs. conventional insemination. However, previous research has suggested that pregnancy rates after cryopreservation of 2 pn embryos are not impacted by fertilisation method [20]. Our retrospective study also did not have a control group, so it is unknown how many patients might have developed OHSS if a fresh transfer had been performed. However, it is reasonable to conclude that some OHSS cases would have been expected from this high-risk population.

In conclusion, this descriptive study finds conservative application of elective embryo cryopreservation to be a useful component of OHSS prophylaxis. High delivery rates are typical among women at risk of OHSS who undergo elective embryo cryopreservation with deferred thaw/transfer, and our outcomes data support this finding as well. While serum oestradiol determinations can be helpful in OHSS surveillance, this report shows that screening based on clinical parameters can also be effective. Further studies are planned to refine specific factors that might be useful in prediction of OHSS risk, with a view to optimise clinical management of this important and potentially dangerous condition.

Competing Interests

The authors declare that they have no competing interests.

Authors' Contributions

ESS and LJM collected data for the study and prepared the original manuscripts; MGG provided design input and statistical analysis; GDC organised the embryology laboratory components and provided data on gametes and reproductive outcome; DJW and APHW supervised the project and directed the research. All authors approved the final manuscript.

References

1. Tiitinen A, Husa LM, Tulppala M, Simberg N, Seppala M: The effect of cryopreservation in prevention of ovarian hyperstimulation syndrome. BJOG 1995, 102:326–9.

2. Delvigne A, Rozenberg S: Systematic review of data concerning etiopathology of ovarian hyperstimulation syndrome. Int J Fertil Womens Med 2002, 47:211–26.

3. Amso NN, Ahuja KK, Morris N, Shaw RW: The management of predicted ovarian hyperstimulation syndrome involving gonadotropin-releasing hormone analog with elective cryopreservation of all pre-embryos. Fertil Steril 1990, 53:1087–90.

4. Sills ES, Schattman GL, Veeck LL, Liu HC, Prasad M, Rosenwaks Z: Characteristics of consecutive in vitro fertilization cycles among patients treated with follicle-stimulating hormone (FSH) and human menopausal gonadotropin versus FSH alone. Fertil Steril 1998, 69:831–5.

5. D'Angelo A, Amso N: Embryo freezing for preventing ovarian hyperstimulation syndrome. Cochrane Database Syst Rev 2007, 18(3):CD002806.

6. Sills ES, Poynor EA, Moomjy M: Ovarian hyperstimulation and oophorectomy following accidental daily clomiphene citrate use over three consecutive months. Reprod Toxicol 2000, 14:541–3.

7. Binder H, Dittrich R, Einhaus F, Krieg J, Müller A, Strauss R, Beckmann MW, Cupisti S: Update on ovarian hyperstimulation syndrome: Part 1 – Incidence and pathogenesis. Int J Fertil Womens Med 2007, 52:11–26.

8. Pattinson HA, Hignett M, Dunphy BC, Fleetham JA: Outcome of thaw embryo transfer after cryopreservation of all embryos in patients at risk of ovarian hyperstimulation syndrome. Fertil Steril 1994, 62:1192–6.

9. Mocanu E, Redmond ML, Hennelly B, Collins C, Harrison R: Odds of ovarian hyperstimulation syndrome (OHSS) – time for reassessment. Hum Fertil (Camb) 2007, 10:175–81.

10. Villasante A, Pacheco A, Ruiz A, Pellicer A, Garcia-Velasco JA: Vascular endothelial cadherin regulates vascular permeability: Implications for ovarian hyperstimulation syndrome. J Clin Endocrinol Metab 2007, 92:314–21.

11. Abramov Y, Schenker JG, Lewin A, Kafka I, Jaffe H, Barak V: Soluble ICAM-1 and E-selectin levels correlate with clinical and biological aspects of severe ovarian hyperstimulation syndrome. Fertil Steril 2001, 76:51–7.

12. Barak V, Elchalal U, Edelstein M, Kalickman I, Lewin A, Abramov Y: Interleukin-18 levels correlate with severe ovarian hyperstimulation syndrome. Fertil Steril 2004, 82:415–20.

13. Wada I, Matson PL, Troup SA, Hughes S, Buck P, Lieberman BA: Outcome of treatment subsequent to the elective cryopreservation of all embryos from women at risk of the ovarian hyperstimulation syndrome. Hum Reprod 1992, 7:962–6.

14. Wada I, Matson PL, Troup SA, Morroll DR, Hunt L, Lieberman BA: Does elective cryopreservation of all embryos from women at risk of ovarian hyperstimulation syndrome reduce the incidence of the condition? BJOG 1993, 100:265–9.

15. Villasante A, Pacheco A, Pau E, Ruiz A, Pellicer A, Garcia-Velasco JA: Soluble vascular endothelial-cadherin levels correlate with clinical and biological aspects of severe ovarian hyperstimulation syndrome. Hum Reprod 2008, 23:662–7.

16. Ferraretti AP, Gianaroli L, Magli C, Fortini D, Selman HA, Feliciani E: Elective cryopreservation of all pronucleate embryos in women at risk of ovarian hyperstimulation syndrome: efficacy and safety. Hum Reprod 1999, 14:1457–60.

17. Veeck LL, Amundson CH, Brothman LJ, DeSciasciolo C, Maloney MK, Muasher SJ, Jones HW Jr: Significantly enhanced pregnancy rates per cycle through cryopreservation and thaw of pronuclear stage oocytes. Fertil Steril 1993, 59:1202–7.

18. Lin YP, Cassidenti DL, Chacon RR, Soubra SS, Rosen GF, Yee B: Successful implantation of frozen sibling embryos is influenced by the outcome of the cycle from which they were derived. Fertil Steril 1995, 63:262–7.

19. Queenan JT Jr, Veeck LL, Toner JP, Oehninger S, Muasher SJ: Cryopreservation of all prezygotes in patients at risk of severe hyperstimulation does not eliminate the symptoms, but the chances of pregnancy are excellent with subsequent frozen-thaw transfers. Hum Reprod 1997, 12:1573–6.

20. Damario MA, Hammitt DG, Galantis TM, Session DR, Dumesic DA: Pronuclear stage cryopreservation after intracytoplasmic sperm injection and conventional IVF: implications for timing of the freeze. Fertil Steril 1999, 72:1049–54.

Neural Differentiation of Embryonic Stem Cells *In Vitro*: A Road Map to Neurogenesis in the Embryo

Elsa Abranches, Margarida Silva, Laurent Pradier,
Herbert Schulz, Oliver Hummel, Domingos Henrique
and Evguenia Bekman

ABSTRACT

Background

The in vitro generation of neurons from embryonic stem (ES) cells is a promising approach to produce cells suitable for neural tissue repair and cell-based replacement therapies of the nervous system. Available methods to promote ES cell differentiation towards neural lineages attempt to replicate, in different ways, the multistep process of embryonic neural development. However, to achieve this aim in an efficient and reproducible way, a better knowledge of the cellular and molecular events that are involved in the process, from the initial specification of neuroepithelial progenitors to their terminal differentiation into neurons and glial cells, is required.

Methodology/Principal Findings

In this work, we characterize the main stages and transitions that occur when ES cells are driven into a neural fate, using an adherent monolayer culture system. We established improved conditions to routinely produce highly homogeneous cultures of neuroepithelial progenitors, which organize into neural tube-like rosettes when they acquire competence for neuronal production. Within rosettes, neuroepithelial progenitors display morphological and functional characteristics of their embryonic counterparts, namely, apico-basal polarity, active Notch signalling, and proper timing of production of neurons and glia. In order to characterize the global gene activity correlated with each particular stage of neural development, the full transcriptome of different cell populations that arise during the in vitro differentiation protocol was determined by microarray analysis. By using embryo-oriented criteria to cluster the differentially expressed genes, we define five gene expression signatures that correlate with successive stages in the path from ES cells to neurons. These include a gene signature for a primitive ectoderm-like stage that appears after ES cells enter differentiation, and three gene signatures for subsequent stages of neural progenitor development, from an early stage that follows neural induction to a final stage preceding terminal differentiation.

Conclusions/Significance

Overall, our work confirms and extends the cellular and molecular parallels between monolayer ES cell neural differentiation and embryonic neural development, revealing in addition novel aspects of the genetic network underlying the multistep process that leads from uncommitted cells to differentiated neurons.

Introduction

Neural induction in vertebrate embryos was first described by Mangold and Spemann in 1924 [1] and results in the establishment of a neuroectodermal primordium from where the nervous system will arise. The molecular signals involved in this crucial event are not yet totally elucidated but it is known that FGF and WNT signalling are required, together with inhibition of BMP signalling activity [2], [3]. In the mouse embryo, the initial population of specified neuroepithelial progenitors (NPs) is known to express various pan-neural genes, like sox1 and sox2 [4], [5]. These NPs will then acquire competence to produce neurons when they become part of the closing neural tube during neurulation, in a process that involves retinoid signalling from adjacent somites and the activity of proneural genes [6].

The embryonic neural tube is composed by a pseudostratified layer of neuroepithelial cells with a clear apico-basal polarity. The apical domain of these cells is located at the luminal surface and is delineated by the presence of apical protein complexes, like the PAR polarity complex [7], as well as by the presence of junctional structures, where N-cadherin and β-catenin accumulate [8]. Centrosomes also localize apically in neuroepithelial cells, which enter mitosis close to the luminal surface due to the characteristic interkinetic nuclear movement (INM) [9]. This particular organization of the neural tube is important for the coordinated production of neurons and glia. Neighbouring neuroepithelial cells signal to each other through Delta/Jagged ligands and Notch receptors, in a process that maintains a population of proliferating NPs and coordinates the timely production of neurons throughout embryonic development (reviewed in [10], [11]). This unique architecture of the embryonic neural tube has transient character and disappears perinatally to give way to definitive CNS structures like the brain and spinal cord.

Several approaches have been used to achieve in vitro neural differentiation starting from embryonic stem (ES) cells, aimed at generating regionally specified neural progenitors and/or differentiated neuronal and glial subtypes. All these methods try to recapitulate, in different ways, the multistep process of neural development that occurs in the embryo, from neural induction to the terminal differentiation of neurons and glial cells. This was initially achieved through embryoid body (EB) formation in the presence of retinoic acid [12] or, alternatively, by co-culture of ES cells with stroma/conditioned medium [13], [14]. However, as ES cells are pluripotential and readily differentiate into almost any cell type, the efficiency of neural conversion is limited and lineage selection is usually needed to ensure homogeneity of the differentiated population [15]. A simpler way to reconstitute neural commitment in vitro and achieve efficient neuronal production relies upon monolayer differentiation of ES cells, a method developed by Ying and co-workers [16]. In this method, ES cells are cultured in defined serum- and feeder-free conditions, in the absence of BMP signals that are known to inhibit neural fate. In these conditions, ES cells undergo neural commitment through a "autocrine" induction mechanism, where FGF signalling plays a pivotal role, as it does in the embryo [17], [18]. This method results in a more efficient neural commitment and differentiation, which likely results from a better mimicry of the events that occur in the embryo. However, a detailed characterization of the cellular and molecular steps involved in promoting ES cell differentiation towards neural lineages is required, not only to enhance our understanding of neurodevelopmental mechanisms but also to develop more rational ES cell-based strategies for treating traumatic injuries and neurodegenerative diseases affecting the human nervous system.

In this work, we describe various aspects of the process that leads from ES cells to differentiated neurons in monolayer cultures. Using improved conditions, we routinely obtain highly homogeneous cultures of NPs that maintain morphological and functional characteristics of their embryonic counterparts, namely apico-basal polarity, active Notch signalling, and proper timing of production of neurons and glia. We show that the transition to neuronal production is accompanied by the organization of NPs into neural tube-like rosettes, where these cells divide and give rise to neurons. Furthermore, we have characterized the global gene expression changes that occur along the path to neural differentiation, from ES cells to neurogenic rosettes. Our results confirm and extend at the molecular level the parallels with embryonic neural development, revealing in addition novel aspects of the genetic network underlying the multistep process that leads from uncommitted cells to differentiated neurons.

Results

Improved Generation of NPs From ES Cells in Defined Serum-Free Media

Commitment of undifferentiated ES cells to neural fate can be achieved with high efficiency in feeder-free adherent monocultures, using the serum-free medium N2B27 [16]. In these conditions, when Sox1-GFP knock-in (46C) ES cells were used, Ying and co-workers reported that cultures with more than 80% of NPs (Sox1-GFP+) can be obtained [16]. Using the same ES cell line, we carried out a comparative study of neural commitment in N2B27 and RHB-A (StemCell-Sciences Inc., UK), a new N2B27-based neural differentiation medium. We monitored cellular growth, the emergence of Sox1-GFP+ NPs and the appearance of various cell-specific markers in these cultures. Our results show that commitment to neural fate in RHB-A occurs faster and produces a higher percentage of Sox1-GFP+ NPs, when compared to N2B27 (Fig. 1A). For instance, three days after ES cell plating in RHB-A, more than 60% of cultured cells are Sox1-GFP+ NPs, while only about 40% of cells became Sox1-GFP+ in N2B27 (p-value = 0.005). The percentage of Sox1-GFP+ NPs in the total population reaches a peak at day 4 in both media, with consistently higher levels in RHB-A (p-value = 0.052). Further culturing for 2 more days results in a sharp increase in the total number of cells (Fig. 1B), but without changes in the percentage of Sox1-GFP+ NPs (Fig. 1A). This suggests that, from day 4 onwards, a "transit-amplifying" population of

Sox1-GFP+ NPs is established and that induction of new NPs contributes little to the growth of this population.

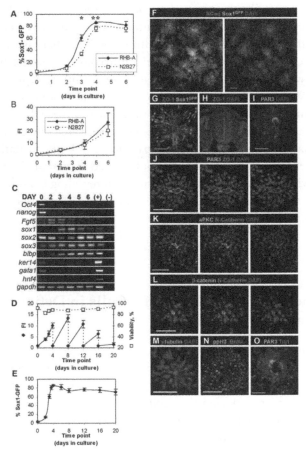

Figure 1. ES-cell derived NPs culture analysis. A) Percentage of GFP+ cells in monolayer cultures grown for 6 days without replating in RHB-A and N2B27 media (* p-value = 0.005; ** p-value = 0.052). B) Fold increase (FI) for monolayer cultures grown for 6 days in RHB-A and N2B27 media. C) Semi-quantitative RT-PCR analysis for selected markers of pluripotency and lineage commitment in day 0–6 RHB-A cultures; mRNA from E10.5 mouse embryos was used as positive control. D) FI (filled squares) and viability (open squares) for RHB-A cultures maintained for 20 days in culture and replated every 4 days at the same initial cell density. E) Percentage of Sox1-GFP+ cells along 20 days in culture in RHB-A, with replating every 4 days. In all graphs data are means±SEM from at least three independent experiments. F) After replating in laminin (day 5), Sox1-GFP+ cells organize in rosettes, with N-Cadherin (in red) present at the centre of these cell clusters. G) ZO-1 accumulates in the cell processes that coalesce at the centre of rosettes, like it does in the apical domain of NPs in the embryonic neural tube (H). I) Anti-PAR3 immunostaining reveals well-defined "apical" domains at the centre of rosettes, where it co-localizes with ZO-1 (J). K) aPKC, another known apical marker is also present at the centre of rosettes and co-localizes with N-Cadherin. L) Adherent junctions' components, ß-catenin and N-Cadherin, co-localize at the central, apical region of rosettes. M) Anti-γ-tubulin staining (in green) shows "apically" localized centrosomes. N) Mitotic figures (ppH3) are localized centrally in rosettes while S-phase nuclei (BrdU) are located at the periphery. O) Differentiating Tuj1+ neurons accumulate at the periphery of rosettes. Nuclei counterstained with DAPI (blue). Scale bar: 50 μm.

RT-PCR analysis confirms that the switch from ES identity (Oct4+, nanog+, sox2+, sox1-, sox3-) to that of NPs (sox1+,2+,3+, blbp+,Oct4- and nanog-) seems to be complete by day 4 (Fig. 1C). Before this, cells pass initially through a primitive ectoderm (PE) stage, as shown by the expression of Fgf5 [19], [20], preceding the appearance of NP markers at day 3. In contrast, markers for endo-dermal (hnf4, gata1) and epidermal (ker14) lineages are rapidly down-regulated during monolayer differentiation (Fig. 1C).

Based on these results, we chose to replate day 4 NPs onto a laminin substrate in the same RHB-A medium, to test their neural differentiation potential. The cultures were maintained until day 20, being replated every 4th day. In these conditions, cell viability remains high (above 90%), although the proliferation rate (shown as fold increase–FI) decreases along time (Fig. 1D). The percentage of Sox1-GFP+ NPs in culture also decreases, stabilizing above 70% around day 8 (Fig. 1E).

NPs Show Proper Apico-Basal Polarity In Vitro and Undergo INM

After replating, we observed that cells grow in tightly packed monolayers resem-bling thick epithelial sheets. However, the distribution of Sox1-GFP+ cells is not uniform in these sheets, being organized in clusters to form rosette-like structures (Fig. 1F). In these clusters, Sox1-GFP+ NPs express the known apical markers of neuroepithelial cells, N-cadherin and ZO-1, which are localized at the centre of rosettes (Fig. 1F,G). This suggests that NPs within these structures are organized with their apical domains coalescing to form a central lumen, like in the embry-onic neural tube (Fig. 1H). This organisation is confirmed by the co-localization of other known neuroepithelial apical markers at the centre of rosettes, like PAR3 (Fig. 1I,J), aPKC (Fig. 1K), β-catenin (Fig. 1L), Numb, Afadin and Occludin (not shown). Furthermore, centrosomes are located close to the central region of the rosettes (Fig. 1M), where mitotic (ppH3+) nuclei are also detected (Fig. 1N). In contrast, S-phase nuclei lie at the periphery of rosettes, as shown by short pulses of BrdU labelling (Fig. 1N). This suggests that the nuclei of NPs within rosettes reproduce the characteristic INM shown by NPs in the embryonic neural tube [9]. To confirm this, we carried out time-lapse imaging of ES cell-derived neural rosettes in culture, revealing that NPs do indeed undergo nuclear move-ments coupled with the cell cycle, like embryonic NPs. Finally, newborn neurons (Tuj1+) localize outside or at the periphery of rosettes (Fig. 1O), resembling also the embryonic neural tube where neurons accumulate outside of the ventricular proliferative zone. Together, these observations reveal that rosettes are remarkably organized like embryonic neural tubes, with ES cell-derived NPs linked by junctional structures at their apical surface and engaged on neurogenesis. This led us to explore whether rosette-like cultures may have other structural and functional similarities with the embryonic neural tube.

Notch Pathway is Active in Rosette Cultures

In the embryonic neuroepithelium, the Notch pathway controls the rate at which proliferating NPs commit to differentiation. When Notch activity is inhibited, precocious neuronal differentiation is usually observed (reviewed in [21]). To test whether Notch signalling is involved in maintaining NPs in cultured neuroepithelial rosettes, as it happens in the embryonic neuroepithelium, we first analysed the expression of various genes known to mediate Notch activity (Fig. 2). RT-PCR data show that Notch1, 2 and 3 are expressed in monolayer cultures, together with various Delta-like and Jagged genes, as well as hes genes known to be involved in embryonic neural development (Fig. 2A). Analysis by in situ hybridization (ISH) reveals that hes5, the main Notch target gene in embryonic NPs [22], is broadly expressed in neuroepithelial rosettes, while Dll1 and hes6, which are normally expressed in newborn neurons [23], [24], show a more scattered expression, consistent with being transcribed in rosette cells singled out for differentiation (Fig. 2B, left panels). To evaluate the functional role of Notch signalling in rosette cultures, its activity was inhibited by treatment with the γ-secretase inhibitor LY411575 [25], resulting in a strong reduction of hes5 expression and the concomitant increase in Dll1 and hes6 expression (Fig. 2B, right panels). These results confirm the efficacy of Notch inhibition and show that rosette progenitors embark on neuronal differentiation in the absence of Notch activity. Indeed, LY411575-treated cultures reveal a significant increase both in Tuj1+ (Fig. 2C) and HuC/D+ (Fig. 2D) neurons, accompanied by striking morphological changes: after 48 h of Notch inhibition, rosette structures disappear and give way to large rounded ganglion-like clusters made up by Tuj1+ and HuC/D+ differentiating neurons, with extensive neurite outgrowths. Quantification of the number of HuC/D+ differentiating neurons reveals that the neurogenic effect due to Notch inhibition is more pronounced in day 8 cultures (n = 3, p-value = 0.002), while day 16 cultures show no increase of neuronal production (Fig. 2D). However, Notch receptors and ligands are still expressed at day 16, albeit at lower levels, making it unlikely that the lack of neurogenic effects is due to the absence of some components of the pathway. An alternative explanation is that, by day 16, NPs have lost most of their neurogenic potential and have switched their competence to gliogenic, as it has been previously described to happen during embryonic neural development and in cultures of isolated cortical NPs [26], [27].

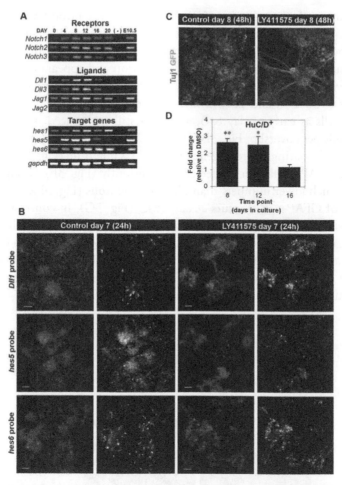

Figure 2. Chemical inhibition of the Notch activity by γ-secretase inhibitor LY411575. A) Expression of Notch pathway genes during monolayer ES cell differentiation, from day 0 to day 20, by RT-PCR analysis. mRNA from E10.5 mouse embryos was used as control. B) Detection by ISH of Dll1, hes5 and hes6 transcripts in control (DMSO-treated) and LY411575-treated rosette cultures. Treatment was done in day 6 cultures for 24 hours. Nuclei counterstained with DAPI. C) After 48 h of LY411575 treatment, starting at day 6, massive neuronal differentiation is observed by Tuj1 immunostaining. D) Notch inhibition with LY411575 at day 8 or 12 of the monolayer protocol results in increased neuronal production, detected by HuC/D imunostaining. No change was detected when inhibition was done in day 16 rosettes. Bars in D represent SEM for the minimum of three independent experiments. * p-value = 0.025; ** p-value = 0.002. Scale bars in B,C: 50 µm.

NPs have both Neurogenic and Gliogenic Potential In Vitro

To test whether NPs in rosette cultures undergo a temporal switch from early, neuron-producing, to late, glia-producing progenitors, we quantified the production of neurons and glia throughout the monolayer differentiation protocol. We found that the number of HuC/D+ neurons in rosette cultures increases up to day 12 and starts to decrease by day 16 (Fig. 3A). On the contrary, GFAP+ glial cells can only be detected from day 16 on, after the third replating, albeit still in reduced numbers (Fig. 3A). We reasoned that 4 days of culture after replating might not be sufficient to allow for glial differentiation and appearance of GFAP immunoreactivity. We therefore extended cultures for 3 additional days without replating (days 8+3, 12+3 and 16+3). In these conditions, we could detect scattered GFAP+ cells as early as day 8+3, although still in reduced numbers (no more than 3 cells per coverslip, Fig. 3B,C), in striking contrast with the number of HuC/D+ differentiating neurons generated at the same time (Fig. 3D). In day 12+3 cultures, GFAP+ cells can be detected consistently (Fig. 3E), though they still appear in much lower numbers than HuC/D+ neurons (Fig. 3F). The maximum number of GFAP+ cells occurs at day 16+3 (Fig. 3G), in contrast to that of HuC/D+ cells which peak at day 8+3 (Fig. 3H). Together, these results reveal that the neurogenic potential of rosette cultures decreases with time, with GFAP+ glial cells appearing consistently after the peak of neuronal production (Fig. 3B), indicating that a switch of progenitor identity, from neurogenic to gliogenic, occurs in these cultures.

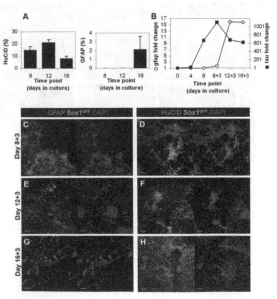

Figure 3. Timing of production of neurons and glia in rosette cultures. A) Percentage of HuC/D+ and GFAP+ cells in rosette cultures, relative to the total number of cells in culture. A decrease in neuronal production is observed at day 16, concomitant with an increase of glial cells. B) Semi-quantitative RT-PCR data showing fold change of expression (relative to day 0) for neuronal (tau) and glial (gfap) markers at successive timepoints of rosette cultures. Data normalized to gapdh. C–H) Rosette cultures at day 8+3, 12+3 and 16+3, labelled with anti-HuC/D and anti-GFAP antibodies to visualize neurons and glial cells, respectively. Few GFAP+ cells appear in day 8+3 cultures (C), with the number increasing at day 12+3 (E) and 16+3 (G). In contrast, a decrease in the number of HuC/D+ neurons is detected at day 16+3 (H). Nuclei counterstained with DAPI (blue). Scale bars: 50 μm.

Neural Stem Cells are Present in Monolayer Cultures

It is known that both embryonic neural tissue and certain regions of the adult vertebrate CNS contain a resident population of progenitor/stem cells [28]. Recent work [29], [30] established conditions for the isolation and clonogenic in vitro propagation of neural stem (NS) cells derived either from ES cells or from embryonic and adult neural tissue. In the present work, using the same experimental conditions, we were able to derive floating aggregates of NS cells from all stages of the in vitro neuroepithelial rosette cultures (days 4, 8, 12, 16 and 20) with similar efficiencies (Fig. 4A,B). When these aggregates are plated en bloc onto laminin substrate, without dissociation, cells migrate out and form neuroepithelial rosettes where all cells are positive for the NS cells markers Sox2 (Fig. 4C) and Nestin (not shown). After several days of culture, these cells develop very long cellular projections similar to those of radial glia (Fig. 4C and data not shown) and are able to differentiate into the all three neural lineages (Fig. 4C), a feature that fits well with the characteristics of NS cells. The constant presence of these cells, both in less proliferative day 20 monolayer cultures as well as in younger day 4 cultures (Fig. 4B), indicates that the floating aggregates are derived from a resident stem cell population, present in neuroepithelial rosette cultures at all time points studied. This, in turn, provides additional evidence for the neuroepithelial identity of these cultures.

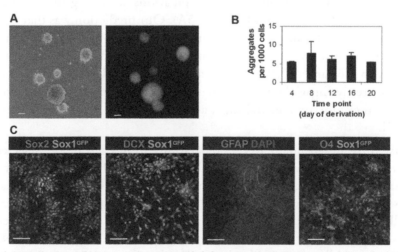

Figure 4. NS cell potential of the in vitro neuroepithelial rosette cultures A) Floating aggregates of NS cells derived from day 4 monolayer cultures (Sox1-GFP 46C cells; phase contrast and GFP fluorescence images). B) Efficiency of derivation of NS cells-derived floating aggregates from several rosette cultures time points (days 4, 8, 12, 16 and 20), expressed as number of aggregates formed per 1000 cells. Bars represent SEM for 3 independent experiments. C) Floating aggregates of NS cells (derived from day 4 monolayer cultures of 46C) were left to attach for 4 days onto laminin substrate and stained for Sox2 (NP marker), Doublecortin (DCX, neuronal marker), GFAP (glia) and O4 (oligodendrocytes). Scale bars: 50 μm.

Transcriptional Profiling of In Vitro Neural Commitment

The results described above indicate that neuroepithelial rosette cultures reca-pitulate several aspects of embryonic neural tube development. In contrast to the scarcity and complexity of cells from early stages of mammalian embryos, these cultures can provide large and highly homogeneous populations of cells at various stages of neural development, with the additional advantage of ob-taining homogeneous populations of Sox1-GFP+ NPs by FACS sorting. This creates a unique opportunity to characterize the transcriptional programs ac-tive at various phases of neural commitment and differentiation, from which it might be possible to predict the molecular pathways regulating these pro-cesses. With this purpose, global gene expression profiling using Affymetrix microarrays (Mouse Genome 430 Version 2.0) was performed at several stages of the monolayer rosette cultures: day 0 (undifferentiated ES cells), day 1, day 3 and day 8. At day 1, ES cells have entered differentiation and our aim was to obtain a gene signature for a population of primitive ectoderm-like cells that is likely to be present, as marked by the up-regulation of Fgf5 expression and down-regulation of nanog. At day 3, a sharp up-regulation of sox1 is de-tected by RT-PCR (Fig. 1C), probably reflecting the emergence of an initial population of NPs after neural induction. To characterize the transcriptional program active in these early NPs, we chose to purify Sox1-GFP+ cells at day 3 by FACS sorting, resulting in two sub-populations according to the levels of GFP expression (GFP+ and GFP++, Fig. 5A). Our prediction was that cells with lower levels of GFP might be at an earlier stage of NP development and that, by separating the two sub-populations of NPs, one could pull out genes associated with the earliest NPs state. Finally, at day 8, NPs are organized in rosettes and already engaged on neurogenesis, in a stage likely to be equivalent to NPs from the embryonic neural tube, after the onset of neuronal differen-tiation [55].

At least three independent RNA preparations from each of the selected time points were processed and hybridized on the arrays. Previous validation of these samples was done by analyzing the expression of Oct4, nanog, hes5 and blbp by semi-quantitative RT-PCR (Fig. 5B). As expected, expression of ES cell genes, Oct4 and nanog, decrease throughout the differentiation process and are no longer detected on day 3. In contrast, expression of NP markers, hes5 and blbp, can only be detected at day 3, increasing significantly at day 8. This pattern of expression was also observed in the microarray profil-ing (Fig. 5C).

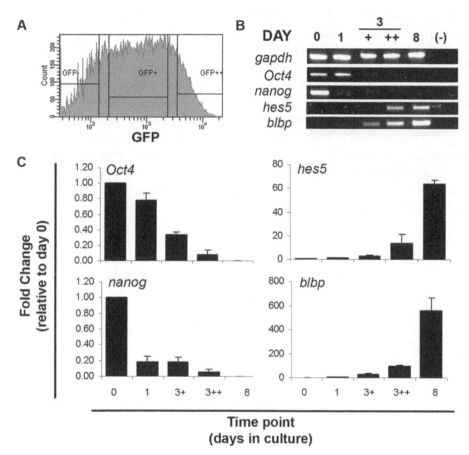

Figure 5. Validation of microarrays results. A) Histogram of sorted Sox1-GFP populations from day 3 monolayers. GFP negative (GFP-) cells were discarded, while two GFP positive populations were collected individually, according to their levels of GFP expression (GFP+ and GFP++). B) RT-PCR analysis of RNA samples collected for microarray analysis for the genes Oct4, nanog, hes5, and blbp. C) Fold changes, relative to day 0, obtained from Affymetrix profiling for the genes Oct4, nanog, hes5, and blbp.

The microarray data were normalized by the log scale robust multi-array analysis [31] and an ANOVA FDR-value of 10-3 (p-value<2.10-4) was used to identify and restrict the number of differentially expressed probe sets to 9456, which correspond to 6563 unique genes. Further analysis of the differentially expressed genes involved their distribution into specific groups, according to the variations in their expression throughout differentiation (Fig. 6). A first group was defined as including genes whose expression peaks at day 0 and is downregulated at all other time points. This includes known pluripotency genes like nanog, rex1 and fbxo15, confirming the ES cell identity of the initial population at day 0. A second group includes genes with a peak of expression at day 1 and might identify a transient

PE population as indicated by the presence of Fgf5 in this group [19], [20]. A third group includes genes that are up-regulated at day 3 but down-regulated in day 8 rosettes, and might characterize a transient population of Sox1-GFP+ NPs (tNPs) that emerge after neural induction. These progenitors will then evolve into neurogenic progenitors (nNPs) competent to initiate neuronal production, identified by a fourth group containing genes that start to be up-regulated at day 3 but continue to be expressed at similar or higher levels in day 8 rosettes. Finally, a fifth group is composed by genes that are only up-regulated in day 8 rosettes and includes genes characteristic of progenitors in the final phase of commitment to differentiation, like the proneural genes ascl1, neuroG1 and neuroG2 [32], as well as genes known to be expressed in early differentiating neurons, like doublecortin and hu/elav [33], [34].

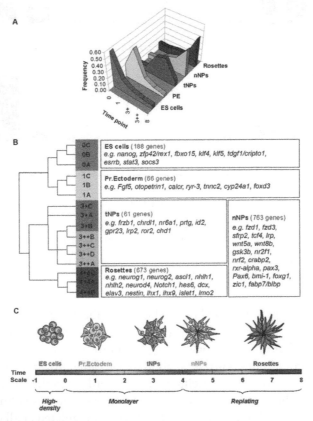

Figure 6. Clustering analysis of differentially expressed genes. A) Frequency distribution of the expression levels of the genes belonging to the five defined groups. B) Dendogram of the relationship of expression of genes belonging to each group (with biological replicates being represented by the letters A, B, C and D) and examples of genes that are present in the five defined groups. C) Schematic representation of the successive cellular states that occur along the path to neural differentiation (see text for definitions of stages).

This distribution of transcriptional profiles depicts, at the molecular level, the successive cellular states that occur along the path to neural differentiation, allowing the identification of gene signatures for each of these states and a better definition of the transitions between them.

Discussion

In this work, we characterize at the cellular and molecular level the processes of neural commitment and differentiation that occur when mouse ES cells are driven into a neural fate, using an improved adherent monolayer protocol [16], [35]. We show that NPs derived from ES cells organize themselves into rosette-like structures, with an apico-basal distribution of polarity proteins similar to that described for neuroepithelial cells in the embryonic neural tube [7], [8]. In addition, ES cell-derived rosette NPs display the characteristic cell cycle-related INM of the embryonic neuroepithelium. We also show that Notch signalling is active in neuroepithelial rosettes and controls the timely production of neurons from ES cell-derived NPs. The intrinsically controlled sequential generation of neurons and glial cells seems to be also preserved in ES cell-derived NPs. Altogether, these results demonstrate that the in vitro generation of neural cells from ES cells, using the monolayer protocol, closely mimics the process of embryonic neural development. Global gene expression analysis, at successive steps of the process that leads ES cells to neurons, provides further support for the similarities between the ES cell-derived rosette culture system and embryonic neural tube development, revealing in addition novel candidate genes that might regulate the processes of neural commitment and differentiation.

Neuroepithelial Rosettes as In Vitro Counterparts of Embryonic Neural Tube

Several methods have been described to achieve neural differentiation of ES cells in vitro (reviewed in [36], [37]), including the treatment of cell aggregates (EBs) with retinoic acid [12] or the co-culture of ES cells with stromal cells that produce uncharacterized neural-inducing factors [14]. The concept of neural induction as a default pathway for differentiation in early vertebrate embryos [2] led to the development of a simple adherent monolayer culture system where ES cells are driven by autocrine signalling into a neural fate, using a defined serum-free media (N2B27). In these conditions, endogenous FGF and Notch signalling seem necessary for ES cells to enter the neural pathway [17], [18], [38], together with the production of endogenous inhibitors of BMP signalling [16]. Using this monolayer protocol, one can routinely obtain an enriched population of NPs (up to 80%) after 4–6 days in culture, although the presence of "contaminating" cells (undifferentiated ES cells and large flatten non-neural differentiating cells) is still observed, probably due to some remaining endogenous BMP signalling. In this work, we report that the use of a new N2B27-derived medium (RHB-A) allows a faster and more efficient production of NPs from ES cells in monolayer culture (Fig. 1A), resulting in highly homogeneous populations of NPs (up to 90%) that can subsequently differentiate into neurons, astrocytes and oligodendrocytes. The observation in the reduction of large flatten non-neural cells in RHB-A cultures correlates with the observed decrease of BMP4 expression, along with an increase in the expression of known BMP-antagonists (e.g. Chordin-like1, Follistatin), in contrast to what has been reported for N2B27 cultures [16].

The transition from ES cells to NPs in RHB-A monolayer cultures is accompanied by the organization of these NPs into characteristic rosette-like structures, in a process that resembles neural tube formation in the embryo. The formation of similar rosettes has been described in other in vitro models of neural differentiation from ES cells [39], [40], suggesting that this is a common behaviour of NPs, associated with their epithelial character. We have studied in detail the organization of these rosettes and show that several proteins normally present in the apical domain of embryonic neuroepithelial cells, like N-Cadherin, ß-catenin, Par3/aPKC and Numb, are localized close to the luminal centre of rosettes, revealing that ES cell-derived NPs are able to acquire a proper apico-basal organization, despite being cultured in a 2D environment. In addition, we show that NPs within rosettes display the characteristic INM observed in the embryonic neuroepithelium, with progenitors entering mitosis when their nuclei are closer to the luminal surface of rosettes. We also noted that differentiating neurons loose contact with the centre of rosettes and migrate to their periphery. Altogether, these findings reveal that NPs in culture are able to self-organize into neural tube-like structures, thus recapitulating the cellular interactions that regulate the process of neuronal production.

Notch activity is a major player in this process (reviewed in [11], [21], [41]) and the parallel between embryonic neurogenesis and in vitro neural differentiation of ES cells is reinforced by the similar dependence on Notch signalling to maintain a population of NPs engaged on neuronal production. Indeed, we show that Notch signalling is active in ES cell derived neuroepithelial rosettes and that chemical inhibition of Notch activity results in massive neuronal differentiation of rosette NPs, similarly to what has been described during CNS embryonic development (reviewed in [11], [21]). Interestingly, this drift to neuronal differentiation in the absence of Notch activity is no longer seen on day 16 cultures, suggesting that neuronal competence decreases with time. This is confirmed by the striking reduction in the number of neurons generated in later cultures, concomitant with an increase in the generation of GFAP+ glial cells, revealing a switch from neurogenic to gliogenic NPs in late monolayer cultures. This switch coincides with the disappearance of rosettes from the cultures: while ES cell-derived NPs grow exclusively in the form of rosettes up to day 12, few rosettes are still present at day 16 as neuronal production is significantly decreased and gliogenesis increases. These observations indicate that the intrinsic temporal regulation of neurogenic vs. gliogenic differentiation, characteristic of embryonic neural tube, is conserved in ES cell-derived rosette cultures. How this temporal regulation occurs in vitro is still unclear but must be independent of extrinsic cues, as previously reported for embryonic NPs [26], [27].

Altogether, our data extend the previous characterization of the monolayer protocol as an efficient and reproducible method to drive ES cells into a neural fate and provide further evidence that the steps involved in the in vitro acquisition of a neural fate closely mimic the events that happen during embryonic neural commitment and differentiation.

Molecular Mechanisms of In Vitro Mammalian Neural Development

The path from ES cells to a neural fate involves various transitions in the potential of the cells, starting with the conversion to a PE-like stage followed by the transition into neuroectoderm and establishment of a population of NPs that will gradually give rise, first, to differentiated neurons and, later, to glial cells. To characterize these cellular states at the molecular level and identify genes that might promote the transitions between successive stages, we have performed global transcriptome analysis of ES cells and their derivatives along the path to a neural fate. This resulted in the identification of a large set of genes (6.563) whose expression significantly changes throughout the monolayer neural differentiation protocol. Analysis of the data involved the clustering of these genes into five groups according to their expression profiles, which we correlated with diverse cell populations that emerge in the course to neural differentiation.

The first group comprises genes with a peak of expression at day 0 and that are rapidly down-regulated as ES cells loose their "stemness" character. This group includes known pluripotency markers of the ES cell state, like nanog, zfp42/rex1, fbxo15, tdgf1/cripto1, socs3, esrrb, klf4 and klf5, and provides an ES cell signature that overlaps extensively with available data on ES cell specific transcripts [42], [43], [44]. Other known "stemness" genes like Pou5f1/Oct4 and sox2 are also strongly expressed in the starting population of ES cells but were excluded from our first gene group as their expression reappears in NPs (sox2) or takes longer to be down-regulated (Pou5f1/Oct4). Together, these data confirm the ES cell identity of the starting population of cells and could also serve to identify novel genes that might be important to maintain the ES cells status.

A second group includes the genes whose expression peaks at day 1 after ES cells have been plated in RHB-A, being subsequently down-regulated in NPs and neural rosettes. This group includes the PE marker Fgf5 and might represent a gene signature for the PE-like stage that emerges after plating of ES cells in the absence of serum and leukaemia inhibitory factor (LIF). Until now, this stage has been characterized by the up-regulation of Fgf5 expression and downregulation of zfp42/rex1, in a population still expressing Oct4 [19], [45], [46], a pattern that is also observed in our data. Although epiblast stem cells (EpiSCs) reveal a similar expression profile [47], [59], the absence of markers of tripotency in our day 1 PE-like cells (e.g. brachyury, otx2, gata4 and gata6), together with the down-regulation of most ES cell genes (included in group I), implies that this population is different from EpiSCs.

A survey of the 66 genes included in this PE-like group reveals the presence of 5 genes involved in calcium homeostasis (calcR, ryr-3, otopetrin1, tnnc2 and cyp24a1) suggesting that calcium signalling plays an important role in the transition of pluripotent stem cells into ectodermal fates. Indeed, an increase in intracellular calcium has been reported to be important for neuralization of ectodermal cells [48]; hence, the activity of the 5 identified genes might contribute to regulate calcium signalling in PE-like cells transiting to a neural fate. It would, therefore, be interesting to test whether these genes, as well as other candidate PE markers present in this group, are expressed in the mouse embryonic PE and what function they have in this tissue.

A third group comprises genes that are up-regulated in NPs (Sox1-GFP+) at day 3 but down-regulated in day 8 neurogenic rosettes. This behaviour indicates that these genes might play a role in the establishment of the initial population of NPs, but are switched-off afterwards to allow the subsequent progression to differentiation. In the embryo, a gene that shows a similar behaviour is sox1, whose down-regulation in NPs seems to be required for their commitment to differentiation, due to its ability to block the neurogenesis-promoting activity of proneural factors [5], [49]. During ES cell differentiation, sox1 expression also peaks in day 3 NPs but does not decrease enough in day 8 rosettes to be included in this group, due to the stringent criteria that was chosen. Still, this group contains various genes that are known to be transiently expressed in embryonic NPs and regulate their generation, like the BMP inhibitor chdl1, the Wnt modulator frzb1 and the orphan nuclear receptor nr6a1 [50], [51], [52], supporting the analogies with embryonic neural development. Of the 61 genes included in this group, 43 are known to be expressed in embryonic NPs (by screening publicly available databases), while there is incomplete or no available data on the expression of the remaining 18 genes. We therefore propose that this group provides a novel gene expression signature for a transient population of NPs (tNPs) that is established following neural induction but that it is not yet competent to enter neurogenesis. This absence of neurogenic competence correlates with the reduced expression of proneural genes in day 3 Sox1-GFP+ NPs and the lack of an increase in Notch activity, as measured by the expression of Notch1 and its targets and effectors hes5 and hes6.

This proposed tNP population is also likely to exist in the mouse embryo but the small number and transient character of tNPs, together with the "dilution" effect due to the presence of several other cell types, has made difficult to pinpoint its existence. The genes we have identified here as markers of the tNP population may now allow the identification of similar progenitors in the mouse embryo and provide an entry point to dissect the genetic circuitry controlling this stage of neural development.

The fourth group comprises genes that are up-regulated in NPs but that, in contrast to tNP genes, continue to be expressed at similar or increased levels in day 8 neurogenic rosettes. Our strategy of separating NP genes into two groups with distinct expression profiles highlights, on one side, genes which are active only during progenitor specification (tNP group) and, on the other side, genes that might also be important for the next stage of NP development (nNP), when competence to enter neurogenesis is acquired. By analogy with embryonic neural development, nNPs are likely to be an in vitro counterpart of the progenitors present in the "transition zone" or "pre-neural tube", located at the caudal open neural plate, rostral to the node but posterior to the level of the first somite [6], [53]. Indeed, a survey of nNP genes reveals that the transition to a proliferative neurogenic population observed in monolayer cultures is accompanied by a significant increase on the expression of genes connected to the retinoic acid signalling (e.g., rxr-alpha, crabp2, nr2f1, nr2f2) and Wnt pathway (e.g., fzd1, fzd3, sfrp2, tcf4, wnt5a, wnt8b, gsk3ß, lrp1), which are known to regulate NP competence in vivo [54], [55]. Together, our data provide an accurate gene signature for two populations of NPs (tNPs and nNPs), with a high degree of confidence that results from the fact that FACS-purified populations of NPs were used in our experiments.

A population of purified Sox1-GFP+ NPs has previously been studied in Sox1GFP transgenic mouse embryos with 15 genes being found to be preferentially expressed in embryonic NPs [56]. Of these, 8 genes are also found in our nNP gene group (sfrp2, lrrn1, sox4, zic1, vim, rtn1, sox11, qk), while 4 other genes (khdrbs3, msi2, hrmtl3, tuba1) show similar expression profile (up-regulated at day 3 NPs and/or day 8 rosettes) but were excluded due to the stringent criteria used to generate the clusters. Concerning the other 3 genes, one is mainly expressed in day 8 rosette NPs (nhlh2), another was not included in the microarrays (Mm.156164) and slc2a1 is not differentially expressed during ES cell differentiation. The fact that none of the tNP genes were found in the embryonic Sox1-GFP+ population might be due to the limited number of genes screened in the embryo (384 in total) and to the expected transient character of tNPs in vivo, which might preclude their isolation from whole mouse embryos at E10.5. Nonetheless, this comparison reveals a strong correlation between the data generated from in vitro neural differentiation of ES cells and the in vivo data obtained from the developing mouse embryo, supporting our proposal that the gene signatures defining NP developmental stages in vitro might serve to identify similar stages during embryonic nervous system development.

The fifth group comprises genes that are up-regulated in day 8 cultures, when NPs are organized in neural tube-like rosettes and actively engaged in neurogenesis. Genes that were already up-regulated in day 3 NPs and that are linked to the previous stages of NP specification and proliferation (included in groups III and IV), were excluded from group V. In this way, this group is enriched in genes linked to the final stages of NP development and commitment to neuronal differentiation, revealing a gene expression profile in day 8 neural rosettes that matches the transcriptional landscape of the embryonic neural tube. For instance, proneural genes like neurog1, neurog2 and ascl1, which are known to promote neuronal commitment, cell cycle exit and entry into differentiation of embryonic NPs, are included in group V. Additionally, genes encoding neuronal determination bHLH proteins, like neurod4, nhlh1 and nhlh2, which are known to be activated by the proneural genes and function in early post-mitotic neurons to implement the neuronal differentiation program, are also present in this group. The similarities between embryonic neural tube and monolayer neural rosettes extend also to the increased transcription of genes of the Notch pathway, which are involved in regulating the balance between NP maintenance and differentiation, both in neural rosettes and in the embryonic neural tube. Other genes up-regulated in day 8 neural rosettes are known to be linked to neuronal type specification, like lhx1, lhx9, islet1, lmo2 and various members of the Brn/Pou family, or associated with the general process of neuronal differentiation, like dcx, elav1, 2, 3 and 4, and neurexin. Altogether, this expression profile provides additional evidence, at the molecular level, of the similarities between the embryonic neural tube and the neural rosettes obtained by monolayer differentiation of ES cells.

A recent study reported the characterization of neural rosettes obtained by differentiating human ES cells through EBs or by co-culture with stromal cells [39]. Exposure of these rosettes to FGF2/EGF signalling resulted in the establishment of NS-like cells similar to those we obtained from mouse neural rosettes with the same growth factors. Gene expression profiling of these human ES cell-derived neural rosettes revealed a group of genes with highly increased expression in rosettes vs. human ES cells. Most of these are also highly expressed in the neural rosettes obtained from mouse ES cells described in this work (for instance, plagl1, dach1, plzf/zbtb16, nr2f1, zic1, fabp7, lhx2, pou3f3), suggesting a conserved general programme of NP/NSC development in mice and humans. Although these genes are highly expressed in rosette cells, our analysis reveals however that they are already up-regulated at day 3 of monolayer culture, before rosette formation, pointing to the existence of evolving populations of NPs/NSCs that emerge at different times of neural development.

To better define these NP populations, we took advantage of the simplicity of the monolayer method and the ability to purify Sox1-GFP+ NPs before rosette formation, to produce an accurate gene profiling dataset at various stages of in vitro neural development. By using embryo-oriented criteria to cluster the differentially expressed genes, our analysis did indeed allow us to pinpoint successive stages in the development of NPs, identified by unique gene signatures. A first signature defines a transient "tNP" population that emerges after neural induction and that gives rise to a subsequent population of "nNPs" with a different gene expression profile and already competent to enter neurogenesis. This is a "transit-amplifying" population of NPs that give rise to a final set of NPs organized in rosettes, expressing proneural genes and committed to exit the cell cycle and enter terminal differentiation. We propose that these stages also exist during embryonic development and future work shall explore whether the gene signatures here defined can serve to identify equivalent NP populations in the mouse embryo.

Materials and Methods

Maintenance and Differentiation of Mouse Es Cells

The ES cell lines used for this study were E14tg2a and two derivatives, 46C (Sox1-GFP, [16]) and S25 (Sox2-βgeo, [15]), all three a gift from Meng Li (MRC Clinical Sciences Centre, Faculty of Medicine, Imperial College, London, UK) and Austin Smith (Wellcome Trust Centre for Stem Cell Research, University of Cambridge, Cambridge UK). ES cells were grown at 37°C in a 5% (v/v) CO_2 incubator in Glasgow Modified Eagles Medium (GMEM, Invitrogen), supplemented with 10% (v/v) fetal bovine serum (FBS) (ES-qualified, Invitrogen), 2 ng/ml LIF and 1 mM 2-mercaptoethanol, on gelatin-coated (0.1% (v/v)) Nunc dishes. Cells were passaged every other day, at constant plating density of 3×10^4 cells/cm2. To start the monolayer protocol, ES cells were plated in serum-free medium ESGRO Complete Clonal Grade medium (Millipore Inc.) at high density (1.5$\times$105 cells/cm2). After 24 hours, ES cells were gently dissociated and plated onto 0.1% (v/v) gelatin-coated tissue culture plastic at 1×104 cells/cm2 in RHB-A or N2B27 media (StemCell Science Inc.), changing media every other day. For replating on day 4, cells were dissociated and plated at 2×104 cells/cm2 onto laminin-coated tissue culture plastic in RHB-A medium supplemented with 5 ng/ml murine bFGF (Peprotech). From this point on, cells were replated in the same conditions every 4th day and the medium was changed every 2nd day, for the total of 20 days in culture. To quantify the number of differentiating neurons at each time point, cells were plated onto laminin-coated glass coverslips in 24-well Nunc plates and, 2 days after plating, medium was changed to a RHB-A:Neurobasal:B27 mixture (1:1:0.02), to allow a better survival of differentiated neurons. To obtain floating aggregates of NS cells, 3×105 cells, dissociated at day 4, 8, 12, 16 and 20 of culture, were plated onto uncoated culture plastic in RHB-A medium supplemented with 10 ng/ml of recombinant murine EGF and bFGF (Peprotech) [57]. Floating aggregates formed within 24 hours and medium was changed after 48 h. After 4 days in suspension culture, aggregates were counted and plated en bloc onto laminin-coated coverslips, being then cultured for 4 days in RHB-A medium (with an intermediate medium change) to allow differentiation. When required, 10 μM BrdU (Sigma) was added to cultures for 5 min immediately before fixation.

Treatment with Γ-Secretase Inhibitor Ly411575

Treatment with LY411575 was done at day 6, 10 or 14 after the beginning of the protocol. At these time points, culture medium was substituted by RHB-A: Neurobasal: B27 (1:1:0.02) medium supplemented either with 0.01% DMSO (control) or with 3 nM LY411575 (in 0.01% DMSO). Cells were fixed in 4% (w/v) paraformaldehyde after 24 h or 48 h of incubation, respectively, for the ISH and for the immunostaining.

Immunocytochemistry

Fixed cells were blocked with 10% (v/v) FBS and 0.05% (v/v) Tween in phosphate buffered saline (PBS) for 1 hour, followed by incubation overnight with primary antibodies. For all double immunostainings (with the exception of those with anti-GFP antibody), monolayer cultures of either S25 or E14tg2a ES cells were used. 46C cells were used in double immunostainings with anti-GFP antibody. Cells were washed 3 times in PBS followed by incubation for 1–2 hours with AlexaFluor-conjugated secondary antibodies (Molecular Probes) and DAPI (1:10000, Sigma). For the detection of BrdU incorporation, cells were treated with 2N HCl for 30 min at 37°C at the beginning of the immunostaining procedure.

Images of fixed cells were obtained with a DM5000B microscope and a DC350F camera (Leica Wetzlar, Germany). Living cells were photographed under an inverted microscope Leica DMIL with a DC200 camera. Images were processed by using Photoshop CS (Adobe, San Jose, CA).

The number of HuC/D and GFAP expressing cells was quantified as a proportion of the total number of cells in culture, counted with the help of ImageJ Cell Counter software. The number of positively labelled cells was quantified by counting 10 to 20 randomly selected fields per coverslip, corresponding to a minimum 5000 cells, counted as DAPI nuclei. Two coverslips were counted per each condition and the analysis was repeated for at least three independent experiments for each of S25 and 46C ES cell lines. Student t-test was used to compare means between groups and p-values lower than 0.05 were considered statistically significant.

In Situ Hybridization

Digoxygenin-labeled RNA probes for hes5, hes6 and Dll1 were synthesized by T7 RNA polymerase from plasmid templates. Whole-mount ISH procedure [24] was adapted to cultured cells with minor modifications. After incubation with AP-conjugated anti-Dig antibody (Roche Diagnostics) coverslips containing cultured cells were incubated with AP substrate FastRed (Roche Diagnostics) for 0.5–1 h at 37°C. Anti-GFP immunostaining was performed after ISH when required.

FACS Analysis

Cells were dissociated and resuspended in 4% (v/v) FBS in PBS. Sox1-GFP analysis was performed on a FACS Calibur cytometer (Becton Dickinson), and all cell sorting experiments were done on a FACS Aria cell sorter (Becton Dickinson). Live cells were gated based on forward scatter and side scatter and/or by propidium iodide dye exclusion. For sorting, the GFP+ and GFP++ NPs populations were collected (the GFP negative cell fraction was discarded) and cell viability at the end of the FACS sorting procedure was determined using trypan blue dye exclusion method. FACS sorted cells were directly processed for RNA extraction.

RNA Extraction and RT-PCR

Total RNA was extracted from 106 cells using High Pure RNA Isolation kit (Roche Diagnostics), with the inclusion of DNAseI treatment according to manufacturer's instructions. The first strand cDNA was synthesized from 0.5 μg of total RNA using SuperscriptII Reverse Transcriptase (Invitrogen) and random hexamers. After synthesis, each cDNA was diluted 5-fold and 5 µl of diluted cDNA used in PCR reaction with gene-specific primers. The absence of contaminating genomic DNA was confirmed for each RNA extraction by PCR amplification of GAPDH-specific product from RT negative samples. The relative amount of each transcript was normalized to the level of GAPDH.

Time-Lapse Movie

Day 4 or day 8 rosette NPs were plated onto laminin-coated MatTek dishes and rosettes were allowed to form for 24–48 h in a conventional CO_2 incubator. Cultures were imaged on an inverted fluorescence Zeiss Axiovert 200M microscope in a chamber kept at 38°C. The chamber stage was buffered with 5% CO2/95% air mix and maintained in a humid environment. Images in bright field were captured using a 40×0.75 NA objective lens (Zeiss EC Plan-Neofluar) with the Hg-arc lamp and acquired with Metamorph software (Molecular Devices). The culture was permanently illuminated and seven focal points were imaged at 2 min intervals, for up to 16 hours. Data was analysed using ImageJ software, by choosing the most focused plane, adjusting brightness and contrast, and after instant time concatenation.

Microarray Sample Preparation and Data Analysis

Total RNA was extracted from day 0 undifferentiated cells, day 1 ectodermal cells, day 3 FACS-purified Sox1-GFP+ and Sox1-GFP++ NPs, and day 8 neuroepithelial rosettes, using High Pure RNA Isolation kit (Roche Diagnostics). The preparation quality was assessed by agarose-formaldehyde gel electrophoresis. Three (or four in the case of 3++ samples) independent preparations (A to D), each containing total RNA from the day 0 (0), day 1 (1) day 3 (3+ and 3++) and day 8 (8) of differentiation were processed at the Max-Delbrück-Centrum für Molekulare Medizin (Berlin, Germany) according to the standardized procedures adopted by all members of the FunGenES European Consortium (http://www.fungenes.org/).

For the synthesis of double-stranded cDNA (from 15 µg of total RNA) the cDNA synthesis system kit (Roche Diagnostics) was used. Biotinylated cRNA were synthesized with Perkin-Elmer nucleotide analogues using the Ambion MEGAScript T7 kit. After fragmenting of the cRNA for target preparation using the standard Affymetrix protocol, 15 µg fragmented cRNA were hybridized for 16 h at 45°C to Mouse Genome 430 Version 2.0 Array (Affymetrix) which includes 45101 probe sets. Following hybridization, arrays were washed and stained with streptavidin-phycoerythrin in the Affymetrix Fluidics Station 450 and further scanned using the Affymetrix GeneChip Scanner 3000 7G. The image data were analyzed with GCOS 1.4 using Affymetrix default analysis settings and global scaling as normalization method. All chips passed quality criteria. Microarray data reported in the manuscript is described in accordance with MIAME guidelines and original datasets have been deposited in the ArrayExpress database for open access (Accession Number E-TABM-717).

After RMA normalization [31], a parametric ANOVA (F-test) and ten pair-wise comparisons using the Student t-test (unpaired, assuming unequal variances) were performed for each time point independently. The false discovery rate of each test-set was calculated using the Benjamini Hochberg procedure [58]. Finally, an ANOVA FDR-value$<10-3$ was used to identify and restrict the number of differentially expressed probe sets (n = 9456). This corresponds to a total of 6563 genes.

To cluster these genes in groups with similar expression profiles along the selected four time points of the monolayer protocol, a cut-off value of 2 for the fold differences in expression levels between time points was imposed. Five groups were defined according to the following criteria: I. "ES cells group"–Expression level on day 0 is at least twice higher than expression in all other time points (days 1, 3 and 8); II. "PE group"–Expression level on day 1 is at least twice higher than expression in all other time points (days 0, 3 and 8); III. "tNPs group"–Expression level on day 3 is at least twice higher than expression in all other time points (days 0, 1 and 8); IV. "nNPs group"–Expression level on day 3 (3+ and/or 3++) is at least twice higher than expression in earlier time points (days 0 and 1), with expression level at day 8 being equal or higher than at day 3 (day (3++)); V. "Rosette group" - Expression level on day 8 is at least twice higher than expression in all other time points (days 0, 1 and 3); in addition, expression levels at day 0, 1 and 3 cannot increase more than twice between them.

Acknowledgements

We would like to thank Austin Smith and Meng Li (Cambridge, UK) for ES cell lines, Jaak Vilo and Raivo Kolde (Egeen Inc., Tartu, Estonia) for bioinformatics support, Ana Luisa Caetano (IMM, Lisbon, Portugal) for FACS sorting, and Tim Alsopp and Lilian Hook (StemCell Sciences Inc) for the opportunity of beta-testing the StemCell Sciences media. We also thank Kate Storey, Catarina Ramos and Filipe Vilas-Boas for critical reading of the manuscript.

Authors' Contributions

Conceived and designed the experiments: EA DH EB. Performed the experiments: EA MS EB. Analyzed the data: EA MS HS DH EB. Contributed reagents/materials/analysis tools: LP HS OH. Wrote the paper: EA DH EB.

References

1. Spemann H, Mangold H (1924) Über die Induktion von Embryonalanlagen durch Implantation artfremder Organisatoren. W Roux' Arch f Entw d Organis u mikrosk Anat 100: 599–638.

2. Muñoz- Sanjuán I, Brivanlou A (2002) NEURAL INDUCTION, THE DEFAULT MODEL AND EMBRYONIC STEM CELLS. Nat Rev Neurosci 3: 271–280.

3. Wilson SI, Edlund T (2001) Neural induction: toward a unifying mechanism. Nat Neurosci 4: Suppl1161–1168.

4. Wood HB, Episkopou V (1999) Comparative expression of the mouse Sox1, Sox2 and Sox3 genes from pre-gastrulation to early somite stages. Mechanisms of Development 86: 197–201.

5. Pevny LH, Sockanathan S, Placzek M, Lovell-Badge R (1998) A role for SOX1 in neural determination. Development 125: 1967–1978.

6. Diez del Corral R, Storey K (2004) Opposing FGF and retinoid pathways: a signalling switch that controls differentiation and patterning onset in the extending vertebrate body axis. Bioessays 26: 857–869.

7. Afonso C, Henrique D (2006) PAR3 acts as a molecular organizer to define the apical domain of chick neuroepithelial cells. Journal of Cell Science 119: 4293–4304.

8. Chenn A, Zhang YA, Chang BT, McConnell SK (1998) Intrinsic polarity of mammalian neuroepithelial cells. Mol Cell Neurosci 11: 183–193.

9. Sauer FC (1935) Mitosis in the neural tube. J Comp Neurol 62: 377–405.

10. Lewis J (1998) Notch signalling and the control of cell fate choices in vertebrates. Semin Cell Dev Biol 9: 583–589.

11. Louvi A, Artavanis-Tsakonas S (2006) Notch signalling in vertebrate neural development. Nat Rev Neurosci 7: 93–102.

12. Bain G, Kitchens D, Yao M, Huettner JE, Gottlieb DI (1995) Embryonic stem cells express neuronal properties in vitro. Developmental Biology 168: 342–357.

13. Perrier A, Tabar V, Barberi T, Rubio M, Bruses J, et al. (2004) Derivation of midbrain dopamine neurons from human embryonic stem cells. Proc Natl Acad Sci USA 101: 12543–12548.

14. Kawasaki H, Mizuseki K, Nishikawa S, Kaneko S, Kuwana Y, et al. (2000) Induction of midbrain dopaminergic neurons from ES cells by stromal cell-derived inducing activity. Neuron 28: 31–40.

15. Li M, Pevny L, Lovell-Badge R, Smith A (1998) Generation of purified neural precursors from embryonic stem cells by lineage selection. Curr Biol 8: 971–974.

16. Ying Q, Stavridis M, Griffiths D, Li M, Smith A (2003) Conversion of embryonic stem cells into neuroectodermal precursors in adherent monoculture. Nat Biotechnol 21: 183–186.

17. Stavridis MP, Lunn JS, Collins BJ, Storey K (2007) A discrete period of FGF-induced Erk1/2 signalling is required for vertebrate neural specification. Development 134: 2889–2894.

18. Kunath T, Saba-El-Leil MK, Almousailleakh M, Wray J, Meloche S, et al. (2007) FGF stimulation of the Erk1/2 signalling cascade triggers transition of pluripotent embryonic stem cells from self-renewal to lineage commitment. Development 134: 2895–2902.

19. Pelton TA, Sharma S, Schulz TC, Rathjen J, Rathjen PD (2002) Transient pluripotent cell populations during primitive ectoderm formation: correlation of in vivo and in vitro pluripotent cell development. J Cell Sci 115: 329–339.

20. Rathjen J, Lake JA, Bettess MD, Washington JM, Chapman G, et al. (1999) Formation of a primitive ectoderm like cell population, EPL cells, from ES cells in response to biologically derived factors. J Cell Sci 112 (Pt 5): 601–612.

21. Yoon K, Gaiano N (2005) Notch signaling in the mammalian central nervous system: insights from mouse mutants. Nat Neurosci 8: 709–715.

22. Ohtsuka T, Ishibashi M, Gradwohl G, Nakanishi S, Guillemot F, et al. (1999) Hes1 and Hes5 as notch effectors in mammalian neuronal differentiation. EMBO J 18: 2196–2207.

23. Bae YK, Shimizu T, Hibi M (2005) Patterning of proneuronal and inter-proneuronal domains by hairy- and enhancer of split-related genes in zebrafish neuroectoderm. Development 132: 1375–1385.

24. Henrique D, Adam J, Myat A, Chitnis A, Lewis J, et al. (1995) Expression of a Delta homologue in prospective neurons in the chick. Nature 375: 787–790.

25. Lanz T (2004) Studies of A Pharmacodynamics in the Brain, Cerebrospinal Fluid, and Plasma in Young (Plaque-Free) Tg2576 Mice Using the -Secretase

Inhibitor N2-[(2S)-2-(3,5-Difluorophenyl)-2-hydrox yethanoyl]-N1-[(7S)-5-methyl-6-oxo-6,7-d ihydro-5H-dibenzo[b,d]azepin-7-yl]-L-alaninamide (LY-411575). Journal of Pharmacology and Experimental Therapeutics 309: 49–55.

26. Shen Q, Wang Y, Dimos JT, Fasano CA, Phoenix T, et al. (2006) The timing of cortical neurogenesis is encoded within lineages of individual progenitor cells. Nat Neurosci 9: 743–751.

27. Qian X, Shen Q, Goderie SK, He W, Capela A, et al. (2000) Timing of CNS cell generation: a programmed sequence of neuron and glial cell production from isolated murine cortical stem cells. Neuron 28: 69–80.

28. Alvarez-Buylla A, Garcia-Verdugo JM (2002) Neurogenesis in adult subventricular zone. J Neurosci 22: 629–634.

29. Conti L, Cattaneo E (2005) Controlling neural stem cell division within the adult subventricular zone: an APPealing job. Trends in Neurosciences 28: 57–59.

30. Pollard S, Conti L, Smith A (2006) Exploitation of adherent neural stem cells in basic and applied neurobiology. Regenerative medicine 1: 111–118.

31. Irizarry RA, Bolstad BM, Collin F, Cope LM, Hobbs B, et al. (2003) Summaries of Affymetrix GeneChip probe level data. Nucleic Acids Res 31: e15–.

32. Guillemot F (2005) Cellular and molecular control of neurogenesis in the mammalian telencephalon. Current Opinion in Cell Biology 17: 639–647.

33. Diez del Corral R, Storey KG (2001) Markers in vertebrate neurogenesis. Nat Rev Neurosci 2: 835–839.

34. Magavi SS, Macklis JD (2008) Immunocytochemical Analysis of Neuronal Differentiation. Neural Stem Cells. pp. 345–352.

35. Ying QL, Smith AG (2003) Defined conditions for neural commitment and differentiation. Methods in Enzymology 365: 327–341.

36. Cai C, Grabel L (2007) Directing the differentiation of embryonic stem cells to neural stem cells. Dev Dyn 236: 3255–3266.

37. Stavridis MP, Smith AG (2003) Neural differentiation of mouse embryonic stem cells. Biochem Soc Trans 31: 45–49.

38. Lowell S, Benchoua A, Heavey B, Smith A (2006) Notch promotes neural lineage entry by pluripotent embryonic stem cells. Plos Biol 4: e121.

39. Elkabetz Y, Panagiotakos G, Al Shamy G, Socci ND, Tabar V, et al. (2008) Human ES cell-derived neural rosettes reveal a functionally distinct early neural stem cell stage. Genes Dev 22: 152–165.

40. Eiraku M, Watanabe K, Matsuo-Takasaki M, Kawada M, Yonemura S, et al. (2008) Self-organized formation of polarized cortical tissues from ESCs and its active manipulation by extrinsic signals. Cell stem cell 3: 519–532.

41. Lewis J (1998) Notch signalling and the control of cell fate choices in vertebrates. Seminars in Cell & Developmental Biology 9: 583–589.

42. Edwards YJ, Bryson K, Jones D (2008) A meta-analysis of microarray gene expression in mouse stem cells: redefining stemness. PLoS ONE 3: e2712.

43. Ivanova NB, Dimos JT, Schaniel C, Hackney JA, Moore KA, et al. (2002) A stem cell molecular signature. Science 298: 601–604.

44. Ramalho-Santos M, Yoon S, Matsuzaki Y, Mulligan RC, Melton DA (2002) "Stemness": transcriptional profiling of embryonic and adult stem cells. Science 298: 597–600.

45. Toyooka Y, Shimosato D, Murakami K, Takahashi K, Niwa H (2008) Identification and characterization of subpopulations in undifferentiated ES cell culture. Development 135: 909–918.

46. Aiba K, Nedorezov T, Piao Y, Nishiyama A, Matoba R, et al. (2009) Defining developmental potency and cell lineage trajectories by expression profiling of differentiating mouse embryonic stem cells. DNA Res 16: 73–80.

47. Tesar P, Chenoweth J, Brook F, Davies T, Evans E, et al. (2007) New cell lines from mouse epiblast share defining features with human embryonic stem cells. Nature 448: 196–199.

48. Webb S, Moreau M, Leclerc C, Miller A (2005) Calcium transients and neural induction in vertebrates. Cell Calcium 37: 375–385.

49. Bylund M, Andersson E, Novitch BG, Muhr J (2003) Vertebrate neurogenesis is counteracted by Sox1-3 activity. Nat Neurosci 6: 1162–1168.

50. Coffinier C, Tran U, Larraín J, De Robertis EM (2001) Neuralin-1 is a novel Chordin-related molecule expressed in the mouse neural plate. Mechanisms of Development 100: 119–122.

51. Hoang BH, Thomas JT, Abdul-Karim FW, Correia KM, Conlon RA, et al. (1998) Expression pattern of two Frizzled-related genes, Frzb-1 and Sfrp-1, during mouse embryogenesis suggests a role for modulating action of Wnt family members. Dev Dyn 212: 364–372.

52. Chung AC, Katz D, Pereira FA, Jackson KJ, DeMayo FJ, et al. (2001) Loss of orphan receptor germ cell nuclear factor function results in ectopic development of the tail bud and a novel posterior truncation. Mol Cell Biol 21: 663–677.

53. Diez del Corral R, Breitkreuz DN, Storey K (2002) Onset of neuronal differentiation is regulated by paraxial mesoderm and requires attenuation of FGF signalling. Development 129: 1681–1691.

54. Diez del Corral R, Olivera-Martinez I, Goriely A, Gale E, Maden M, et al. (2003) Opposing FGF and retinoid pathways control ventral neural pattern, neuronal differentiation, and segmentation during body axis extension. Neuron 40: 65–79.

55. Olivera-Martinez I, Storey K (2007) Wnt signals provide a timing mechanism for the FGF-retinoid differentiation switch during vertebrate body axis extension. Development 134: 2125–2135.

56. Aubert J, Stavridis MP, Tweedie S, O'Reilly M, Vierlinger K, et al. (2003) Screening for mammalian neural genes via fluorescence-activated cell sorter purification of neural precursors from Sox1-gfp knock-in mice. Proc Natl Acad Sci USA 100: Suppl 111836–11841.

57. Conti L, Pollard SM, Gorba T, Reitano E, Toselli M, et al. (2005) Niche-Independent Symmetrical Self-Renewal of a Mammalian Tissue Stem Cell. PLoS Biology 3: 1594–1606.

58. Benjamini Y, Hochberg Y (1995) Controlling the false discovery rate: a practical and powerful approach to multiple testing. J Roy Statist Soc Ser 57: 289–300.

59. Brons IG, Smithers L, Trotter MW, Rugg-Gunn P, Sun B, et al. (2007) Derivation of pluripotent epiblast stem cells from mammalian embryos. Nature 448: 191–195.

Defining Human Embryo Phenotypes by Cohort-Specific Prognostic Factors

Sunny H. Jun, Bokyung Choi, Lora Shahine, Lynn M. Westphal,
Barry Behr, Renee A. Reijo Pera, Wing H. Wong and
Mylene W. M. Yao

ABSTRACT

Background

Hundreds of thousands of human embryos are cultured yearly at in vitro fertilization (IVF) centers worldwide, yet the vast majority fail to develop in culture or following transfer to the uterus. However, human embryo phenotypes have not been formally defined, and current criteria for embryo transfer largely focus on characteristics of individual embryos. We hypothesized that embryo cohort-specific variables describing sibling embryos as a group may predict developmental competence as measured by IVF cycle outcomes and serve to define human embryo phenotypes.

Methodology/Principal Findings

We retrieved data for all 1117 IVF cycles performed in 2005 at Stanford University Medical Center, and further analyzed clinical data from the 665 fresh IVF, non-donor cycles and their associated 4144 embryos. Thirty variables representing patient characteristics, clinical diagnoses, treatment protocol, and embryo parameters were analyzed in an unbiased manner by regression tree models, based on dichotomous pregnancy outcomes defined by positive serum ß-human chorionic gonadotropin (ß-hCG). IVF cycle outcomes were most accurately predicted at ~70% by four non-redundant, embryo cohort-specific variables that, remarkably, were more informative than any measures of individual, transferred embryos: Total number of embryos, number of 8-cell embryos, rate (percentage) of cleavage arrest in the cohort and day 3 follicle stimulating hormone (FSH) level. While three of these variables captured the effects of other significant variables, only the rate of cleavage arrest was independent of any known variables.

Conclusions/Significance

Our findings support defining human embryo phenotypes by non-redundant, prognostic variables that are specific to sibling embryos in a cohort.

Introduction

Developmental arrest of human embryos cultured in vitro is common and presents a major obstacle to achieving pregnancy through IVF, as well as a major obstacle to research in human embryonic stem cell (hESC) biology [1], [2], [3], [4], [5], [6], [7]. While the culture of embryos to blastocyst stage for subsequent transfer yields high pregnancy rates and minimizes the risk of multiple gestation, the availability of blastocysts is limited even in the best IVF clinics because of the high rates of attrition in in vitro embryo culture [1], [3], [4], [8], [9], [10], [11].

Although developmental defects such as cleavage arrest, polyploidy, and fragmentation are commonly encountered and have been used for scoring individual embryos in IVF [11], [12], [13], [14], [15], [16], [17], [18], [19], [20], [21], the lack of well-defined human embryo phenotypes has hindered translational research and mechanistic investigations. One key challenge to defining human embryo phenotypes relates to the unclear and often highly interactive relationships amongst variables pertaining to patient characteristics, clinical infertility diagnoses, IVF treatment protocols, and observed embryo characteristics. Further, since any single couple may typically produce a few oocytes or embryos that are abnormal merely by chance, it is difficult to determine whether sibling embryos as a group, or an embryo cohort, is "normal". (Note that "embryo cohort" refers to an embryo sibling group from the same couple within the same IVF treatment.)

Nonetheless, we envision that the identification of cohort-specific parameters to define human embryo phenotypes is a necessary step towards translational investigations of molecular determinants of developmental competence. Thus, we sought to test the hypothesis that embryo cohort-specific variables have prognostic value in measuring IVF cycle outcomes by identifying non-redundant, prognostic variables in an unbiased manner using regression tree models.

Results

Of all 1117 IVF treatments performed at Stanford University in 2005, 822 were fresh IVF cycles that used the patients' own oocytes (Figure 1A). Based on our exclusion criteria, 157 cycles were excluded for a variety of medical and non-medical reasons. Clinical and embryology data on the remaining 665 cycles that satisfied inclusion and exclusion criteria, and their 4144 embryos, respectively, were analyzed to test the hypothesis that cohort-specific variables predict IVF cycle outcomes (Figure 1). Of those 4144 embryos, the number of blastomeres or cells on day 3 was recorded for 4002 embryos (96.6%). Overall, 38.8% had 8 cells, the developmentally appropriate cell number, while 18.2% of embryos had ≤4 cells, and 33.6% had 5–7 cells (Figure 2).

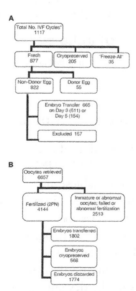

Figure 1. Source of data. A) IVF cycles performed in 2005. B) Utilization of oocytes and embryos in 665 fresh, non-donor IVF cycles. * All numbers in Panel A indicate the number of cycles and numbers in Panel B indicate the number of oocytes or embryos. Fresh cycles are defined by ovarian stimulation of gonadotropins and embryo transfer performed within the same cycle; cryopreserved cycles utilize embryos that were obtained and cryopreserved from a previous cycle; "freeze-all" are cycles in which ovarian stimulation was performed, but embryos were cryopreserved instead of being transferred back within the same cycle for medical or non-medical reasons. 157 cycles were removed from analysis for a variety of medical and non-medical reasons that did not result in fresh embryo transfer (see SI Text for details).

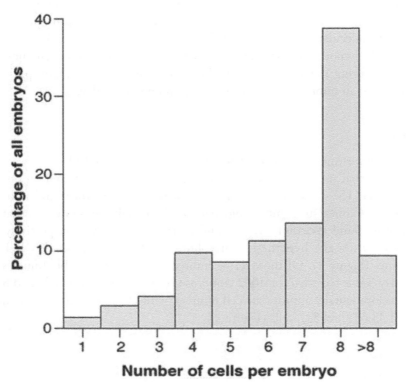

Figure 2. Distribution of all embryos from 665 fresh, non-donor IVF cases according to their cell number on Day 3.

Prognostic Significance and Correlation of Variables

We systematically examined the association of each variable with IVF outcomes, and the correlation of each pair of variables. Pair-wise logistic regression tests confirmed many known prognostic variables, including female age, day 3 FSH, and the number of 8-cell embryos. However, in addition to these known prognostic variables, we observed that cohort-specific variables such as fertilization rate and the rate of cleavage arrest were also significantly associated with IVF cycle outcome (p<0.001; Table 1). In contrast, except for male factor infertility (p<0.05), none of the conventional clinical infertility diagnoses were significantly associated with IVF outcomes. Notably, despite a high degree of correlation between many variables and age or day 3 FSH level, which estimates ovarian aging, neither age nor day 3 FSH level was correlated to cohort-specific embryo parameters. Collectively, these results suggest that determinants other than age-related mechanisms and clinical diagnoses impact cohort-specific embryo developmental competence.

Table 1. Association of each variable with pregnancy outcome.

Variables	Estimate*	S.E.	P-Value
Patient Characteristics and Clinical Diagnoses[†]			
Age	−0.10	0.02	2.16E-007
Maximum Day 3 FSH level	−0.08	0.03	1.70E-003
Gravidity	0.036	0.066	5.86E-001
Male Factor (infertility diagnosis)	0.50	0.24	3.71E-002
IVF Cycle Characteristics			
Microdose lupron (flare) protocol	−1.14	0.24	2.53E-006
Antagonist protocol	−0.74	0.19	9.98E-005
Performance of ICSI	−0.15	0.16	3.47E-001
No. of oocytes	0.08	0.01	1.58E-009
Embryo Cohort Parameters			
Fertilization rate	1.24	0.36	5.37E-004
No. of embryos	0.14	0.02	2.67E-012
Average cell no. of embryos	0.29	0.06	6.34E-006
No. of 8-cell embryos	0.26	0.04	2.88E-012
Percentage of 8-cell embryos	0.76	0.28	5.75E-003
Cleavage arrest rate[‡]	−1.28	0.35	2.76E-004
Average grade of embryos	−0.091	0.17	5.88E-001
Parameters of Transferred Embryos			
Day 5 embryo transfer[§]	1.40	0.19	7.51E-013
No. of embryos transferred	0.0058	0.053	9.12E-001
Average cell no. of embryos transferred	0.47	0.07	2.19E-010
Percentage of transferred embryos at the 8-cell stage	1.33	0.21	5.35E-010
No. of 8-cell embryos transferred	0.41	0.08	4.40E-008
No. of embryos with ≤4 cells transferred	−2.14	0.49	1.06E-005
Average grade of embryos transferred	−0.52	0.17	2.61E-003

*Positive and negative estimates indicate association with positive and negative pregnancy outcomes, respectively.

[†]Clinical infertility diagnoses that were not significantly associated with pregnancy outcome (p-value >0.05) were not listed: uterine factor, polycystic ovarian syndrome, endometriosis, tubal ligation, tubal disease, hydrosalpinges, unexplained infertility, and "other diagnoses". Each IVF case may have more than one clinical infertility diagnosis.

[‡]Cleavage arrest rate is defined as the percentage of embryos with 4 or fewer cells on Day 3 of *in vitro* culture.

[§]Day 5 embryo transfer is arbitrarily listed under Parameters of Transferred Embryos. It can also be considered an Embryo Cohort Parameter, as it depends on the total number of embryos and the number of 8-cell embryos.

Thresholds of Non-Redundant, Prognostic Variables Defining Human Embryo Cohort Phenotypes

Sequential Multiple Additive Regression Tree (MART®) and Classification and Regression Tree (CART) analyses of all 30 variables (listed in Table 1 and its legend) determined that IVF cycle outcomes were most accurately predicted at ~70% by using only four non-redundant variables: total number of embryos, rate of cleavage arrest in an embryo cohort, the number of 8-cell embryos in a cohort, and day 3 FSH level. Remarkably, these four variables all describe the embryo cohort rather than individual embryos, and were more informative than age, clinical diagnoses, or any measures of the transferred embryos. Interestingly, the total number of embryos, day 3 FSH, and the number of 8-cell embryos depended on and thus captured the effects of many other variables. In contrast, the rate of cleavage arrest was independent of any of those known variables.

Of the prognostic thresholds identified, the most robust phenotypes are A1 and A2, and B1 and B2 (Table 2). Number of embryos <6 or ≥6 is used by all 5 top CART models, defines all other phenotypes (B to F), and can be applied to all cases. Specifically, the phenotype defined by having fewer than 6 embryos, has an odds ratio of 3.9 for no pregnancy compared to cases with ≥6 embryos (95% Confidence Interval [CI], 2.8 to 5.5). Similarly, the next most robust phenotypes are defined by the number of embryos and cleavage arrest rate, such that for cases with ≥6 embryos, those with cleavage arrest rate >14.6% are 3.0 times more likely to result in no pregnancy than those with cleavage arrest rate ≤14.6% (95% CI, 1.9 to 4.9).

Table 2. Prognostic thresholds defining cohort-specific phenotypes.

	Embryos (No.)*	Cleavage Arrest (%)*	8-cell embryo (No.)*	FSH (mIU/mL)*	Pregnancy -No. (%)†	No Pregnancy -No. (%) ‡	Applicable Cases-No. (%) §	No. Trees ¶	Reference Condition‖	Odds Ratio	95% Confidence Interval (C.I.)
A1**	≥6				177 (57.7)	130 (42.3)	307 (46.2)	5			
A2	<6				92 (25.7)	266 (74.3)	358 (53.8)	5	A1	3.9	(2.8, 5.5)
B1	≥6	≤14.6			112 (70.4)	47 (29.6)	159 (23.9)	4			
B2	≥6	>14.6			65 (43.9)	83 (56.1)	148 (22.3)	4	B1	3.0	(1.9, 4.9)
B3	≥6	14.6–52.8			62 (47.3)	69 (52.7)	131 (19.7)	1	B1	2.6	(1.6, 4.3)
B4	≥6	≤52.8			174 (60.0)	116 (40.0)	290 (43.6)	n/a			
B5	≥6	>52.8			3 (17.6)	14 (82.3)	17 (2.6)	1	B1	10.6	(3.2, 49.6)
									B4	6.7	(2.1, 30.9)
C1	≥6		≥2		157 (63.6)	90 (36.4)	247 (37.1)	1			
C2	≥6		<2		20 (33.3)	40 (66.7)	60 (9.2)	1	C1	3.5	(1.9, 6.4)
D1	≥6	>14.6	≥2		51 (53.1)	45 (46.9)	96 (14.4)	1			
D2	≥6	>14.6	<2		14 (26.9)	38 (73.1)	52 (7.8)	1	D1	3.0	(1.5, 6.5)
E1	≥6	>14.6	≥2	≤4.6	14 (82.4)	3 (17.6)	17 (2.6)	1			
E2	≥6	>14.6	≥2	>4.6	34 (46.9)	37 (53.1)	71 (12.2)	1	E1	4.8	(1.4, 23.4)

*Cohort phenotypes defined by thresholds of non-redundant prognostic variables. Each set of conditions (A–E) use "AND" as the operator where more than one condition is listed.

†No. of cases that satisfy the threshold conditions and have pregnancy outcome. This percentage is calculated by using the No. Applicable Cases as denominator. In general, conditions that discriminate between pregnancy and no pregnancy outcomes more highly are more robust and are expected to be more useful in both clinical management and translational research.

‡No. of cases that satisfy the threshold conditions and have no pregnancy outcome. This percentage is calculated by using the No. Applicable Cases as denominator.

§The No. Applicable Cases is the total number of cases that satisfy the threshold conditions. This percentage is calculated by using the total number of cycles (665) as the denominator. In general, the larger the number of applicable cases, the more useful the set of conditions are for clinical management and counseling. However, for the purpose of translational research, conditions that define a smaller number of cases may have more specific correlates on a molecular level.

¶No. Trees shows the number of CART trees that utilize each set of conditions. There are a total of 5 trees. (See Supplemental Results.) Increased utilization indicates "usefulness" or "robustness" of that particular set of conditions.

‖Reference condition against which the Odds Ratio and 95% C.I. for having no pregnancy is calculated.

**Conditions A–E are listed from most robust and "useful" to least "useful" based on: the number of trees that utilize each set of conditions, the number of applicable cases, and the odds ratio and 95% CI.

In contrast, the rest of the thresholds listed in Table 2 are used by only 1 CART model each, and is applicable to fewer cases. However, as some of those phenotypes describe very specific subset of cases and have odds ratios that are highly discriminatory, they may be extremely useful depending on the clinical or translational research context. For example, for cases with ≥6 embryos, having cleavage arrest rates of 14.6–52.8% and >52.8% increase the odds of no pregnancy by 2.6 (95% CI 1.6 to 4.3) and 10.6 (95% CI 3.2 to 49.6), respectively, when compared to cases with cleavage rates of ≤14.6%.

Discussion

Since the introduction of IVF in the 1970s, the major challenges of assisted reproductive technologies (ART) have been the high attrition rates of embryos cultured in vitro [1], [3], [4], [8], [9], [10], [11], [22], [23], the limited value of embryo morphology in predicting developmental competence [24], [25], [26], and finding criteria to help determine the number of embryos to transfer [22], [23]. In addition, the benefit of aneuploidy screening by preimplantation genetic screening (PGS) has recently been refuted [27]. Thus, there is a need to reassess factors that determine human embryo quality.

Our findings represent a first step towards this goal by using regression tree models, MART® and CART, as unbiased methods to analyze IVF and embryo data. These methods allowed us to consider and control for a large number of variables, even if only a few of them have significant impact on outcomes. This feature is critical for the analysis of the highly interactive and multicollinear IVF and human embryo data, as arbitrary selection of variables may compromise completeness of data and introduce bias, while including all of them would cause the conventional multivariate regression to breakdown (see SI Text). Indeed, such application of CART analysis was taken by Guzick et al. to define semen parameters that predicted male infertility [28]. In our study, we further used MART®, a more powerful statistical method that "boosts" or increases accuracy in the CART method [29], [30], [31], [32].

We identified four non-redundant variables that predict outcomes in the current IVF cycle with ~70% accuracy. Most remarkably, these variables–total number of embryos, cleavage arrest rate, number of 8-cell embryos, and day 3 FSH (in order of relative importance)–describe the entire embryo cohort, and are more predictive than any measures of the transferred embryos. In addition, we show that most prognostic information carried by highly interacting and multicollinear conventional variables such as age and clinical diagnoses, is captured by three of the four variables.

Previous reports mainly focused on the prognostic value of individual embryo scores, in which the relative weighting of score components was determined arbitrarily rather than by objective or statistical methods [15], [18]. Further, although individual variables that were significantly related to IVF cycle outcomes were reported, there has been no attempt to compare their relative prognostic value, or to identify redundancy amongst variables [12], [14], [17], [21]. For example, age, serum FSH, number of oocytes and number of embryos were each reported to be significantly related to IVF outcomes [17]. However, as shown by our analyses, the prognostic value of age and number of oocytes was captured by three of the four non-redundant variables. Similarly, the total number of embryos and the number of 8-cell embryos have been advocated for use in selecting patients for blastocyst transfer in some IVF clinics to minimize the risk of having no embryos to transfer due to failed blastocyst development [21], [33], [34]. However, the prognostic value of these two variables has not been compared to that of others, and their ability to capture prognostic information from most other variables were not known.

Indeed, cleavage arrest rate is the only variable that is independent of the others, which suggests that it may be linked to biological mechanisms that are not currently recognized in the management of clinical infertility or hESC biology. Encountered in ~18% of human embryos cultured in vitro overall, its underlying defects are likely diverse, and may be due to suboptimal in vitro culture environment, biological mechanisms underlying infertility, a generally poor reproductive fitness of our species or all of these factors. Although cleavage arrest coincides with the maternal-embryonic transition during which maternal transcripts are degraded and the embryonic genome is activated [35], gene expression analyses of arrested single human embryos did not show failure in embryonic genome activation, and no specific molecular defects have been identified [8], [11].

Our study has some limitations. Although we took advantage of the power of regression tree models to analyze a very comprehensive range of variables, we did not include cryopreservation of sibling embryos and assisted hatching as variables. In addition, it would also be valuable to analyze blastocyst development rate of sibling embryos, because this variable has been shown to correlate with positive pregnancy outcomes [36]. Those variables are now being investigated in a larger study that encompasses four years of data. As the goal of this current study was to explore new paradigms in human embryo development in IVF, and not to arrive at recommendations to change clinical practice, we used positive serum hCG status as the surrogate outcome measure to identify nonredundant predictors of IVF cycles in which at least one embryo attaches to the endometrium and secretes hCG, from those in which no embryo attaches. In the future, we will use later endpoints, such as clinical pregnancy or live birth, to address clinical questions.

In spite of over 30 years of ART, many challenges remain. Ongoing and future investigations may incorporate approaches common to genetics and developmental biology, in order to reassess defective human embryo development in terms of phenotypes that can be diagnosed, defined, and translated into improved clinical practices. Collectively, our results indicate that embryos from a cohort share as yet undefined genetic or epigenetic determinants of developmental competence, which is consistent with the greater increase in implantation relative to pregnancy rates conferred by blastocyst transfer [37]. The concept of cohort-specific determinants suggest a paradigm shift from strictly focusing research efforts on selecting the "best" embryos to identifying methods that would improve the quality of the entire cohort. In addition, it raises the question of whether quality of the entire cohort is intrinsic due to the shared origins of the embryos, or if it is merely a result of group culture in vitro, especially since the benefits of group culture have been reported in animal and human embryos [38], [39], [40], [41]. While embryo-specific parameters may help to identify embryos that would maximize the immediate pregnancy outcome for each couple, in the long term, understanding cohort-specific parameters is critical in counseling patients, improving treatment, and ultimately in developing mechanism-specific and more customized treatments.

We reason that well-defined criteria for embryo cohort phenotypes in selecting abnormal embryos for molecular analyses would maximize the chance of finding non-random genetic or epigenetic molecular defects that are consistent in an embryo cohort. For example, we are applying our findings to analyze arrested embryos from embryo cohorts in which the number of embryos are ≥6 and cleavage arrest rate is >52.8% (see Condition B5 in Table 2). Overall, ~2.5% of fresh, non-donor IVF cases (or ~17 cases per year, at our center) are expected to fulfill these criteria. This approach should allow for objective interpretation and comparison of data both internally and amongst research groups.

We are also applying this research strategy to investigate predictors of pregnancy outcomes in subsequent IVF cycles to contrast couple- versus embryo cohort-specific prognostics variables. More importantly, new hypotheses that are generated by this investigation can be further tested as additional years of data become available. For example, our findings indicate that a low day 3 FSH (<4.6 mIU/mL) confer high pregnancy rates in a very small and specific subset of patients (see Condition E in Table 2), and offer new perspectives on this controversial entity. While abnormally high levels of day 3 FSH have been associated with ovarian aging, poor ovarian response in IVF, and poor IVF cycle outcomes, many studies have cautioned against its use in clinical management due to its low sensitivity, especially in women under 40 [42], [43], [44], [45]. However, the clinical utility of this test may be improved by determining appropriate thresholds and conditions [46].

Similar to the implications for ART, our results also raise questions about the effects of cohort-specific determinants on the success rate of hESC line derivation, the quality of hESC lines, and most importantly, embryo cohort selection for hESC line derivation, or oocyte cohort selection for somatic cell nuclear transfer. Currently, most scientific reports on successful derivation of hESC lines do not include information on embryo cohort characteristics, clinical information or IVF outcomes of sibling embryos. Our findings suggest that correlation of clinical IVF data and hESC line characteristics may provide valuable insight that would move both the fields of reproductive medicine and hESC research forward. We envision that dissection of human embryo phenotypes and their corresponding molecular correlates is not only a necessary step towards improving the treatment of clinical infertility, but will also contribute significantly to research efforts in the hESC field.

Materials and Methods

Data Collection, Inclusion and Exclusion Criteria

Data related to clinical diagnoses, IVF treatment protocol and monitoring, embryology data and treatment outcomes for all IVF cycles performed between January 1, 2005 and December 31, 2005 at Stanford University Medical Center were retrieved from BabySentryPro (BabySentry Ltd, Limassol, Cyprus), a widely used fertility database management system, or obtained from medical and embryology records as necessary. Retrospective data collection, de-identification, and analysis were performed according to a Stanford University Institutional Review Board-approved protocol. The inclusion criteria for data analysis were fresh, stimulated, non-donor oocyte IVF cycles. We excluded cycles that did not result in embryo transfer for any reason, cycles performed for women aged over 45, and those performed for preimplantation genetic screening.

Assessment of Embryo Development

Our standard clinical protocols for ART treatment, fertilization, embryo culture, embryo assessment, cryopreservation criteria, and clinical outcomes are described in methods in SI Text. The normal progression of human embryo development in vitro is characterized by the appearance of 2 pronuclei at 16–20 hours after insemination as evidence of fertilization on Day 1, with Day 0 as the day of oocyte retrieval. By late Day 1, embryo development has reached the 2-cell stage, followed by the 4-cell and 8-cell stages on Days 2 and 3, respectively. On Days 4 and 5, embryo development is characterized by the establishment of the morula and blastocyst stages, respectively. All embryos were available for evaluation on Day 3. The day of embryo transfer was determined by the number of blastomeres on Day 3. In general, if 4 or more 8-cell embryos were present, we would recommend extended embryo culture until Day 5, when blastocyst transfer, which has been associated with higher pregnancy rates, would be performed. If fewer than four 8-cell embryos were present, embryo transfer would be performed on Day 3.

Patient, IVF Cycle, and Embryo Parameters

We analyzed 30 variables for association with IVF treatment outcomes, as listed in Table 1, under four main categories: patient characteristics and clinical diagnoses, IVF cycle characteristics, embryo cohort parameters, and parameters of transferred embryos. The cleavage arrest rate was defined as the percentage of embryos within a cohort with 4 or fewer cells on Day 3 of in vitro culture. All other variables were self-explanatory.

Statistical Analysis

Since some patients underwent more than one IVF cycle during the study period, the analyses were performed based on treatment cycles rather than patients. Statistical analyses were performed based on the dichotomous outcomes of no pregnancy, as defined by negative serum ß-hCG, and pregnancy, as defined by positive serum ß-hCG, and included biochemical pregnancy, clinical pregnancy, spontaneous abortion, and ectopic pregnancy. We performed pair-wise logistic regression of each variable to the outcome and determined the Pearson correlation coefficient between each pair of continuous variables.

For the main analyses, boosted classification trees were constructed by MART® to identify non-redundant prognostic variables, which were then further analyzed by CART to identify thresholds that would define them as categorical variables. MART® is a robust method used to identify interactive structure of variables that are predictive of outcomes [29], [30], [31], [32]. The use of cross-validation and boosting in parameter selection and model assessment in MART® also preserve parsimony and prevent over-fitting [31]. In the MART® tree constructions, the whole data set is divided into 10 subsets to achieve 10 fold cross validation for model assessment. The same 10 fold cross validation was repeated 1000 times to perform a robust prediction rate estimation and identify tree models with the highest prediction rates in the CART. While MART® is powerful in selecting non-redundant prognostic variables from a large set of highly interactive variables, CART analysis results in simple algorithms, and more easily understood "decision trees", that are used in the medical literature [28]. Thus non-redundant, prognostic variables identified by MART® to confer prediction were analyzed by CART to further define prognostic thresholds.

Acknowledgements

We thank the Embryology staff and L. Morcom for their efforts on data entry, L. Wu and C. Mak for valuable discussion, M. Zamah for help with preliminary data collection.

Authors' Contributions

Conceived and designed the experiments: WW MY SJ BC. Performed the experiments: BC. Analyzed the data: WW BC. Contributed reagents/materials/analysis tools: WW LW BB. Wrote the paper: WW MY SJ BC RR LS. Other: Acquired data: MY SJ LW BB. Interpreted data analysis: SJ MY. Interpreted findings: BB RRP LW.

References

1. Behr B (1999) Blastocyst culture and transfer. Hum Reprod 14: 5–6.

2. Cowan CA, Klimanskaya I, McMahon J, Atienza J, Witmyer J, et al. (2004) Derivation of embryonic stem-cell lines from human blastocysts. N Engl J Med 350: 1353–1356.

3. Gardner DK, Lane M, Schoolcraft WB (2000) Culture and transfer of viable blastocysts: a feasible proposition for human IVF. Hum Reprod 15: Suppl 69–23.

4. Milki AA, Hinckley MD, Fisch JD, Dasig D, Behr B (2000) Comparison of blastocyst transfer with day 3 embryo transfer in similar patient populations. Fertil Steril 73: 126–129.

5. Chen H, Qian K, Hu J, Liu D, Lu W, et al. (2005) The derivation of two additional human embryonic stem cell lines from day 3 embryos with low morphological scores. Hum Reprod 20: 2201–2206.

6. Strelchenko N, Verlinsky O, Kukharenko V, Verlinsky Y (2004) Morula-derived human embryonic stem cells. 9: 623–629.

7. Mitalipova M, Calhoun J, Shin S, Wininger D, Schulz T, et al. (2003) Human embryonic stem cell lines derived from discarded embryos. Stem Cells 21: 521–526.

8. Artley JK, Braude PR, Johnson MH (1992) Gene activity and cleavage arrest in human pre-embryos. Hum Reprod 7: 1014–1021.

9. Barratt CL, St John JC, Afnan M (2004) Clinical challenges in providing embryos for stem-cell initiatives. Lancet 364: 115–118.

10. Hardy K, Spanos S, Becker D, Iannelli P, Winston RM, et al. (2001) From cell death to embryo arrest: mathematical models of human preimplantation embryo development. Proc Natl Acad Sci USA 98: 1655–1660.

11. Dobson AT, Raja R, Abeyta MJ, Taylor T, Shen S, et al. (2004) The unique transcriptome through day 3 of human preimplantation development. Hum Mol Genet 13: 1461–1470.

12. Alikani M, Cohen J, Tomkin G, Garrisi GJ, Mack C, et al. (1999) Human embryo fragmentation in vitro and its implications for pregnancy and implantation. Fertil Steril 71: 836–842.

13. Ebner T, Moser M, Sommergruber M, Tews G (2003) Selection based on morphological assessment of oocytes and embryos at different stages of preimplantation development: a review. Hum Reprod Update 9: 251–262.

14. Ebner T, Yaman C, Moser M, Sommergruber M, Polz W, et al. (2001) Embryo fragmentation in vitro and its impact on treatment and pregnancy outcome. Fertil Steril 76: 281–285.

15. Giorgetti C, Terriou P, Auquier P, Hans E, Spach JL, et al. (1995) Embryo score to predict implantation after in-vitro fertilization: based on 957 single embryo transfers. Hum Reprod 10: 2427–2431.

16. Jun SH, O'Leary T, Jackson KV, Racowsky C (2006) Benefit of intracytoplasmic sperm injection in patients with a high incidence of triploidy in a prior in vitro fertilization cycle. Fertil Steril 86: 825–829.

17. Keltz MD, Skorupski JC, Bradley K, Stein D (2006) Predictors of embryo fragmentation and outcome after fragment removal in in vitro fertilization. Fertil Steril 86: 321–324.

18. Lan KC, Huang FJ, Lin YC, Kung FT, Hsieh CH, et al. (2003) The predictive value of using a combined Z-score and day 3 embryo morphology score in the assessment of embryo survival on day 5. Hum Reprod 18: 1299–1306.

19. Nagy ZP, Dozortsev D, Diamond M, Rienzi L, Ubaldi F, et al. (2003) Pronuclear morphology evaluation with subsequent evaluation of embryo morphology significantly increases implantation rates. Fertil Steril 80: 67–74.

20. Stone BA, Greene J, Vargyas JM, Ringler GE, Marrs RP (2005) Embryo fragmentation as a determinant of blastocyst development in vitro and pregnancy outcomes following embryo transfer. Am J Obstet Gynecol 192: 2014–2019; discussion 2019–2020.

21. Volpes A, Sammartano F, Coffaro F, Mistretta V, Scaglione P, et al. (2004) Number of good quality embryos on day 3 is predictive for both pregnancy and implantation rates in in vitro fertilization/intracytoplasmic sperm injection cycles. Fertil Steril 82: 1330–1336.

22. (2006) Guidelines on number of embryos transferred. Fertil Steril 86: 5 SupplS51–52.

23. Racowsky C (2002) High rates of embryonic loss, yet high incidence of multiple births in human ART: is this paradoxical? Theriogenology 57: 87–96.

24. Milki AA, Hinckley MD, Gebhardt J, Dasig D, Westphal LM, et al. (2002) Accuracy of day 3 criteria for selecting the best embryos. Fertil Steril 77: 1191–1195.

25. Neuber E, Mahutte NG, Arici A, Sakkas D (2006) Sequential embryo assessment outperforms investigator-driven morphological assessment at selecting a good quality blastocyst. Fertil Steril 85: 794–796.

26. Rijnders PM, Jansen CA (1998) The predictive value of day 3 embryo morphology regarding blastocyst formation, pregnancy and implantation rate after day 5 transfer following in-vitro fertilization or intracytoplasmic sperm injection. Hum Reprod 13: 2869–2873.

27. Mastenbroek S, Twisk M, van Echten-Arends J, Sikkema-Raddatz B, Korevaar JC, et al. (2007) In vitro fertilization with preimplantation genetic screening. N Engl J Med 357: 9–17.

28. Guzick DS, Overstreet JW, Factor-Litvak P, Brazil CK, Nakajima ST, et al. (2001) Sperm morphology, motility, and concentration in fertile and infertile men. N Engl J Med 345: 1388–1393.

29. Friedman J (1999) Greedy function approximation: A stochastic boosting machine. Department of Statistics, Stanford University. Technical Report.

30. Friedman J (1999) Stochastic gradient boosting. Department of Statistics, Stanford University. Technical Report.

31. Friedman J (2002) Tutorial: Getting started with MART in R. Department of Statistics, Stanford University.

32. Friedman J, Meulmann JJ (2003) Multiple additive regression trees with application in epidemiology. Statistics in medicine 22: 1365–1381.

33. Gardner DK, Schoolcraft WB, Wagley L, Schlenker T, Stevens J, et al. (1998) A prospective randomized trial of blastocyst culture and transfer in in-vitro fertilization. Hum Reprod 13: 3434–3440.

34. Racowsky C, Jackson KV, Cekleniak NA, Fox JH, Hornstein MD, et al. (2000) The number of eight-cell embryos is a key determinant for selecting day 3 or day 5 transfer. Fertil Steril 73: 558–564.

35. Braude P, Bolton V, Moore S (1988) Human gene expression first occurs between the four- and eight-cell stages of preimplantation development. Nature 332: 459–461.

36. Fisch JD, Milki AA, Behr B (1999) Sibling embryo blastocyst development correlates with the in vitro fertilization day 3 embryo transfer pregnancy rate in patients under age 40. Fertil Steril 71: 750–752.

37. Gardner DK, Surrey E, Minjarez D, Leitz A, Stevens J, et al. (2004) Single blastocyst transfer: a prospective randomized trial. Fertil Steril 81: 551–555.

38. Lane M, Gardner DK (1992) Effect of incubation volume and embryo density on the development and viability of mouse embryos in vitro. Hum Reprod 7: 558–562.

39. Rijinders PM, Jansen CAM (1998) The predictive value of day 3 embryo morphology regarding blastocyst formation, pregnancy, and implantation rate after day 5 transfer following in vitro fertilization or intracytoplasmic sperm injection. Hum Reprod 13: 2869–2873.

40. Spyropoulou I, Karamalegos C, Bolton VN (1999) A prospective randomized study comparing the outcome of in-vitro fertilization and embryo transfer following culture of human embryos individually or in groups before embryo transfer on day 2. Hum Reprod 14: 76–79.

41. Paria BC, Dey SK (1990) Preimplantation embryo development in vitro: cooperative interactions among embryos and role of growth factors. Proc Natl Acad Sci USA 87: 4756–4760.

42. Broekmans FJ, Kwee J, Hendriks DJ, Mol BW, Lambalk CB (2006) A systematic review of tests predicting ovarian reserve and IVF outcome. Hum Reprod Update 12: 685–718.

43. Jain T, Soules MR, Collins JA (2004) Comparison of basal follicle-stimulating hormone versus the clomiphene citrate challenge test for ovarian reserve screening. Fertil Steril 82: 180–185.

44. Srouji SS, Mark A, Levine Z, Betensky RA, Hornstein MD, et al. (2005) Predicting in vitro fertilization live birth using stimulation day 6 estradiol, age, and follicle-stimulating hormone. Fertil Steril 84: 795–797.

45. Watt AH, Legedza AT, Ginsburg ES, Barbieri RL, Clarke RN, et al. (2000) The prognostic value of age and follicle-stimulating hormone levels in women over forty years of age undergoing in vitro fertilization. J Assist Reprod Genet 17: 264–268.

46. Scott RT Jr, Elkind-Hirsch KE, Styne-Gross A, Miller KA, Frattarelli JL (2007) The predictive value for in vitro fertility delivery rates is greatly impacted by the method used to select the threshold between normal and elevated basal follicle-stimulating hormone. Fertil Steril.

Search for the Genes Involved in Oocyte Maturation and Early Embryo Development in the Hen

Sebastien Elis, Florence Batellier, Isabelle Couty,
Sandrine Balzergue, Marie-Laure Martin-Magniette,
Philippe Monget, Elisabeth Blesbois and Marina S. Govoroun

ABSTRACT

Background

The initial stages of development depend on mRNA and proteins accumulated in the oocyte, and during these stages, certain genes are essential for fertilization, first cleavage and embryonic genome activation. The aim of this study was first to search for avian oocyte-specific genes using an in silico and a microarray approaches, then to investigate the temporal and spatial dynamics of the expression of some of these genes during follicular maturation and early embryogenesis.

Results

The in silico approach allowed us to identify 18 chicken homologs of mouse potential oocyte genes found by digital differential display. Using the chicken Affymetrix microarray, we identified 461 genes overexpressed in granulosa cells (GCs) and 250 genes overexpressed in the germinal disc (GD) of the hen oocyte. Six genes were identified using both in silico and microarray approaches. Based on GO annotations, GC and GD genes were differentially involved in biological processes, reflecting different physiological destinations of these two cell layers. Finally we studied the spatial and temporal dynamics of the expression of 21 chicken genes. According to their expression patterns all these genes are involved in different stages of final follicular maturation and/or early embryogenesis in the chicken. Among them, 8 genes (btg4, chkmos, wee, zpA, dazL, cvh, zar1 and ktfn) were preferentially expressed in the maturing oocyte and cvh, zar1 and ktfn were also highly expressed in the early embryo.

Conclusion

We showed that in silico and Affymetrix microarray approaches were relevant and complementary in order to find new avian genes potentially involved in oocyte maturation and/or early embryo development, and allowed the discovery of new potential chicken mature oocyte and chicken granulosa cell markers for future studies. Moreover, detailed study of the expression of some of these genes revealed promising candidates for maternal effect genes in the chicken. Finally, the finding concerning the different state of rRNA compared to that of mRNA during the postovulatory period shed light on some mechanisms through which oocyte to embryo transition occurs in the hen.

Background

The activation of molecular pathways underlying oocyte to embryo transition (OET) depends exclusively on maternal RNAs and proteins accumulated during growth of the oocyte [1]. During OET and preimplantation development in mice, the embryo becomes almost autonomous, and may gradually eliminate maternal components. Indeed, by the two cell stage, the major pathways regulated by maternal mRNA are targeted protein degradation, translational control and chromatin remodelling [2]. The recruitment of maternal mRNA for translation has long been recognized as a widespread mechanism to generate newly synthesized proteins in maturing oocytes and fertilized eggs [3]. Conversely, RNA that is no longer needed is actively degraded in the early embryo [4]. Moreover, careful regulation of proteolysis during the same period is likely to be important in oocytes, which are predominantly transcriptionally inactive and must often wait for long periods before fertilization in different species such as Drosophila, Xenopus, Caenorhabditis and Zebrafish [5]. Maternal transcripts that are present in the early pre-implantation embryo can be subdivided into two classes according to whether they are re-synthesized soon after embryonic genome activation or not. The first is common to the oocyte and early embryo and is replenished after activation of the zygotic genome. The second consists of oocyte-specific mRNA that is not subsequently transcribed from zygotic genes in the embryo. This class of mRNA may be detrimental to early post-fertilization development [6].

Maternal effect genes have been found in several species ranging from invertebrates to mammals. Wide screening of mutants has been performed in invertebrates as Drosophila melanogaster [7] and Caenorhabditis elegans [8] where several mutations lead to arrest of early embryo development. Although females bearing this type of mutation are viable and appear to be normal, the development and survival of their embryos are compromised [9]. Maternal effect mutations have also been described in other vertebrates such as Danio rerio for the nebel gene [10], and Xenopus laevis for the af gene [11]. Despite the fact that maternal effect mutations are well known in lower organisms, only a few examples have been reported in mammals. All of them are based on knock-out experiments and concern three murine genes, i.e. Dnmt1, Hsf1 and Mater [9]. Mater (Maternal antigen that embryos require) is a single-copy gene that is transcribed in growing oocytes. Although its transcripts are degraded during meiotic maturation, MATER protein persists into the blastocyst. Female mice lacking this 125 kDa cytoplasmic protein produce no offspring because of an embryonic block at the early cleavage stage. Thus, Mater is one of few documented genes for maternal effect in mammalian development [12]. Mater has been found in bovine models but there is no report in the literature on maternal effect genes conserved between species.

No information has been available to date on maternal effect genes in birds. However, birds represent a good model to observe progressive accumulation of mRNA in the oocyte before ovulation. The embryonic genome of a model bird, i.e. the chicken, is activated when the embryo contains 30,000–50,000 cells [13] 24 h after fertilization. Proteins and mRNA, accumulated as the chicken oocyte matures, are essential not only for fertilization and first cleavage but also for supporting a high number of embryonic cell divisions before genome activation. By comparison, the embryonic genome is activated at the 8-cell stage in bovines [14] and at the 2-cell stage in the mouse [15]. The avian oocyte consists of a large amount of yolk and a structure called the germinal disc (GD) [16]. The GD is a white plaque of about 3–4 mm diameter on the top of the oocyte. It contains the nucleus and 99% of oocyte organelles although it occupies less than 1% of the cell volume [17]. Structurally, and therefore functionally, the GD is mostly equivalent to the mammalian oocyte. The ovary of the reproductively active hen consists of small pre-hierarchical follicles and maturing preovulatory follicles showing a hierarchy according to size (F6 to F1) [18].

Only a few studies have reported on gene expression in the oocyte and during early embryo development in the chicken. The dynamics of the overall RNA profile of the chicken oocyte through different maturation stages has been described by Olzanska et al. [13,19-22]. Chicken vasa homolog protein (CVH) was hypothesized to be maternally inherited in the chicken embryo, since it has been localized in chicken oocytes and during first cleavage [23]. Another protein, Epidermal Growth Factor, was found in F2 GD and its potential role in follicular development has also been investigated [24].

Since oocyte-specific genes expressed during follicular maturation and after ovulation are potentially involved in the fertilization process and in early embryo development, and almost no information is available on these genes in birds, the aim of this study was to identify avian oocyte-specific genes and then to investigate the temporal and spatial dynamics of their expression during follicular maturation and early embryogenesis. We chose initially to focus on oocyte-specific genes because the accumulation of their transcripts in the oocyte should have greater consequences on fertilization and OET. Two different strategies were used to identify avian genes potentially involved in oocyte developmental competence. The first was based on a candidate gene approach and consisted of a search for avian homologs of murine oocyte genes, previously identified by digital differential display [25]. The second strategy involved a global transcriptomic approach based on chicken Affymetrix microarray. We report here several novel chicken genes with potential maternal effect identified using these two strategies. We also describe the spatial and temporal dynamics of the expression of some of these genes as well as some potential mechanisms in which they could be involved. We also compare chicken and murine orthologs in terms of their tissue specificity and their potential involvement in oocyte developmental competence and/or early embryogenesis.

Results

In Silico Search For Chicken Homologs of Murine Oocyte Genes

Differential digital display analysis performed on murine tissues provided a list of 101 potentially oocyte-specific murine genes [25-27]. Bioinformatic analyses were performed on this list of genes in order to find potentially oocyte-specific chicken orthologs. Genes with a blast score higher than 100 were localized using mapview [28] and blatsearch [29] tools. The syntenic regions were checked: chromosome localization of murine genes and chicken homologs were compared, in order to obtain the correct chicken ortholog of mice genes. Only genes with sufficient homology or whose localization was in accordance with syntenic regions were selected. Forty-one chicken genes were eliminated because of their poor homology with murine genes and 32 other genes were eliminated because they were localized outside the syntenic regions. Among the remaining 28 chicken genes the transcript of only 18 genes could be correctly amplified using real time or classic RT-PCR, of which the detailed study of two genes bmp15 and gdf9 has previously been reported [30]. Thus 16 avian genes were finally retained (Table 1). The homology with murine genes was strong for 11 of these genes (blast score between 288 and 2149) and was weak for 5 (blast score between 104 and 132). The last five chicken genes were nevertheless considered as potential orthologs of murine genes and kept as candidate genes because of their correct localizations with respect to the syntenic region. Eleven of the selected genes were localized in the expected syntenic region (btg4, chkmos, msh4, mtprd, mcmip, znfingerRIZ, discs5, trans fact 20, wee, zar1 and ktfn). Three other genes were localized in the vicinity of the expected syntenic region (dazL, fbox and mark3). Two genes were localized in the unexpected syntenic region, but they were identified with the same name as murine genes (zpA and zpC).

Table 1. Accession numbers of murine sequences used and of homolog chicken sequences found. Bold text represents chicken genes whose syntenic regions are conserved with the appropriate murine homologs.

Murine genes				Chicken genes				
Accession number	Name	Localisation		Accession number	Name	tblastn score	Localisation	abbreviation
		Chromosome	Position (kb)				Chromosome	Position (kb)	
XM_205433.2	Oog2-like	4 E1	142170	XM_417634	similar to zinc finger protein RIZ, partial	100	21	4843	znfingerRIZ
XM_357175.2	TRAF-interacting protein	2A1	6081	XM_416727	similar to mtprd protein – mouse	102	1	100734	mtprd
AK054339.1	FBXO12A	9F2	109192	XM_419103	similar to hypothetical protein	104	2	91887	fbox
AY351591.1	Msh4	3H3	154505.7	XM_422549	similar to MutS homolog 4	107	8	29983	msh4
XM_138939.3	Speer-like	14A3	20635	XM_421604	similar to Discs. large homolog 5	132	6	12497.5	discs5
AK018361.1	Zfp393	4D2	116103.6	XM_422416	similar to Kruppel-like transcription factor neptune	288	8	20700.7	ktfn
NM_172481.1	Nalp9E	7A1	5473.6	XM_420951	similar to mast cell maturation inducible protein 1	473	5	3008	mcmip
BC066811.1	Btg4	9A5	51279	XM_417919	similar to p30 B9.10	599	24	5212.7	btg4
XM_355960.1	PAR-1Alike	7A3	11760.3	XM_421385	similar to MAP/microtubule affinity-regulating kinase 3 long isoform	643	5	47042.5	mark3
AY191415.1	Zar1	5	72968	XM_424318	similar to zygote arrest 1	643	4	68326.1	zar1
XM_139155.2		14D3	71005	XM_416218	similar to transcription factor 20 isoform 1	688	1	46012.9	transfact20
NM_010021.2	Dazl	17B1	48475	NM_204218	deleted in azoospermia-like	863	2	33589.4	dazL
NM_020021.1	Mos	4A1	3798.4	M19412	Chicken c-mos proto-oncogene	955	2	110418	chkmos
NM_011775.2	ZP2	7F2	107432.5	XM_424608	similar to zona pellucida A	1146	6	15797.3	zpA
NM_201370.1	WEE1hu	6B1	40383	XM_425491	similar to Wee1A kinase	1218	1	54018.4	wee
NM_011776.1	ZP3	5G2	133429.5	D89097	zona pellucida C protein	2149	10	188.600	zpC

Comparing Oocyte and Granulosa Cells Transcription Profiles at Final Maturation Steps Using Chicken Affymetrix Microarrays

The samples studied were: F1 GCs, corresponding to granulosa cells (GCs) of the largest follicles before ovulation (F1); F1 GDR, corresponding to the germinal disc region (GDR) of F1 follicles, and Ov GDR, corresponding to the germinal disc region of ovulated oocytes (Ov) (Fig. 1). Apart from the stage of maturation the main difference between these samples involved the presence of granulosa cells. They were not present in GDR from ovulated oocytes, they were slightly present in F1 GDR, and they constituted a major component of F1 GCs (see Materials and Methods). Statistical analysis of data obtained after Affymetrix microarray hybridization provided lists of genes differentially expressed in three comparisons: F1 GDR and Ov GDR, F1 GDR and F1 GCs, and Ov GDR and F1 GCs (accession number GSE7805). There were only a few differentially expressed genes in each comparison (fewer than 500 out of 28000 genes on the Chip) (Table 2). Indeed, the first comparison, between the F1 GDR and Ov GDR, showed 92 genes over-expressed at the F1 stage, including one of our in silico identified genes (zpC). In the second comparison, between F1 GDR and F1 GCs, 342 differentially expressed genes were identified. These genes involved 104 genes over-expressed in F1 GCs and 238 genes over-expressed in F1 GDR. Five of our in silico identified genes btg4, chkmos, dazL, zpA and ktfn were found among the latter genes over-expressed in F1 GDR (Table 2). The third analysis compared the expression of genes between F1 GCs and Ov GDR. We obtained a set of 448 genes that were differentially expressed between F1 GCs and Ov GDR, of which 392 genes were over-expressed in F1 GCs and 56 genes were over-expressed in Ov GDR. We found 1 of our in silico identified genes among these genes, (btg4) (Table 2). The Venn diagram (Fig. 2) shows overlapping differentially expressed genes between different dataset comparisons. Eighty-five differentially expressed genes were common for two comparisons (F1 GCs and Ov GDR, and F1 GCs and F1 GDR). Only one gene was common for the comparisons between F1 GDR and F1 GCs, and F1 GDR and Ov GDR and 1 gene was common for all three comparisons. F1 GDR samples contained a quantity of granulosa cells in contrast to Ov GDR samples which were free of granulosa cells. Analysis of the redundancy of over-expressed genes between different comparisons was therefore performed in order to distinguish between genes found over-expressed in F1 GDR samples due to granulosa cell contamination and those really over-expressed in the oocyte (Fig. 3). This analysis revealed that 85 of 92 genes over-expressed in F1 GDR were also over-expressed in F1 GCs, both compared to Ov GDR, indicating that these genes characterized granulosa cell expression rather than variation in the expression between F1 GD and Ov GD. Thus only 7 genes should be considered as overexpressed in F1 GD compared to Ov GD. Moreover, of the 104 genes overexpressed in F1 GCs compared to F1 GDR, 36 were also overexpressed in F1 GCs compared to Ov GDR. Consequently, taking into account the latter redundancy, a total of 460 genes was overexpressed in granulosa cells compared to oocytes. On the other hand 49 genes were redundant among genes overexpressed in the Ov GDR and F1 GCs, and F1 GDR and F1 GCs comparisons.

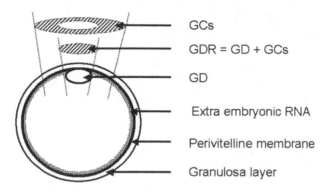

Figure 1. Schematic representation of avian oocyte. The oocyte is delimited by the perivitelline membrane, and the overlying layer of granulosa cells (GCs). Inside the oocyte, the perivitelline membrane is covered by extra-embryonic RNA. At the top of the oocyte, the germinal disc (GD) is visible. The GD with the overlying GCs constituted the germinal disc region (GDR). Two hatched area represent two different samples used in this study. The GDR comprises the GD and the lowest possible number of GCs, and GCs comprises by the GCs located in the vicinity of the GDR. Both samples are localized on the apical part of the oocyte.

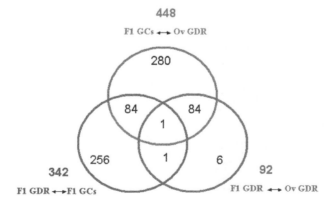

Figure 2. The relationships between differentially expressed genes in different comparisons. Diagram shows the overlap of differentially expressed genes in different comparisons. Each circle represents the total number of differentially expressed genes in one comparison. The overlapping areas represent differentially expressed genes common for different comparisons.

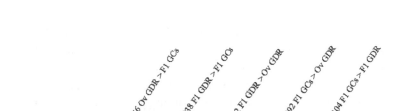

Figure 3. The relationships between overexpressed genes in different comparisons. Diagram shows the overlap of overexpressed genes in different comparisons. Each circle represents the total number of overexpressed genes in one comparison. The overlapping areas represent overexpressed genes common for different comparisons. Genes identified by the in silico approach and found to be overexpressed in different comparisons on the chips are indicated.

Table 2. Genes differentially expressed in Affymetrix experiment. Bold text represents genes that were further studied using real time RT-PCR

Affymetrix reference	Accession number	Gene name found by Blast search	Abbreviation of genes studied	Fold change	Comparison	Number of overexpressed genes
Ggs.305.1.S1_at	NM_204783.1	Gallus gallus wingless-type MMTV integration site family, member 4		150	F1 GCs and Ov GDR	56 genes overexpressed in Ov GDR
GgaAffx.13126.1.S1_at	AJ721023	Gallus gallus mRNA for hypothetical protein, clone 33l2		129		
Ggs.5341.1.S1_at	BU365377	Finished cDNA, clone ChEST746h6		105		
Ggs.7980.1.S1_at	CD730241	similar to Interferon regulatory factor 6		102		
Ggs.4536.2.S1_at	BX934646.1	aldo-keto reductase family 1, member D1 (delta 4-3-ketosteroid-5-beta-reductase)		94		
Ggs.9136.2.S1_a_at	CD740066	Gallus gallus tumor-associated calcium signal transducer 1		92		
GgaAffx.11525.1.S1_s_at	AJ719422	Gallus gallus claudin 1		91		
Ggs.20379.1.S1_at	CR524234.1	Finished cDNA, clone ChEST914e3		90		
Ggs.9936.1.S1_at	BX936026.2	Amyotrophic lateral sclerosis 2 chromosomal region candidate gene protein 7		83		
Ggs.5597.1.S1_at	BU450115	B-cell translocation gene 4	btg4	52		
Ggs.14454.1.S1_at	NM_213576.1	zona pellucida protein D	zpD	890	F1 GCs and Ov GDR	392 genes overexpressed in F1 GCs
Ggs.13391.1.S1_at	BX935169.2	Gallus gallus similar to adrenodoxin homolog		718		
Ggs.1824.1.S1_at	BU350625	Gallus gallus finished cDNA, clone ChEST974b18		703		
Ggs.6358.1.S1_at	BX265773	Gallus gallus similar to chromosome 9 open reading frame 61		687		
Ggs.596.1.S2_at	NM_205118.1	Gallus gallus 3beta-hydroxysteroid dehydrogenase/delta5-delta4 isomerase		667		
Ggs.17706.1.S1_at	CR388473.1	Gallus gallus finished cDNA, clone ChEST591g11		579		
Ggs.3095.1.S1_a_at	BX932425.2	similar to hypothetical protein FLJ22662		561		
Ggs.13065.1.S1_at	BU424424	Gallus gallus finished cDNA, clone ChEST537h21		559		
GgaAffx.5954.1.S1_at	ENSGALT00000015374.1	Gallus gallus similar to LRT5841		557		
GgaAffx.24194.1.S1_at	ENSGALT00000021874.1	Gallus gallus similar to CG8947-PA		534		
Ggs.305.1.S1_at	NM_204783.1	Gallus gallus wingless-type MMTV integration site family, member 4		90	F1 GDR and F1 GCs	238 genes overexpressed in F1 GDR
Ggs.7980.1.S1_at	CD730241	similar to Interferon regulatory factor 6		80		
Ggs.6103.1.S1_x_at	BU242707	Gallus gallus similar to CG3613-PA		75		
Ggs.4901.1.S1_at	BU424477	Gallus gallus similar to carbonic anhydrase 9		72		
Ggs.9136.2.S1_a_at	CD740066	Gallus gallus tumor-associated calcium signal transducer 1		66		
Ggs.5133.1.S1_at	NM_204218.1	Gallus gallus deleted in azoospermia-like	dazL	34		
Ggs.5597.1.S1_at	BU450115	B-cell translocation gene 4	btg4	33		
Ggs.8089.1.S1_at	BU258896	Gallus gallus similar to Kruppel-like transcription factor neptune	ktfn	18		
Ggs.5714.1.S1_at	BX268842	Gallus gallus similar to zona pellucida A	zpA	12		
GgaAffx.9819.1.S1_at	ENSGALT00000024851.1	oocyte maturation factor Mos	chkmos	10		
Ggs.9254.1.S1_at	CF384921	Gallus gallus similar to CG8947-PA		120	F1 GDR And F1 GCs	104 genes overexpressed in F1 GCs
GgaAffx.96.1.S1_s_at	ENSGALT00000000192.1	similar to relaxin 3 preproprotein		68		
Ggs.12454.1.S1_at	BU435007	similar to relaxin 3 preproprotein		68		
Ggs.11031.1.S1_at	BU450054	Gallus gallus finished cDNA, clone ChEST699k2		56		
GgaAffx.24194.1.S1_at	ENSGALT00000021874.1	Gallus gallus similar to CG8947-PA		31		
Ggs.572.1.S1_at	NM_205078.1	Gallus gallus nuclear receptor subfamily 5, group A, member 2		27		
Ggs.10434.1.S1_at	BX933855.1	Gallus gallus similar to chromosome 9 open reading frame 61		19		
Ggs.4510.1.S1_a_at	NM_205361.1	Gallus gallus finished cDNA, clone ChEST159o8		17		
Ggs.7212.1.S1_at	BU124346	Gallus gallus similar to Ephx1 protein		14		
Ggs.3667.1.S1_at	NM_204839.1	Gallus gallus reversion-induced LIM protein		13		
Ggs.14454.1.S1_at	NM_213576.1	Gallus gallus zona pellucida protein D	zpD	602	F1 GDR And Ov GDR	92 genes overexpressed in F1 GDR
Ggs.13391.1.S1_at	BX935169.2	similar to adrenodoxin homolog - chicken		321		
Ggs.596.1.S2_at	NM_205118.1	Gallus gallus 3beta-hydroxysteroid dehydrogenase/delta5-delta4 isomerase		317		
Ggs.17706.1.S1_at	CR388473.1	Finished cDNA, clone ChEST591g11		307		
GgaAffx.5954.1.S1_at	ENSGALT00000015374.1	similar to LRT5841		247		
Ggs.6358.1.S1_at	BX265773	weak similarity to HUMAN Putative protein X123		240		
Ggs.1824.1.S1_at	BU350625	Finished cDNA, clone ChEST974b18		213		
Ggs.13065.1.S1_at	BU424424	Finished cDNA, clone ChEST537h21		195		
Ggs.3095.1.S1_a_at	BX932425.2	similar to hypothetical protein FLJ22662		183		
Ggs.7210.1.S1_at	NM_204389.1	Gallus gallus zona pellucida glycoprotein 3	zpC	64		

Based on GO annotation the genes upregulated in GCs were mostly related to metabolic processes, transport, proteolysis, regulation of transcription, immune response and cell adhesion, whereas genes preferentially expressed in GDR were preferentially involved in cell cycle, chromosome organization, phosphorylation of proteins, regulation of transcription, multicellular organism development and DNA metabolic processes (Fig. 4).

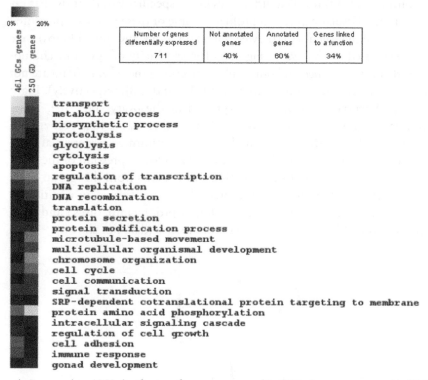

Number of genes differentially expressed	Not annotated genes	Annotated genes	Genes linked to a function
711	40%	60%	34%

transport
metabolic process
biosynthetic process
proteolysis
glycolysis
cytolysis
apoptosis
regulation of transcription
DNA replication
DNA recombination
translation
protein secretion
protein modification process
microtubule-based movement
multicellular organismal development
chromosome organization
cell cycle
cell communication
signal transduction
SRP-dependent cotranslational protein targeting to membrane
protein amino acid phosphorylation
intracellular signaling cascade
regulation of cell growth
cell adhesion
immune response
gonad development

Figure 4. Gene ontology (GO) classification of genes overexpressed in GDR of mature oocytes and in F1 GCs identified using Affymetrix microarray analysis. The rows represent percentages of genes linked to a function calculated from a total number of genes linked to any function. The columns present sets of genes overexpressed in different cell layers. GO annotations were found with Netaffx software, as described in Materials and Methods.

Tissular Pattern of Gene Expression

On the basis of the in silico and microarray approaches, 17 genes were retained for further study. Among these, 16 genes were found using the in silico approach as described above (Table 1), of which 6 genes (chkmos, dazL, btg4, zpA, ktfn and zpC) were also found using the microarray approache (Table 2) and 1 gene (zpD) was found only in the microarray approach. The latter gene was chosen because of its strong involvement in the fertilization process [31]. In contrast to zpA and zpC, the zpD gene was not found in the mouse [32], explaining why it was absent from the list of murine potentially oocyte-specific genes. Four other genes already known for their involvement in final follicular maturation or early embryo development in the chicken or in other vertebrates were added (foxL2 [33], igf2 [34,35], hsf1 [36], and cvh [23]) (Table 3). Finally 21 chicken genes were further studied. Real time RT-PCR performed on 11 adult tissue samples (total ovary, spleen, intestine, gizzard, liver, heart, skin, brain, pectoralis muscle, lung and pituitary gland) revealed differences in the specificity of their tissular expression patterns. Tissular expression profiles of some of these genes are presented on Fig. 4. Seven of these genes (dazL, wee, zar1, zpA, btg4, zpC and chkmos) were specifically expressed in the ovary (Fig 5A). The specificity of zpC and chkmos has previously been described and our results concerning these genes (data are not shown) were in accordance with the literature ([37] and [38], respectively). Eight genes were preferentially and strongly expressed in the ovary (Fig. 5B). Three genes from last group were slightly expressed in another tissue: fbox and zpD in the pituitary gland and ktfn in the muscle. We also confirmed preferential ovarian expression of cvh [23] and foxL2 [33], as well as a low expression of the latter in the pituitary gland (data not shown). Three other genes from this group (igf2, mark3 and znfingerRIZ) were slightly expressed in other tissues, in addition to the ovary. The last 6 genes (trans fact 20, msh4, mtprd, mcmip, discs5 and hsf1) were expressed as highly in the ovary as in other tissues (data not shown).

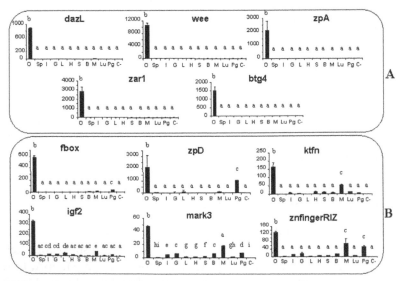

Figure 5. Real-time PCR analysis of the expression of candidate genes in hen tissues. Total RNA was isolated from whole ovary (O), spleen (Sp), intestine (I), gizzard (G), liver (L), heart (H), skin (S), brain (B), pectoralis muscle (M), lung (Lu) and pituitary gland (Pg) and real time PCR was performed as described in the Materials and Methods. Ribosomic RNA 18S was used as a reporter gene. The negative control (water) is indicated as (C-). The results represent the means ± SEM. The same letters indicate that differences were not significant. Different letters indicate that differences were significant (p < 0.05).

Table 3. Accession number of chicken genes selected on the basis of the literature RNA state during follicular maturation and early embryo development

Bibliographic reference	Accession number	Name	Abbreviation
Govoroun et al. 2004	NM_001012612	forkhead box L2	foxL2
Aegerter et al. 2005, Heck et al. 2005	XM_421026	Insulin growth factor 2	Igf 2
Tsunekawa et al. 2000	NM_204708	Gallus gallus DEAD (Asp-Glu-Ala-Asp) box polypeptide 4 (DDX4)	cvh
Hsf1 Anckar et al. 2007	L06098	chicken heat shock factor protein 1	hsf1

Analysis of total RNA, assessed with Agilent RNA nano chips (Fig. 6A), showed an atypical state of RNA from GDR between ovulation and oviposition. In our conditions, rRNA seemed degraded from ovulation until oviposition. With three different RNA extraction methods (tri reagent (Euromedex), RNeasy kit (QIAGEN) and MasterPure™ RNA Purification Kit (Epicentre Biotechnologies)), the RNA profile of GDR from the ovulation stage and from the following embryonic stages remained degraded (data not shown). Moreover, we assessed rRNA 18S and 28S by real time PCR in order to confirm this atypical RNA state (Fig. 6B and 6C). These results confirmed our previous observations; 18S and 28S rRNA expression showed a huge decrease from ovulation untill the oviposition. We then performed labelled reverse transcription to investigate whether mRNA was also degraded. We compared three samples, i.e. the RNA from GDR of F1 stage, from a whole adult ovary that had a normal rRNA profile and GDR of ovulation stage with had degraded rRNA. Electrophoresis profiles of labelled cDNA in denaturing agarose gel were almost identical for the three samples investigated and the smear corresponding to the reverse transcribed mRNA was still present in all these samples (Fig. 6D).

Gene Expression During Follicular Maturation and Early Embryo Development

A detailed study of the expression profiles during follicular maturation and early embryo development using real time RT-PCR was performed for 21 genes, for which tissue specificity of the expression was characterized. First unsupervised hierarchical clustering of our data was performed in order to confirm the biological appropriateness of the selected genes and samples (Fig 7). Two other previously studied genes (gdf9 and bmp15) were added to this analysis in order to facilitate the clustering process because of their already known oocyte-preferential localization [30]. This analysis discriminated two major groups of samples. The first included all granulosa samples from F6 to F1 (correlation threshold 0.58). The second group corresponded to GDR from F6 to the ovulation stage and to all embryo stages (correlation threshold 0.60). Several subgroups could be distinguished within each group (correlation threshold 0.69–0.94). Each subgroup corresponded to a different physiological state, indicated on Fig 6A, suggesting that the genes selected were pertinent. Moreover, unsupervised clustering arranged samples in the perfect chronological order.

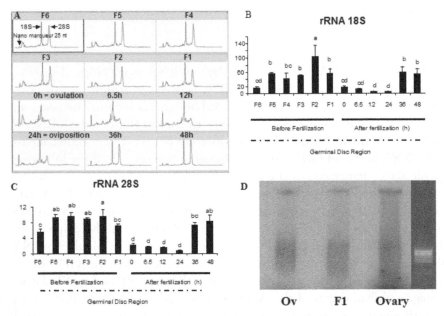

Figure 6. Ribosomic and messenger RNA profiles during oocyte maturation and early embryo development. Total RNA was extracted from the GDR of an oocyte or from the embryo. RNA quality was assessed using nanochips (Agilent technologies) as described in the Materials and Methods. This analysis represents rRNA 18S and 28S subunit profiles in the samples (A). Ribosomic RNA 18S and 28S subunit profiles were then confirmed by real time PCR analysis (B and C, respectively) using the TaqMan kit (Eurogentec) as described in the Materials and Methods. Labelled 32P reverse transcription was performed to investigate the quality of mRNA extracted from F1 GDR, Ov GDR and from the ovary. The radioactive signal for samples and ethidium bromide staining for the ladder are shown (D).

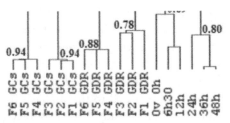

Figure 7. Unsupervised hierarchical clustering of candidate genes during follicular maturation and early embryo development. Real time PCR analysis was performed in duplicate on samples at 18 different stages with two biological replicates. Total GDR and GC RNA from different preovulatory follicles (F1 to F6), and total GDR RNA from just ovulated oocytes and early embryos at 6.5 h, 12 h, 24 h, 36 h and 48 h post ovulation were extracted as described in the Materials and Methods. Unsupervised hierarchical clustering of biological samples was performed using Cluster 3.0 software as described in the Materials and Methods. Node correlation thresholds are indicated.

In order to group genes with similar expression patterns supervised clustering was performed with samples arranged first according to the nature of the sample (GCs or GDRs) and then according to chronological order (Fig. 8). Five clusters (threshold > 0.64) were identified (C1 to C5). A representative example of the expression profile characterizing each cluster is shown on Fig. 9. The chkmos, btg4, wee, zpA, dazL, hsf1, fbox and bmp15 genes, forming cluster C1, were preferentially expressed in GDR. These genes showed a significant increase in mRNA expression in GDR during follicular maturation and a steady decrease after ovulation, becoming nearly undetectable at 36 h after fertilization. The cvh, ktfn and zar1 genes included in Cluster C2 were also preferentially expressed in GDR but, in contrast to genes from cluster C1, they were constantly expressed in the early embryo. Moreover, ktfn and zar1 genes displayed a significant increase in expression from 24 to 48 hours after fertilization. The genes from cluster C3 were expressed in GCs, in GDRs and in the embryo. Moreover, mark3, igf2, gdf9 and transfact20 genes showed a significant decrease in expression during the last stages of follicular maturation in both GDR and GCs. Cluster C4, including discs5, zn-fingerRIZ, foxL2 and mtprd genes, was characterized by expression that was fairly similar to that of cluster C3. The difference consisted of less pronounced variations in gene expression in GCs and more pronounced variations in gene expression in GDR for cluster C4 compared to genes in cluster C3 through the stages investigated. Genes belonging to cluster C5 were expressed in GCs and GDR but, in contrast to the clusters C3 and C4, their expression dropped dramatically at ovulation, especially for zpC and zpD whose transcripts showed increasing expression in GCs during the last stages of follicular maturation and a less marked increase in the expression in GDR. In contrast, the levels of mcmip decreased progressively during the same period. The last gene (msh4) was not clustered. Its expression profile was fairly similar to that of genes belonging to cluster C3, but its expression increased significantly 36 h after fertilization.

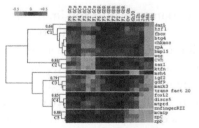

Figure 8. Supervised hierarchical clustering of candidate genes during follicular maturation and early embryo development. Real time PCR analysis was performed in duplicate on samples at 18 different stages with two biological replicates. Total GDR and GC RNA from different preovulatory follicles (F1 to F6), and total GDR RNA from just ovulated oocytes and early embryos at 6.5 h, 12 h, 24 h, 36 h and 48 h post ovulation were extracted as described in the Materials and Methods. Supervised hierarchical clustering of genes was performed using Cluster 3.0 software as described in the Materials and Methods. Five clusters are shown (C1 to C5). Node correlation thresholds are indicated for each cluster.

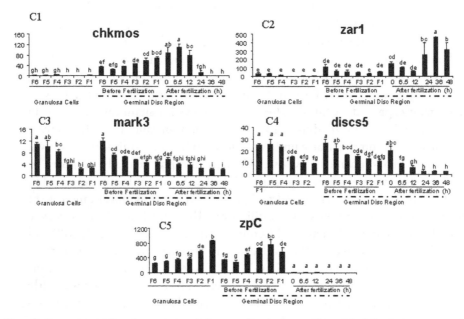

Figure 9. Expression profiles of some representative genes from 5 clusters (C1 to C5), defined using supervised hierarchical clustering, during follicular maturation and early embryogenesis. Real time PCR analyses were performed in duplicate on samples at 18 different stages with two biological replicates. Total mRNA of GDR and GCs from different preovulatory follicles (F1 to F6), and from GDR of just ovulated oocytes and early embryos at 6.5 h, 12 h, 24 h, 36 h and 48 h post ovulation were extracted as described in the Materials and Methods. Three different reporter genes were used (β actin, ef1 α and gapdh) because their expression was stable during oogenesis and early embryo development (data not shown). Results represent means ± SEM. The same letters indicate that differences were not significant. Different letters indicate that differences were significant ($p < 0.05$).

Localization of Gene Expression in the Ovary by in Situ Hybridization

We performed in situ hybridization on mature and immature ovarian sections. All the probes were assessed on the two stages, but only the most significant results are shown on Fig. 10. For all the genes studied we detected homogeneous signals of mRNA expression in oocytes similar to those we have previously described for bmp15 and gdf9 [30]. The mRNA of four of eight genes localized by in situ hybridization (btg4, dazL, cvh and fbox) were detected with a high intensity signal in the oocytes from follicles of 300 μm – 600 μm from both immature (Fig. 10A,B,C and 10D) and mature (data not shown) ovaries. No significant expression was detected in somatic cells. For four other genes (chkmos, zpA, zpC and zpD) a signal was found in oocytes from the largest follicles of 500 μm-6 mm of mature ovaries (Fig. 10E,F,G and 10H). The signal was particularly high for chkmos and zpA (Fig. 10E and 10F), whereas zpC and especially zpD had weaker signals in both oocyte and somatic cells (Fig. 10G and 10H). No significant signal was detected with the corresponding sense probes.

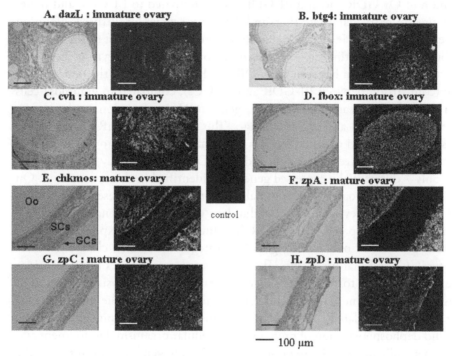

Figure 10. Localization of candidate genes mRNA in the hen ovary. Localization of dazL(A), btg4 (B), vasa (C), fbox (D), chkmos (E), zpA (F), zpC (G) and zpD (H) by in situ hybridization in immature ovaries (A, B, C and D) or in mature ovaries (E, F, G and H) as described in the Materials and Methods. Bright fields are on the left, and dark fields are on the right. The control sections were hybridized with sense probes. The hybridization with only one sense probes is represented because no signal was observed after hybridization with any sense probes. Oocytes (Oo), granulosa cells (GCs) and somatic cells (SCs) are indicated. Scale bar = 100 μm.

Discussion

In the present study we identified and characterized for the first time several genes expressed in the chicken oocyte during follicular maturation and/or in early embryo. Moreover, we showed that our candidate gene approach and microarray approach were complementary in finding new avian genes potentially involved in follicular maturation and/or early embryo development. Five genes preferentially and highly expressed in the oocyte (btg4, chkmos, dazL, zpA and ktfn) were identified using both microarray analysis and digital differential display on murine genes. Moreover, 2 genes (zpC and zpD), identified as overexpressed in GCs by microarray analysis, were also confirmed by real time PCR analysis to be highly preferentially expressed in chicken GCs compared to GDR. Microarray analysis identified a total of 245 genes upregulated in the hen F1 and ovulated oocytes both compared to F1 GCs. Among these, 49 overexpressed genes were common to Ov GDR and and F1 GDR, both compared to F1 GCs, and therefore represent particular interesting candidate oocyte genes for further exploration of their potential role in oocyte maturation, fertilization and OET. The fact that we found almost five times fewer genes overexpressed in the oocyte at the ovulation stage (comparison between Ov GDR and F1 GCs) than in the oocyte at F1 stage (comparison between F1 GDR and F1 GCs), compared to granulosa cells from F1 follicles, means that for some genes mRNA expression in the oocyte decreased between F1 and ovulation stages. This change in the mRNA expression levels between F1 and ovulated oocytes is probably insufficient to be detected in the comparison between Ov GDR and F1 GDR by microarray hybridization, since only 7 differentially expressed genes were identified in this comparison. On the other hand microarray analysis enabled us to detect the presence of GCs in F1 GDR samples, revealed by the redundant overexpressed genes in F1 GDR and in F1 GCs compared to Ov GDR. The functions of overexpressed genes in the various comparisons according to GO categories revealed clear differences between GCs and mature oocytes. For GCs overexpressed genes these functions were mostly related to metabolic processes, transport, proteolysis, regulation of transcription, immune response and cell adhesion, whilst for the mature oocyte they were mostly related to cell cycle, chromosome organization, phosphorylation and dephosphorylation of proteins, multicellular organism development and DNA metabolic process. These presumed functions of genes overexpressed in the oocyte are consistent with the physiological processes that it must undergo: i.e. fertilization, cleavage, chromatin remodeling, and supporting early embryo development.

The use of both bioinformatics and microarray approaches provided information on the molecular mechanisms through which OET is driven in the hen. The expression of several oocyte-specific genes increased during final follicular maturation, suggesting that transcription was still effective. After ovulation, despite the fact that 18S and 28S ribosomal RNA subunits were degraded, we showed by both labelled reverse transcription and microarray analysis that the integrity of mRNA was almost unaffected. In fact, mRNA levels were nearly the same for many genes because only 92 of 28000 genes were differentially expressed between GDRs before and after ovulation, of which only 7 really corresponded to the oocyte genes. The high number of replicates performed and the different extraction methods used strongly indicate that the difference in quality between oocyte rRNA and mRNA after ovulation is not the artifact of the experiment but reflects a real physiological feature of chicken OET, consisting probably of the arrest and degradation of the oocyte translational machinery. It could thus be hypothesized that the maternal translational system has to be replaced by the embryonic translational system. Indeed, the maternal ribosome in the embryo must be degraded before activation of the genome, in other words, before the beginning of transcription, and translation, when new embryonic ribosomes are required. This suggests that, because there are no maternal ribosomes at the stage between ovulation and oviposition, there is probably no translation or only translation of a few specific genes. If this is the case, maternal proteins should be the major essential components that support early embryo development after fertilization. This is supported by the fact that, based on GO annotation, a considerable number of the genes overexpressed in the mature oocyte are related to protein phosphorylation. Further investigation is required onto whether accumulated proteins have such an important role during these early stages of development in birds or if de novo protein synthesis still occurs and is dependent on the oocyte pool of ribosomes as in mammals [39].

On the basis of in silico and microarray approaches and analysis of the literature, 21 chicken genes were chosen in this study for further investigation of their expression using real time PCR and in situ hybridization. All these genes showed state-specific and/or cell type-specific expression patterns throughout the period, beginning from the first stages of final follicular maturation until embryonic genome activation. This suggests that these genes have different functions and have a role at the different physiological stages investigated. The observed decrease in the mRNA expression of almost all genes studied between ovulation and oviposition, which corresponds to late genome activation in chicken embryo [13], is consistent with a potential arrest of transcription and progressive maternal mRNA degradation occurring during meiotic maturation [40]. However, the rate of maternal mRNA degradation in the chicken seems to be considerably lower than that of rRNA, as demonstrated by the present study. Five genes (chkmos, btg4, wee, dazL and zpA) belonging to cluster C1 (Fig. 6B) were no longer expressed after activation of the embryonic genome, and thus transcripts of these genes are only maternally inherited. Moreover, chkmos, btg4 and wee genes are known to be involved in the cell cycle in other species. In our study these maternally inherited genes were increasingly expressed during follicular maturation and thus are probably used during the last stages of final follicular maturation and/ or early embryo development. Chkmos is the chicken homolog of mos [38], protein kinase required for meiotic maturation in vertebrates [41,42] and for mitosis in Xenopus laevis [43]. Meiotic maturation is brought about by steroids using redundant pathways involving synthesis of Mos, which regulates the activity of MPF (M-phase promoting factor). The Mos-MAPK pathway has long been implicated in the arrest of mitosis in vertebrate eggs [43]. The B cell translocation gene 4 (btg4) belongs to a family of cell-cycle inhibitors. In the mouse and bovine, btg4 is preferentially expressed in the oocyte [44,45] where it exerts a marked antiproliferative activity, [46]. Wee is a conserved gene from invertebrates to mammals and regulates meiotic maturation during oocyte development [38,41-43]. The transcripts of dazL are also maternally inherited in the medaka embryo [47], and in adult medaka fish the expression of dazL was detected exclusively in the ovary and in the testis [48].

ZpA, zpC and zpD, that belong to the ZP (zona pellucida) gene family, are known to be involved in oogenesis, fertilization and preimplantation development [49]. In our study the expression pattern of zpA was different from that of zpD and zpC using both real time PCR analysis, where they were distributed in different clusters (C1 for zpA and C5 for zpD and zpC respectively), and in situ hybridization analysis. In contrast to chicken zpD and zpC, which were expressed in oocytes and somatic cells, chicken zpA was found to be specific to the oocyte, as is the case in the mouse [50,51], and expressed earlier than zpD and zpC. In the mouse the expression of zpA also precedes that of zpC [50,51]. Indeed in our in situ hybridization experiments zpA expression detected in small follicles of the mature ovary was oocyte-specific, whereas that of zpC and zpD was weaker and found in oocytes and in the somatic cells of the same follicles. Real time RT-PCR, showed increasing expression of zpC and zpD in both GCs and GDR from F6 to F1. Our finding on the dynamics of zpC in GCs are in accordance with a previously reported study [52]. Moreover at the F1 stage both zpC and zpD were significantly more highly expressed in GCs than in GD and this was consistent with our results for microarray hybridization, but expression decreased dramatically after ovulation. In contrast to the cellular expression pattern found for chicken zpC, in several mammals (murines, bovines and porcines) zpC (zp3) is specifically expressed in the oocyte. However, in the equine species, ZPC protein synthesis is completely taken over by cumulus cells [53]. These findings indicate species specificity of zpC distribution inside the follicle. ZPC protein plays a crucial role in the fertilization process in mammals and birds, [31,37,49].

As zpA other genes (zar1, ktfn and cvh) were preferentially expressed in the oocyte and might play role in fertility (zar1, ktfn) or in germ cell specification (cvh). Both zar1 and ktfn were expressed at higher levels after activation of the embryonic genome. We could therefore hypothesize that these genes might be involved not only in oocyte maturation, but also in early embryo development, just after maternally inherited genes. Of these 2 genes, only zar1 has been studied in reproduction. It is one of the few known oocyte-specific maternal-effect genes essential for OET in mice. In mammals and humans it is hypothesized to be involved in the initiation of embryo development and fertility control [54,55]. CVH protein has been previously proposed to be a part of the mechanism for germ cell specification in birds [16,23]. Our results concerning the spatio-temporal expression of cvh mRNA during follicular maturation and early embryo development are consistent with previously reported studies on the CVH protein.

The genes belonging to clusters C3 and C4 were all preferentially expressed in the ovary in both GCs and GDR and had quite similar expression patterns. Except for foxL2, their expression declined during follicular maturation in GCs and less in GDR and persisted at low levels in the early embryo. This suggests that they are especially involved in the first stages of final follicular maturation as well as in oocyte maturation. The chicken homolog of the mouse par-1a-like gene, i.e. mark3, is required for oocyte differentiation and microtubule organization in the Drosophila [56], and its role in cell polarity and Wnt signaling is conserved from invertebrates to mammals [57,58]. The expression of another gene (msh4), which did not belong to any cluster, also decreased during follicular maturation in GCs and in GD but, as for zar1 and ktfn, it showed a significant increase after embryonic genome activation. Msh4 is known to be involved in mediating recombination of homologous chromosomes and DNA mismatch repair in the mouse [59]. These events occur during the meiotic prophase, the stage where oocytes are blocked for a long time before meiotic maturation.

Conclusion

In conclusion, the findings of the present study on spatio-temporal expression of 8 chicken oocyte specific genes (chkmos, btg4, wee, zpA, dazL, cvh, zar1 and ktfn) were consistent with our hypothesis that oocyte-specific genes in the chicken should play a major role in oocyte maturation, fertilization and early embryo development as in the mouse [12]. Other genes, whose mRNA expression were found in our study to be more specific for GCs or detected in both GCs and GDR depending on the stage, seem to be involved in follicular maturation (foxl2, transfac 20, mark3, ...) and fertilization (zpD and zpC) rather than in early embryo development. Moreover, the microarray approach provided allowed the discovery of a set of new potential chicken mature oocyte and chicken granulosa cell markers for future studies. Interestingly, 40% of these genes had no homologs in the gene databases and some of them probably correspond to specific chicken mechanisms such as hierarchical follicular maturation or rapid yolk accumulation.

Methods

Animals

Laying breed hens aged 60–70 weeks (ISA Brown, egg layer type, Institut de Selection Animale, Saint Brieuc, France) were housed individually in laying batteries with free access to feed and water and were exposed to a 15L:9D photoperiod, with lights-on at 8.00 pm. Individual laying patterns were monitored daily. For in situ hybridization, these hens and younger ones of (10 weeks old) were used to provide mature and immature ovaries, in order to study follicles at each stage. Hens used to provide fertilized eggs were bred in the same conditions and insemi-nated once a week.

Collection of Tissues, Oocytes and Embryos

Hens aged 60 weeks were used. Tissue samples were collected from the ovary, spleen, intestine, gizzard, liver, heart, skin, brain, pectoralis muscle, lung and pitu-itary gland. Germinal disc regions (GDR) and granulosa cells (GCs) surrounding the GDR (Fig. 1) were collected from different preovulatory follicles (F1 to F6), just ovulated oocytes and early embryos at 6.5 h, 12 h, 24 h, 36 h and 48 h post ovulation. The GDR and GCs surrounding the GDR were carefully dissected in the same way under a binocular microscope using fine forceps and scissors (World Precision Instruments) as previously described for quail oocytes [60]. After wash-ing in phosphate buffer saline (PBS, Gibco, Cergy Pontoise, France) GDR and GCs were frozen in liquid nitrogen and then were stored at -80°C until use. For the last two stages, eggs were incubated at 37.8°C for 12 and 24 h, respectively. The 6.5 h, 12 h, and 24 h stages of embryo development correspond to stages I, V and X of the Eyal-Giladi and Kochav classification, respectively [61]. The 36 h and 48 h stages correspond to stages 3 and 6 of the Hamburger-Hamilton classifi-cation, respectively [62]. During follicular maturation the germinal disc is closely associated with its overlying granulosa cells (GCs) and forms a structure called the germinal disc region (GDR) (Figure 1). The GDR from F6-F1 follicles used for these studies consisted not only of the germinal disc but also of the overlying layer of GCs, because GD and overlying GCs cannot be completely separated [60,63,64] and the number of GCs in GDR preparations could not be counted.

Bioinformatic Analysis

A differential digital display analysis has already been performed with mouse ESTs [25-27], providing a list of murine oocyte-specific genes. Using this murine gene list, we systematically searched for chicken orthologs of these genes in internation-al public databases pubmed [65]. Blast bit scores higher than 100 were retained. Moreover, the physical localization of genes identified on chicken chromosomes was retrieved from both mapview [28] and from blat search [29]. We also verified that chicken homologs were localized in the syntenic genomic regions conserved with that of mouse species to have a better chance that true orthologs were studied with ensembl [66].

Rna Isolation and Microarray Analysis

Total RNA was extracted from GDR of F1 and ovulated oocytes, and from GCs of F1 follicles as described above. We thus had 3 samples with a biological replicate of each sample. The RNeasy Mini Kit (QIAGEN, Hilden, Germany) was used according to the manufacturer's instructions. The tissues (GDR or GC) from 25 hens were pooled for each stage investigated to obtain enough RNA for probe synthesis. Two such a pools were constituted for each sample in order to achieve two biological replicates of microarray hybridization. All RNA samples were checked for their integrity on the Agilent 2100 bioanalyzer according to Agilent Technologies guidelines (Waldbroon, Germany). Two micrograms of total RNA were reverse transcribed with the One-cycle cDNA synthesis kit (Affymetrix, Santa Clara, CA), according to the manufacturer's procedure. Clean up of the double-stranded cDNA was performed with Sample Cleanup Module (Affymetrix, Santa Clara, CA) followed by in vitro transcription (IVT) in the presence of biotin-labelled UTP using GeneChip® IVT labelling Kit (Affymetrix). The quantity of the cRNA labelled with RiboGreen® RNA Quantification Reagent (Turner Biosystems, Sunnyvale, CA) was determined after cleanup by the Sample Cleanup Module (Affymetrix). Fragmentation of 15 µg of labelled-cRNA was carried out for 35 minutes at 94°C, followed by hybridization for 16 hours at 45°C to Affymetrix GeneChip® Chicken Genome Array, representing approximately 32,773 transcripts, corresponding to over different 28,000 Gallus gallus genes. After hybridization, the arrays were washed with 2 different buffers (stringent: 6X SSPE, 0.01% Tween-20 and non-stringent: 100 mM MES, 0.1 M [Na+], 0.01% Tween-20) and stained with a complex solution including Streptavidin R-Phycoerythrin conjugate (Invitrogen/molecular probes, Carlsbad, CA) and anti Streptavidin biotinylated antibody (Vectors laboratories, Burlingame, CA). The washing and staining steps were performed in a GeneChip® Fluidics Station 450 (Affymetrix). The Affymetrix GeneChip® Chicken Genome Arrays were finally scanned with the GeneChip® Scanner 3000 7G piloted by the GeneChip® Operating Software (GCOS).

All these Steps were Performed on Affymetrix Equipement at Inra-Urgv, Evry, France

The raw CEL files were imported in R software for data analysis. All raw and normalized data are available from the Gene Expression Omnibus (GEO) repository at the National Center for Biotechnology Information (NCBI) [67], accession number GSE7805. Gene Ontology annotations were performed with NetAffx.

RNA Extraction and Reverse Transcription

Total RNA was extracted from whole adult tissues (ovary, spleen, intestine, gizzard, liver, heart, skin, brain, pectoralis muscle, lung and pituitary gland) using Tri-reagent (Euromedex, Mundolsheim, France) according to the manufacturer's procedure. RNA quality and quantity were then assessed by using RNA nano chips and Agilent RNA 6000 nano reagents (Agilent Technologies, Waldbronn, Germany) according to the manufacturer's instructions. Samples were stored at -80°C until use. Reverse transcription (RT) was performed to test the expression of candidate genes in different tissues and at different stages of follicular maturation and embryo development using polymerase chain reactions (PCR). One microgram of total RNA extracted from tissues or GDR was digested by RQ1 DNase (Promega, Madison, WI, USA) and reverse transcribed to first-strand cDNA using Moloney Murine Leukemia Virus reverse transcriptase I with an oligo dT-random primer mix (Promega, Madison, WI, USA) according manufacturer's instructions.

Labelled RT was performed in order to assess mRNA quality in GDR of F1 stage and just ovulated oocytes. Ten μCi αP32 dCTP was added to the reverse transcription mix in order to label cDNA. Labelled cDNA was then separated on 1.2% denaturing agarose gel (50 mM NAOH, 1 mM EDTA) by electrophoresis. A storage phosphor screen (Amersham Biosciences, Bucks, UK) was placed on the gel in an exposure cassette (Amersham Biosciences, Bucks, UK). The signal was detected one hour later with a STORM 840 (Molecular Dynamics), a phosphor screen imaging system.

Table 4. Oligonucleotide primer sequences

	Abbreviation	Forward	Reverse	Efficiency	Length (bp)
Real time PCR	btg4	TTGGGTGTTTTTGGGAGG	AGTGCTTCAACTGCTTCTCAGACC	1.93	187
	chkmos	TACTCGTGTGACATCGTGACTGGC	TTGCTGGCAAACATGGTGGC	2.09	177
	dazL	TACCCATTCGTCAACAACCTGC	CCCTTTGGAAACACCAGTTCTG	1.88	194
	fbox	ACCTGTGCTGGATGATGTTGACC	CAACAAGAGGTATGTGCTTTGCG	1.81	197
	hsf1	ACCCCTATTTCATTCCGTGGC	AGTCCATGCTCTCCTGCTTTCC	1.80	165
	mah4	GATTCTCGGAATGGTCACACGC	AGCATCAACAAGTGGCTCCAGG	2.16	105
	mrgrd	TGAAGATCAAGGTCCAACAACTGC	TTGCTTCCTGAAACCTTTGGC	1.80	180
	mcmip	CGTGATGGCAGGAGAGAAAAGC	ATGAAGTGAGGAAGGATGGGTGAC	1.80	165
			C		
	zn finger RiZ	GAGCAAAAGAGTACATCAGAGG	CCATTTGATTCACCTCTTGC	1.80	89
	mark3	AGTGAAAGAACCACTGCTGATAGG	TTGAAGCGACAGGCGTTCTTT	1.83	91
		C			
	dixcx5	ATGACTGTATGGTGGTTGAGACTG	CGGCCTTTTGAAAGTATCAGGG	1.81	176
	trans fact 20	TCAAAACTCAGCACCAGCCC	GTGCAGGTTTCTCTTGTCCACC	1.92	178
	wee	GGAGAAGAGGGTGAAGACAGAAG	TGAGCTTCCTGTGAGGAGTTGC	1.90	121
		C			
	zar1	GTTTGTTTAGGGCTCTTCCAGGG	TTTACTCGCAGCTTTCCCAGC	2.20	92
	kdfn	ACTATTTTTCTCCTCCTGCCTCTGC	AATGCCAAATACAAGCGGGG	2.05	186
	zpA	CCTTAAATCCAACAACTCCACAGC	CAGCAAAAATCCCAACAAGAGG	1.92	131
	zpC	ACTAGCTCTGCCTTCATCACACCC	GGCAGGTGATGTAGATCAGGTTCC	1.82	109
	foxL2	Govorrun et al. 2004		1.83	
	igf 2	Heck et al. 2003		1.85	
	cvh	AGTTCCTGGCATCTTTGGGC	AGCGTCCTTTGAGAACTCCTGC	2.00	131
	zpD	TCATTGAGACAGGGAGGGAAGC	TCTTCACCACCTGCTCATAGGC	1.90	102
	gapdh	TGCTGCCCAGAACATCATCC	ATCAGCAGCAGCCTTCACTACC	2.05	199
	beta actin	CAGATGTGGATCAGCAAGCAGG	TTTCATCACAGGGGTGTGGG	1.88	107
	efl alpha	AGC AGA CTT TGT GAC CTT GCC	TGA CAT GAG ACA GAC GGT TGC	2.00	90
in situ hybridization	btg4	ACGGTCTTCTTCATCACGAGGC	TCTGTAGCACCAGCCTTCATCC		729
	chkmos	CCCTGGCAAAGATGGAAAAGC	CAGAGGGTGATGGCAAAGGAGT		733
	dazL	TGTTTTTAAGTGTGCGGGCG	ACTATTACCAGACATCTGTGTGGG		749
			C		
	fbox	TGTGTTCCTTTACCCCGATTGC	AACTGCTACACTGCTTTCAGTCAG		794
			G		
	zpA	TCATCGCTCCTCTCTTTGTTGG	TTTGCATGTGGGATCTCTGAGC		773
	zpC	TACCGCACGCTCATCAACTACG	ATCAGCTGCAACCTCTTTCCCG		744
	cvh	AGTTCCTGGCATCTTTGGGC	ATATCGAGCATGCGGTCTGC		792
	zpD	TATTTGCTGCTGTTCTCTGCCC	TGGTGCTGCCCTTCTATCTTCA		740

Real Time RT-PCR

Specific sets of primer pairs (Sigma Genosis), designed using Vecteur NTI software to amplify fragments of 21 different transcripts, are shown in Table 4. Real time PCR reactions were carried out in 25 µl containing primers at a final concentration of 150 nm of each, 5 µl of the RT reaction diluted 1/30 and qPCR Mastermix Plus for Sybr Green I (Eurogentec) according to the manufacturer's instructions. Real time PCR was performed using an ABI Prism 7000 (Applied Biosystems). After incubation at 50°C for 2 min, and 95°C for 10 min, the thermal cycling protocol was as follows: 40 cycles at 95°C for 15 sec and 60°C for 1 min. The amounts of 18S rRNA and 28S rRNA were measured in the RT reactions diluted 1/5000 using 28S rRNA and 18S rRNA control kits (Eurogentec), respectively, according to the manufacturer's instructions. PCR amplification without cDNA was performed systematically as a negative control. 18S rRNA was used as a reporter gene in the study of the mRNA tissular expression pattern, whereas β actin, gapdh and ef1α were used as reporter genes in the mRNA expression study during follicular maturation and early embryo development, the expression of these three genes being similar in GCs and in GDR (data not shown). In both cases (tissular expression study or temporal dynamic study) the samples were analyzed in duplicate on the same plate for a given gene. Real time RT-PCR was performed in the temporal dynamic study with two biological replicates of each sample. Melting curve analysis was systematically performed for all genes in order to verify the specificity of the PCR product. Real time PCR efficiency (E) was measured in duplicate on serial dilutions of cDNA (pool of reverse-transcribed RNA from GDR and GC samples at the same stages as used for the temporal dynamic study) for each primer pair and was calculated using the following equation: E = (101/slope)-1. In the tissular expression study the relative amounts of gene transcripts (R) were calculated according to the equation:

$$R = \frac{E_{gene}^{-Ct} gene}{E_{18S}^{-Ct} \; 18S \; rRNA} \\ rRNA$$

where Ct is a cycle threshold. In the temporal dynamic study R was calculated using following equation:

$$R = \frac{E_{gene}^{-Ct} gene}{Mean(Rc_{ef1\alpha}, Rc_{\beta actin}, Rc_{gapdh})}$$

where Rc is corrected relative reporter gene expression calculated as explained below. In order to take into account only fold changes in the expression levels of reporter genes between samples but not the differences in the expression levels of reporter genes in the sample, the expression levels of reporter genes in each sample (samplei) was adjusted against the relative amount of ef1α in the F1GDR sample according to the equation:

$$Rc_{reporter} = E_{ef1\alpha}^{-Ct_{ef1\alpha \text{ in F1GDR}}} \times E_{reporter}^{\Delta Ctreporter}$$

where $\Delta Ct_{reporter} = Ct_{reporter \text{ in F1GDR}} - Ct_{reporter \text{ in sample}_i}$.

Hierarchical Clustering

The hierarchical classification of data obtained using real time RT-PCR was performed with the Cluster 3.0 program using unsupervised single linkage or supervised complete linkage clustering in order to classify biological samples or to group together genes with a similar expression pattern, respectively [68].

In Situ Hybridization

Female chickens were sacrificed at different stages of sexual development. Two types of tissue were used, i.e. mature ovaries, containing follicles of different sizes (50 μm-7 mm) from 60-week-old hens (most follicles being larger than 300 μm) and immature ovaries, containing a majority of small follicles (25–500 μm) from 10-week-old hens (most follicles being smaller than 100 μm). Mature and immature ovaries were then collected and included in Tissue-Tek (Sakura Finetek Europe BV, Zoeterwoude, The Netherlands). Frozen ovaries were serially sectioned with a cryostat (thickness 10 μm) to perform in situ hybridization experiments using 35S-labeled chicken gene cRNA. The gene antisense and sense constructs used for in situ hybridization were generated by inserting 700 – 800 bp fragments of chicken gene cDNA into the pGEM-T vector (Promega, Madison, WI, USA), and selecting a clone with the appropriate antisense or sense orientation. The gene cDNA fragments were generated by RT-PCR from chicken ovary mRNA using forward and reverse primers (Table 4). The in situ hybridization was performed as previously described [69]. Hybridization specificity was assessed by comparing signals obtained with the cRNA antisense probe and the corresponding cRNA sense probe.

Statistical Analysis

Data obtained after Affymetrix microarray hybridization analysis were normalized with the gcrma algorithm [70], available in the Bioconductor package [71].

Differential analysis was performed with the varmixt package of R [72]. A double-sided, unpaired t-test was computed for each gene between the two conditions. Variance of the difference in gene expression was split between subgroups of genes with homogeneous variance [72]. The raw P values were adjusted by the Bonferroni method, which controls the Family Wise Error Rate (FWER) [73]. A gene is declared differentially expressed if the **Bonferroni-Corrected P-Value Is Less Than 0.05.**

All other experimental data are presented as means ± SEM. One-way analysis of variance (ANOVA) was used to test differences. If ANOVA revealed significant effects, the means were compared by Fisher's test, with P < 0.05 considered significant. Different letters indicate significant differences.

Authors' Contributions

SE performed the experiences, the sequence alignment and the microarray analysis. FB participated in the design of the study and coordinated oocytes and embryos collection. IC participated in the experiences. SB carried out the microarrays hybridization. MLMM performed the statistical analysis of microarray data. PM participated in the design of the study and provided the list of murine oocyte genes identified in silico. EB participated in the design of the study. MSG conceived of the study, and participated in its design and coordination. SE and MSG wrote and revised the manuscript. All authors read and approved the final manuscript.

Acknowledgements

We thank Svétlana Uzbekova and Rozenn Dalbies-Tran for helpful discussion, Sonia Métayer for igf2 primers and Frederic Mercerand and Jean-Didier Terlot-Bryssine for expert animal care. This study was supported by the "Institut National de la Recherche Agronomique." S Elis was supported by a fellowship from the Institut National de la Recherche Agronomique and "Région Centre."

References

1. Evsikov AV, Graber JH, Brockman JM, Hampl A, Holbrook AE, Singh P, Eppig JJ, Solter D, Knowles BB: Cracking the egg: molecular dynamics and evolutionary aspects of the transition from the fully grown oocyte to embryo. Genes Dev 2006, 20:2713–2727.

2. Evsikov AV, de Vries WN, Peaston AE, Radford EE, Fancher KS, Chen FH, Blake JA, Bult CJ, Latham KE, Solter D, Knowles BB: Systems biology of the 2-cell mouse embryo. Cytogenet Genome Res 2004, 105:240–250.

3. Hake LE, Richter JD: Translational regulation of maternal mRNA. Biochimica et Biophysica Acta (BBA) - Reviews on Cancer 1997, 1332:M31–M38.

4. Bashirullah A, Cooperstock RL, Lipshitz HD: Spatial and temporal control of RNA stability. Proc Natl Acad Sci USA 2001, 98:7025–7028.

5. DeRenzo C, Seydoux G: A clean start: degradation of maternal proteins at the oocyte-to-embryo transition. Trends Cell Biol 2004, 14:420–426.

6. Alizadeh Z, Kageyama S, Aoki F: Degradation of maternal mRNA in mouse embryos: selective degradation of specific mRNAs after fertilization. Mol Reprod Dev 2005, 72:281–290.

7. Schupbach T, Wieschaus E: Female sterile mutations on the second chromosome of Drosophila melanogaster. I. Maternal effect mutations. Genetics 1989, 121:101–117.

8. Golden A, Sadler PL, Wallenfang MR, Schumacher JM, Hamill DR, Bates G, Bowerman B, Seydoux G, Shakes DC: Metaphase to anaphase (mat) transition-defective mutants in Caenorhabditis elegans. J Cell Biol 2000, 151:1469–1482.

9. Christians ES: [When the mother further impacts the destiny of her offspring: maternal effect mutations]. Med Sci (Paris) 2003, 19:459–464.

10. Pelegri F, Knaut H, Maischein HM, Schulte-Merker S, Nusslein-Volhard C: A mutation in the zebrafish maternal-effect gene nebel affects furrow formation and vasa RNA localization. Curr Biol 1999, 9:1431–1440.

11. Kubota HY, Itoh K, Asada-Kubota M: Cytological and biochemical analyses of the maternal-effect mutant embryos with abnormal cleavage furrow formation in Xenopus laevis. Dev Biol 1991, 144:145–151.

12. Dean J: Oocyte-specific genes regulate follicle formation, fertility and early mouse development. J Reprod Immunol 2002, 53:171–180.

13. Zagris N, Kalantzis K, Guialis A: Activation of embryonic genome in chick. Zygote 1998, 6:227–231.

14. Meirelles FV, Caetano AR, Watanabe YF, Ripamonte P, Carambula SF, Merighe GK, Garcia SM: Genome activation and developmental block in bovine embryos. Animal Reproduction Science Research and Practice III 15th International Congress on Animal Reproduction 2004, 82–83:13–20.

15. Eric M. Thompson EL Jean-Paul Renard,: Mouse embryos do not wait for the MBT: Chromatin and RNA polymerase remodeling in genome activation at the onset of development. Developmental Genetics 1998, 22:31–42.

16. Callebaut M: Origin, fate, and function of the components of the avian germ disc region and early blastoderm: role of ooplasmic determinants. Dev Dyn 2005, 233:1194–1216.

17. Yao HH, Bahr JM: Chicken granulosa cells show differential expression of epidermal growth factor (EGF) and luteinizing hormone (LH) receptor messenger RNA and differential responsiveness to EGF and LH dependent upon location of granulosa cells to the germinal disc. Biol Reprod 2001, 64:1790–1796.

18. Etches RJ, Petitte JN: Reptilian and avian follicular hierarchies: models for the study of ovarian development. J Exp Zool Suppl 1990, 4:112–122.

19. Olszanska B: [Role of polyadenylic segments and RNA polyadenylation in embryonic development]. Postepy Biochem 1985, 31:365–384.

20. Olszanska B, Borgul A: Quantitation of nanogram amounts of nucleic acids in the presence of proteins by the ethidium bromide staining technique. Acta Biochim Pol 1990, 37:59–63.

21. Olszanska B, Borgul A: Maternal RNA content in oocytes of several mammalian and avian species. J Exp Zool 1993, 265:317–320.

22. Olszanska B, Kludkiewicz B, Lassota Z: Transcription and polyadenylation processes during early development of quail embryo. J Embryol Exp Morphol 1984, 79:11–24.

23. Tsunekawa N, Naito M, Sakai Y, Nishida T, Noce T: Isolation of chicken vasa homolog gene and tracing the origin of primordial germ cells. Development 2000, 127:2741–2750.

24. Wang Y, Li J, Ying Wang C, Yan Kwok AH, Leung FC: Epidermal growth factor (EGF) receptor ligands in the chicken ovary: I. Evidence for heparin-binding EGF-like growth factor (HB-EGF) as a potential oocyte-derived signal to control granulosa cell proliferation and HB-EGF and kit ligand expression. Endocrinology 2007, 148:3426–3440.

25. Dade S, Callebaut I, Mermillod P, Monget P: Identification of a new expanding family of genes characterized by atypical LRR domains. Localization of a cluster preferentially expressed in oocyte. FEBS Lett 2003, 555:533–538.

26. Dade S, Callebaut I, Paillisson A, Bontoux M, Dalbies-Tran R, Monget P: In silico identification and structural features of six new genes similar to MATER specifically expressed in the oocyte. Biochem Biophys Res Commun 2004, 324:547–553.

27. Paillisson A, Dade S, Callebaut I, Bontoux M, Dalbies-Tran R, Vaiman D, Monget P: Identification, characterization and metagenome analysis of oocyte-specific genes organized in clusters in the mouse genome. BMC Genomics 2005, 6:76.

28. Mapview [http://www.ncbi.nlm.nih.gov/projects/mapview/].

29. blat search [http://www.genome.ucsc.edu/cgi-bin/hgBlat].

30. Elis S, Dupont J, Couty I, Persani L, Govoroun M, Blesbois E, Batellier F, Monget P: Expression and biological effects of bone morphogenetic protein-15 in the hen ovary. J Endocrinol 2007, 194:485–497.

31. Okumura H, Kohno Y, Iwata Y, Mori H, Aoki N, Sato C, Kitajima K, Nadano D, Matsuda T: A newly identified zona pellucida glycoprotein, ZPD, and dimeric ZP1 of chicken egg envelope are involved in sperm activation on sperm-egg interaction. Biochem J 2004, 384:191–199.

32. Goudet G, Mugnier S, Callebaut I, Monget P: Phylogenetic Analysis and Identification of Pseudogenes Reveal a Progressive Loss of Zona Pellucida Genes During Evolution of Vertebrates. Biol Reprod 2007.

33. Marina S Govoroun MP Eric Pailhoux, Julie Cocquet, Jean-Pierre Brillard, Isabelle Couty, Florence Batellier, Corinne Cotinot,: Isolation of chicken homolog of the FOXL2 gene and comparison of its expression patterns with those of aromatase during ovarian development. Developmental Dynamics 2004, 231:859–870.

34. Aegerter S, Jalabert B, Bobe J: Large scale real-time PCR analysis of mRNA abundance in rainbow trout eggs in relationship with egg quality and post-ovulatory ageing. Mol Reprod Dev 2005, 72:377–385.

35. Heck A, Metayer S, Onagbesan OM, Williams J: mRNA expression of components of the IGF system and of GH and insulin receptors in ovaries of broiler breeder hens fed ad libitum or restricted from 4 to 16 weeks of age. Domest Anim Endocrinol 2003, 25:287–294.

36. Anckar J, Sistonen L: Heat shock factor 1 as a coordinator of stress and developmental pathways. Adv Exp Med Biol 2007, 594:78–88.

37. Okumura H, Aoki N, Sato C, Nadano D, Matsuda T: Heterocomplex Formation and Cell-Surface Accumulation of Hen's Serum Zona Pellucida B1 (ZPB1) with ZPC Expressed by a Mammalian Cell Line (COS-7): A Possible Initiating Step of Egg-Envelope Matrix Construction. Biol Reprod 2007, 76:9–18.

38. Schmidt M, Oskarsson MK, Dunn JK, Blair DG, Hughes S, Propst F, Vande Woude GF: Chicken homolog of the mos proto-oncogene. Mol Cell Biol 1988, 8:923–929.

39. Maddox-Hyttel P, Svarcova O, Laurincik J: Ribosomal RNA and nucleolar proteins from the oocyte are to some degree used for embryonic nucleolar formation in cattle and pig. Theriogenology 2007, 68 Suppl 1:S63–70.

40. De La Fuente R, Viveiros MM, Burns KH, Adashi EY, Matzuk MM, Eppig JJ: Major chromatin remodeling in the germinal vesicle (GV) of mammalian oocytes is dispensable for global transcriptional silencing but required for centromeric heterochromatin function. Dev Biol 2004, 275:447–458.

41. Haccard O, Jessus C: Oocyte maturation, Mos and cyclins--a matter of synthesis: two functionally redundant ways to induce meiotic maturation. Cell Cycle 2006, 5:1152–1159.

42. Inoue D, Ohe M, Kanemori Y, Nobui T, Sagata N: A direct link of the Mos-MAPK pathway to Erp1/Emi2 in meiotic arrest of Xenopus laevis eggs. Nature 2007, 446:1100–1104.

43. Yue J, Ferrell JE Jr: Mechanistic Studies of the Mitotic Activation of Mos. Mol Cell Biol 2006, 26:5300–5309.

44. Buanne P, Corrente G, Micheli L, Palena A, Lavia P, Spadafora C, Lakshmana MK, Rinaldi A, Banfi S, Quarto M, Bulfone A, Tirone F: Cloning of PC3B, a novel member of the PC3/BTG/TOB family of growth inhibitory genes, highly expressed in the olfactory epithelium. Genomics 2000, 68:253–263.

45. Pennetier S, Uzbekova S, Guyader-Joly C, Humblot P, Mermillod P, Dalbies-Tran R: Genes Preferentially Expressed in Bovine Oocytes Revealed by Subtractive and Suppressive Hybridization. Biol Reprod 2005, 73:713–720.

46. Vallee M, Gravel C, Palin MF, Reghenas H, Stothard P, Wishart DS, Sirard MA: Identification of Novel and Known Oocyte-Specific Genes Using Complementary DNA Subtraction and Microarray Analysis in Three Different Species. Biol Reprod 2005, 73:63–71.

47. Aizawa K, Shimada A, Naruse K, Mitani H, Shima A: The medaka midblastula transition as revealed by the expression of the paternal genome. Gene Expr Patterns 2003, 3:43–47.

48. Xu H, Li M, Gui J, Hong Y: Cloning and expression of medaka dazl during embryogenesis and gametogenesis. Gene Expression Patterns 2007, 7:332–338.

49. Wassarman PM, Jovine L, Litscher ES: Mouse zona pellucida genes and glycoproteins. Cytogenet Genome Res 2004, 105:228–234.

50. Zeng F, Schultz RM: Gene expression in mouse oocytes and preimplantation embryos: use of suppression subtractive hybridization to identify oocyte- and embryo-specific genes. Biol Reprod 2003, 68:31–39.

51. Epifano O, Liang LF, Familari M, Moos MC Jr., Dean J: Coordinate expression of the three zona pellucida genes during mouse oogenesis. Development 1995, 121:1947–1956.

52. Takeuchi Y, Nishimura K, Aoki N, Adachi T, Sato C, Kitajima K, Matsuda T: A 42-kDa glycoprotein from chicken egg-envelope, an avian homolog of the ZPC family glycoproteins in mammalian zona pellucida. Its first identification, cDNA cloning and granulosa cell-specific expression. European Journal of Biochemistry 1999, 260:736–742.

53. Kolle S, Dubois CS, Caillaud M, Lahuec C, Sinowatz F, Goudet G: Equine zona protein synthesis and ZP structure during folliculogenesis, oocyte maturation, and embryogenesis. Mol Reprod Dev 2007, 74:851–859.

54. Wu X, Viveiros MM, Eppig JJ, Bai Y, Fitzpatrick SL, Matzuk MM: Zygote arrest 1 (Zar1) is a novel maternal-effect gene critical for the oocyte-to-embryo transition. Nat Genet 2003, 33:187–191.

55. Uzbekova S, Roy-Sabau M, Dalbies-Tran R, Perreau C, Papillier P, Mompart F, Thelie A, Pennetier S, Cognie J, Cadoret V, Royere D, Monget P, Mermillod P: Zygote arrest 1 gene in pig, cattle and human: evidence of different transcript variants in male and female germ cells. Reprod Biol Endocrinol 2006, 4:12.

56. Cox DN, Lu B, Sun TQ, Williams LT, Jan YN: Drosophila par-1 is required for oocyte differentiation and microtubule organization. Curr Biol 2001, 11:75–87.

57. Ossipova O, He X, Green J: Molecular cloning and developmental expression of Par-1/MARK homologues XPar-1A and XPar-1B from Xenopus laevis. Mech Dev 2002, 119 Suppl 1:S143–8.

58. Ossipova O, Dhawan S, Sokol S, Green JB: Distinct PAR-1 proteins function in different branches of Wnt signaling during vertebrate development. Dev Cell 2005, 8:829–841.

59. Acevedo N, Smith GD: Oocyte-specific gene signaling and its regulation of mammalian reproductive potential. Front Biosci 2005, 10:2335–2345.

60. Malewska A, Olszanska B: Accumulation and localisation of maternal RNA in oocytes of Japanese quail. Zygote 1999, 7:51–59.

61. Eyal-Giladi H, Kochav S: From cleavage to primitive streak formation: a complementary normal table and a new look at the first stages of the development of the chick. I. General morphology. Dev Biol 1976, 49:321–337.

62. Hamburger V, Hamilton HL: A series of normal stages in the development of the chick embryo. 1951. Dev Dyn 1992, 195:231–272.

63. Perry MM, Gilbert AB, Evans AJ: The structure of the germinal disc region of the hen's ovarian follicle during the rapid growth phase. J Anat 1978, 127:379–392.

64. Tischkau SA, Bahr JM: Avian germinal disc region secretes factors that stimulate proliferation and inhibit progesterone production by granulosa cells. Biol Reprod 1996, 54:865–870.

65. pubmed [http://www.ncbi.nlm.nih.gov/BLAST/Blast.cgi].

66. ensembl [http://www.ensembl.org/Gallus_gallus/syntenyview].

67. Wheeler DL, Barrett T, Benson DA, Bryant SH, Canese K, Chetvernin V, Church DM, DiCuccio M, Edgar R, Federhen S, Geer LY, Helmberg W, Kapustin Y, Kenton DL, Khovayko O, Lipman DJ, Madden TL, Maglott DR, Ostell J, Pruitt KD, Schuler GD, Schriml LM, Sequeira E, Sherry ST, Sirotkin K, Souvorov A, Starchenko G, Suzek TO, Tatusov R, Tatusova TA, Wagner L, Yaschenko E: Database resources of the National Center for Biotechnology Information. Nucleic Acids Res 2006, 34:D173–80.

68. Eisen MB, Spellman PT, Brown PO, Botstein D: Cluster analysis and display of genome-wide expression patterns. Proc Natl Acad Sci USA 1998, 95:14863–14868.

69. Pierre A, Pisselet C, Dupont J, Bontoux M, Monget P: Bone morphogenetic protein 5 expression in the rat ovary: biological effects on granulosa cell proliferation and steroidogenesis. Biol Reprod 2005, 73:1102–1108.

70. Irizarry RA, Hobbs B, Collin F, Beazer-Barclay YD, Antonellis KJ, Scherf U, Speed TP: Exploration, normalization, and summaries of high density oligonucleotide array probe level data. Biostatistics 2003, 4:249–264.

71. Gentleman R, Carey V: Bioconductor. RNews 2002, 2:1116.

72. Delmar P, Robin S, Daudin JJ: VarMixt: efficient variance modelling for the differential analysis of replicated gene expression data. Bioinformatics 2005, 21:502–508.

73. Ge Y, Dudoit S, Speed TP: Resampling-based multiple testing for microarray data analysis. TEST 12 2003, 1–44.

Ectopic Pregnancy Rates with Day 3 Versus Day 5 Embryo Transfer: A Retrospective Analysis

Amin A. Milki and Sunny H. Jun

ABSTRACT

Background

Blastocyst transfer may theoretically decrease the incidence of ectopic pregnancy following IVF-ET in view of the decreased uterine contractility reported on day 5. The purpose of our study is to specifically compare the tubal pregnancy rates between day 3 and day 5 transfers.

Methods

A retrospective analysis of all clinical pregnancies conceived in our IVF program since 1998 was performed. The ectopic pregnancy rates were compared for day 3 and day 5 transfers.

Results

There were 623 clinical pregnancies resulting from day 3 transfers of which 22 were ectopic (3.5%). In day 5 transfers, there were 13 ectopic pregnancies out of 333 clinical pregnancies (3.9%). The difference between these rates is not statistically significant (P = 0.8).

Conclusions

Our data suggests that the ectopic pregnancy rate is not reduced following blastocyst transfer compared to day 3 transfer. While there may be several benefits to extended culture in IVF, the decision to offer blastocyst transfer should be made independently from the issue of ectopic pregnancy risk.

Keywords: Blastocyst, Ectopic pregnancy, Embryo transfer, IVF

Background

Ectopic pregnancy has been reported to occur in approximately 2–5% of all clinical pregnancies after IVF-ET [1-4]. Although the direct injection of transfer media with embryos into the fallopian tubes may account for the development of tubal pregnancies after IVF, migration of embryos to the tubes by reflux expulsion from uterine contractions has been proposed as another possible explanation. [3,5]

Uterine junctional zone activity has been shown to decrease with increasing time after oocyte retrieval. [6]. When comparing day 2 to day 3 transfers, Lesny et al. [3] showed a trend for a lower ectopic pregnancy rate in the day 3 transfer group which they attributed to the decreased uterine contractility further along in the luteal phase. Fanchin et al. [7] reported a significant reduction in retrograde uterine contractility, from the cervix to the fundus, 7 days after hCG administration compared to both 4 days after and the day of hCG injection. These findings suggest that blastocyst transfer should be associated with a lower incidence of ectopic pregnancy compared to cleavage stage transfer. The larger diameter of the blastocyst was proposed as an additional factor in reducing the rate of tubal pregnancies after day 5 transfer. [8]

Despite these theoretical considerations, large series that specifically compare the incidence of ectopic pregnancy with blastocyst versus cleavage stage transfers are lacking in the literature. The purpose of our study is to shed light on this issue by examining the ectopic pregnancy rates after day 3 transfer compared to day 5 transfer in our program over a 5 year period.

Methods

We reviewed all clinical pregnancies conceived in our IVF program since 1998 when blastocyst transfer was introduced to our center. The incidence of ectopic pregnancy was compared between day 3 and day 5 transfers in the same time period.

The controlled ovarian hyperstimulation protocol consisted of pretreatment with oral contraceptive pills with overlapping GnRH agonist down-regulation followed by FSH/hHMG and hCG, microdose flare or antagonist protocols. Oocytes were inseminated conventionally or by ICSI 3–4 hours after retrieval. Embryos were cultured in groups under mineral oil in 150 μL droplets of P1 medium (Irvine Scientific, Santa Ana, CA, USA) with 10% Serum Substitute Supplement (SSS) at 37 degrees Celsius in a 5% O_2, 5% CO_2 and 90% N_2 environment for 72 hours. For the blastocyst transfer group, the embryos were moved on day 3 into Blastocyst medium (Irvine Scientific) with 10% SSS and cultured for 48 hours before transfer. Additional blastocysts were cryopreserved on day 5 or day 6.

All transfers were performed using a Tefcat catheter (Cook Ob/Gyn, Spencer, IN, USA) 1 to 1.5 cm short of the fundus under transabdominal ultrasound guidance. The transfer volume was 20–30 μL.

Clinical pregnancies were defined by seeing a gestational sac on transvaginal ultrasound or by diagnosing an ectopic pregnancy. Ectopic pregnancies were diagnosed by ultrasound or by laparoscopic visualization of a gestational sac in the fallopian tube or by the absence of an intrauterine gestational sac and rising βhCG levels following the failure of suction D&C to reveal products of conception.

The rate of ectopic pregnancies for day 3 and day 5 transfers was compared. Chi-square testing was used for statistical analysis. Significance was set at $P <$ 0.05. Institutional review board approval was obtained for chart review.

Results

There were 623 clinical pregnancies resulting from day 3 transfer of which 22 were ectopic (3.5%). In day 5 transfers, there were 13 ectopic pregnancies out of 333 clinical pregnancies (3.9%). The difference between these rates is not statistically significant ($P = 0.8$). Of the 22 ectopic pregnancies with day 3 transfer, 9 were in patients with tubal disease compared to 5 out of the 13 ectopic pregnancies with day 5 transfer ($P = 0.9$). More importantly, the incidence of tubal disease was similar in the day 3 transfer and the day 5 transfer groups, 22 and 24%, respectively ($P = 0.4$). The mean ages were 37.7(± 4.9) years in the day 3 group and 35.3 (± 4.7) years in the day 5 group (P < 0.01). The mean BMI was similar in both groups. (Table 1)

Table 1. Comparison of ectopic pregnancy rates in day 3 versus day 5 embryo transfers

	Day 3-ET	Day 5 Transfer	P-value
EctopicPregnancy/Clinical Pregnancy	22/623 (3.5%)	13/333 (3.9%)	NS
Ectopic Pregnancy/Clinical Pregnancy (Excluding Frozen Embryo transfers)	22/615 (3.5%)	9/271(3.3%)	NS
Mean Age (yrs)	37.7 ± 4.9	35.3 ± 4.7	<0.01
Mean BMI	22.9 ± 3.4	23.2 ± 3.5	NS
%Tubal Disease	22%	24%	NS

In our program, we primarily perform cryopreservation at the blastocyst stage. [9] Accordingly, the day 5 transfer group includes the vast majority of the thaw embryo transfers. When these pregnancies are excluded and only fresh transfers are considered, the ectopic pregnancy rate remains similar for day 3 and day 5 at 3.5% (22/615) and 3.3% (9/271), respectively. (P = 0.8)

Discussion

Studies that showed decreased uterine contractility further along in the luteal phase [6,7] would imply that the ectopic pregnancy rates should be reduced after a day 5 transfer compared to a cleavage stage transfer. It has also been postulated that the larger size of the blastocyst may decrease the chances of the day 5 embryo from migrating to the fallopian tube. [8] Despite these theoretical mechanisms which suggest that day 5 transfer is associated with a lowered ectopic pregnancy risk, our study, which examined close to a thousand pregnancies, failed to show such a trend. It is possible that when a blastocyst is transferred, it does indeed have a lower probability of entering the fallopian tube. However, the blastocyst that does reach the tube may have a higher tendency to implant there while the day 3 embryo has 2 additional days, compared to the day 5 embryo, to migrate back into the uterine cavity.

A potential source of bias in our study is the fact that blastocyst transfer was offered to patients with more than 3 eight cell embryos on day 3 which is likely to occur in patients with a higher number of oocytes and higher estrogen levels. Specific data on oocyte number is not available for our study, and we do not routinely measure estradiol levels in our program. However, when specifically analyzed in previous studies in the literature [1,2], these parameters were not found to affect the incidence of tubal pregnancy. Although the patients in the blastocyst transfer group were on the average 2 years younger than those in the day 3 group, it is unlikely that this small difference could have had an impact on increasing the ectopic pregnancy rate in the day 5 group. If anything, the rate of ectopic pregnancy has been reported to increase with age [10]. Another confounding factor could be the prevalence of tubal disease in the two patient populations studied, as tubal pathology has been shown to be a major risk factor for the development of an ectopic pregnancy with IVF [11,12] The incidence of tubal disease is unlikely to be a source of bias in our study since it was similar in our day 3 and day 5 transfer groups.

Although a comparison of the incidence of ectopic pregnancy between cleavage stage transfer and blastocyst transfer has not been the specific subject of any prior study in the literature, information on this issue can be found in a report by Marek et al. [13] In their study, the authors compared the pregnancy rates in their program when they switched from day 3 to day 5 transfer for all patients. The ectopic pregnancy rates can be extrapolated from their tabulated data as being 1% (2/199) with day 3 and 1.3% (2/159) with day 5 transfers. The findings of this smaller series confirm the absence of a decrease in ectopic rates after blastocyst transfer.

The literature contains 2 additional studies that incidentally report data allowing the computation of the ectopic pregnancy rate with blastocyst transfer without any information on day 3 transfers. In one study, Pantos et al. [14] examined the influence of age on the pregnancy rate after blastocyst transfer and mentioned 4 ectopic pregnancies out of a total of 99 pregnancies (4%). In the other study, Tarlatzis et al. [15] looked at monozygotic twinning with blastocyst transfer after ICSI and conventional IVF and noted an ectopic pregnancy rate of 2 out of 48 pregnancies (4.2%). Although the purpose of these studies was not related to the issue of ectopic pregnancy and they lacked day 3 controls, the rates of about 4% are in line with what has been reported with day 3 transfers and suggest that blastocyst transfer does not reduce the likelihood of tubal pregnancy.

Conclusion

We believe that blastocyst transfer is a valuable tool that has enabled IVF programs to more accurately select the embryos with the highest potential for implantation [16-18] allowing for a good pregnancy rate while avoiding high order multiple gestations [19,20]. In our program, we offer blastocyst transfer to patients of any age [21] if they have more than 3 eight cell embryos. Although some authors have advocated routine blastocyst transfer in all patients [13,22], offering extended culture when there are no eight cell embryos on day 3 has been reported to be detrimental [23]. We suggest that programs establish the criteria that work for them for offering blastocyst culture and transfer. However, based on the results of this study, the presence of risk factors favoring ectopic pregnancy should not be taken into account in the decision making for choosing to transfer on day 3 or day 5.

Competing Interests

None declared.

Authors' Contributions

AAM, faculty attending physician, treated the patients, conceived of the study and co-wrote the manuscript. SHJ, resident, collected the data and co-wrote the manuscript.

References

1. Marcus SF, Brinsden PR: Analysis of the incidence and risk factors associated with ectopic pregnancy following in-vitro fertilization and embryo transfer. Hum Reprod 1995, 10:199–203.

2. Strandell A, Thorburn J, Hamberger L: Risk factors for ectopic pregnancy in assisted reproduction. Fertil Steril 1999, 71:282–286.

3. Lesny P, Killick SR, Robinson J, Maguiness SD: Transcervical embryo transfer as a risk factor for ectopic pregnancy. Fertil Steril 1999, 72:305–309.

4. American Society for Reproductive Medicine and Society for Assisted Reproductive Technology: Assisted reproductive technology in the United States: 1999 results generated from the American Society for Reproductive Medicine/Society for Assisted Reproductive Technology Registry. Fertil Steril 2002, 78:918–931.

5. Russell JB: The etiology of ectopic pregnancy. Clin Obstet Gynecol 1998, 30:181–190.

6. Lesny P, Killick SR, Tetlow RL, Robinson J, Maguiness SD: Uterine junctional zone contractions during assisted reproduction cycles. Hum Reprod Update 1998, 4:440–445.

7. Fanchin R, Ayoubi JM, Righini C, Olivennes F, Schonauer LM, Frydman R: Uterine contractility decreases at the time of blastocyst transfers. Hum Reprod 2001, 16:1115–1119.

8. Schoolcraft WB, Surrey ES, Gardner DK: Embryo transfer: techniques and variables affecting success. Fertil Steril 2001, 76:863–870.

9. Behr B, Gebhardt J, Lyon J, Milki AA: Factors relating to a successful cryopreserved blastocyst transfer program. Fertil Steril 2002, 77:697–699.

10. Bouyer J, Coste J, Shojaei T, Pouly JL, Fernandez H, Gerbaud L, Job-Spira N: Risk factors for ectopic pregnancy: a comprehensive analysis based on a large case-control, population-based study in France. Am J Epidemiol 2003, 157:185–194.

11. Herman A, Ron-El R, Golan A, Weinraub Z, Bukovsky I, Caspi E: The role of tubal pathology and other parameters in ectopic pregnancies occurring in in vitro fertilization and embryo transfer. Fertil Steril 1990, 54:864–868.

12. Verhulst G, Camus M, Bollen N, Van Steiterghem A, Devroey P: Analysis of risk factors with regard to the occurrence of ectopic pregnancy after medically assisted procreation. Hum Reprod 1993, 8:1284–1287.

13. Marek D, Langley M, Gardner DK, Confer N, Doody KM, Doody KJ: Introduction of blastocyst culture and transfer for all patients in an in vitro fertilization program. Fertil Steril 1999, 72:1035–1040.

14. Pantos K, Athanasiou V, Stefanidis K, Stavrou D, Vaxevanoglou T, Chronopoulou M: Influence of advanced age on the blastocyst development rate and pregnancy rate in assisted reproductive technology. Fertil Steril 1999, 71:1144–1146.

15. Tarlatzis BC, Qublan HS, Sanopoulou T, Zepiridis L, Grimbizis G, Bontis J: Increase in the monozygotic twinning rate after intracytoplasmic sperm injection and blastocyst stage embryo transfer. Fertil Steril 2002, 77:196–198.

16. Rijnders PM, Jansen CA: The predictive value of day 3 embryo morphology regarding blastocyst formation, pregnancy and implantation rate after day 5 transfer following in-vitro fertilization or intracytoplasmic sperm injection. Hum Reprod 1998, 13:2869–2873.

17. Graham J, Han T, Porter R, Levy M, Stillman R, Tucker MJ: Day 3 morphology is a poor predictor of blastocyst quality in extended culture. Fertil Steril 2000, 74:495–497.

18. Milki AA, Hinckley MD, Behr B: Comparison of blastocyst transfer to day 3 transfer with assisted hatching in the older patient. Fertil Steril 2002, 78:1244–1247.

19. Gardner DK, Schoolcraft WB, Wagley L, Schlenker T, Stevens J, Hesla J: A prospective randomized trial of blastocyst culture and transfer in in-vitro fertilization. Hum Reprod 1998, 13:3434–3440.

20. Milki AA, Fisch JD, Behr B: Two-blastocyst transfer has similar pregnancy rates and a decreased multiple gestation rate compared with three-blastocyst transfer. Fertil Steril 1999, 72:225–228.

21. Milki AA, Hinckley MD, Gebhardt J, Dasig D, Westphal LM, Behr B: Accuracy of day 3 criteria for selecting the best embryos. Fertil Steril 2002, 77:1191–1195.

22. Wilson M, Hartke K, Kiehl M, Rodgers J, Brabec C, Lyles R: Integration of blastocyst transfer for all patients. Fertil Steril 2002, 77:693–696.

23. Racowsky C, Jackson KV, Cekleniak NA, Fox JH, Hornstein MD, Ginsburg ES: The number of eight-cell embryos is a key determinant for selecting day 3 or day 5 transfer. Fertil Steril 2000, 73:558–564.

Transcriptome Analysis of Mouse Stem Cells and Early Embryos

Alexei A. Sharov, Yulan Piao, Ryo Matoba,
Dawood B. Dudekula, Yong Qian, Vincent VanBuren,
Geppino Falco, Patrick R. Martin, Carole A. Stagg,
Uwem C. Bassey, Yuxia Wang, Mark G. Carter,
Toshio Hamatani, Kazuhiro Aiba, Hidenori Akutsu,
Lioudmila Sharova, Tetsuya S. Tanaka, Wendy L. Kimber,
Toshiyuki Yoshikawa, Saied A. Jaradat, Serafino Pantano,
Ramaiah Nagaraja, Kenneth R. Boheler, Dennis Taub,
Richard J. Hodes, Dan L. Longo, David Schlessinger,
Jonathan Keller, Emily Klotz, Garnett Kelsoe,
Akihiro Umezawa, Angelo L. Vescovi, Janet Rossant,
Tilo Kunath, Brigid L. M. Hogan, Anna Curci, Michele D'Urso,
Janet Kelso, Winston Hide and Minoru S. H. Ko

ABSTRACT

*Understanding and harnessing cellular potency are fundamental in biology
and are also critical to the future therapeutic use of stem cells. Transcriptome*

analysis of these pluripotent cells is a first step towards such goals. Starting with sources that include oocytes, blastocysts, and embryonic and adult stem cells, we obtained 249,200 high-quality EST sequences and clustered them with public sequences to produce an index of approximately 30,000 total mouse genes that includes 977 previously unidentified genes. Analysis of gene expression levels by EST frequency identifies genes that characterize preimplantation embryos, embryonic stem cells, and adult stem cells, thus providing potential markers as well as clues to the functional features of these cells. Principal component analysis identified a set of 88 genes whose average expression levels decrease from oocytes to blastocysts, stem cells, postimplantation embryos, and finally to newborn tissues. This can be a first step towards a possible definition of a molecular scale of cellular potency. The sequences and cDNA clones recovered in this work provide a comprehensive resource for genes functioning in early mouse embryos and stem cells. The nonrestricted community access to the resource can accelerate a wide range of research, particularly in reproductive and regenerative medicine.

Introduction

With the derivation of pluripotent human embryonic stem (ES) (Thomson et al. 1998) and embryonic germ (EG) (Shamblott et al. 1998) cells that can differentiate into many different cell types, excitement has increased for the prospect of replacing dysfunctional or failing cells and organs. Very little is known, however, about critical molecular mechanisms that can harness or manipulate the potential of cells to foster therapeutic applications targeted to specific tissues.

A related fundamental problem is the molecular definition of developmental potential. Traditionally, potential has been operationally defined as "the total of all fates of a cell or tissue region which can be achieved by any environmental manipulation" (Slack 1991). Developmental potential has thus been likened to potential energy, represented by Waddington's epigenetic landscape (Waddington 1957), as development naturally progresses from "totipotent" fertilized eggs with unlimited differentiation potential to terminally differentiated cells, analogous to a ball moving from high to low points on a slope. Converting differentiated cells to pluripotent cells, a key problem for the future of any stem cell-based therapy, would thus be an "up-hill battle," opposite the usual direction of cell differentiation. The only current way to do this is by nuclear transplantation into enucleated oocytes, but the success rate gradually decreases according to developmental stages of donor cells, providing yet another operational definition of developmental potential (Hochedlinger and Jaenisch 2002; Yanagimachi 2002).

What molecular determinants underlie or accompany the potential of cells? Can the differential activities of genes provide the distinction between totipotent cells, pluripotent cells, and terminally differentiated cells? Systematic genomic methodologies (Ko 2001) provide a powerful approach to these questions. One of these methods, cDNA microarray/chip technology, is providing useful information (Ivanova et al. 2002; Ramalho-Santos et al. 2002; Tanaka et al. 2002), although analyses have been restricted to a limited number of genes and cell types. To obtain a broader understanding of these problems, it is important to analyze all transcripts/genes in a wide selection of cell types, including totipotent fertilized eggs, pluripotent embryonic cells, a variety of ES and adult stem cells, and terminally differentiated cells. Despite the collection of a large number of expressed sequence tags (ESTs) (Adams et al. 1991; Marra et al. 1999) and full-insert cDNA sequences (Okazaki et al. 2002), systematic collection of ESTs on these hard-to-obtain cells and tissues has been done previously only on a limited scale (Sasaki et al. 1998; Ko et al. 2000; Solter et al. 2002).

Accordingly, we have attempted to (i) complement other public collections of mouse gene catalogs and cDNA clones by obtaining and indexing the transcriptome of mouse early embryos and stem cells and (ii) search for molecular differences among these cell types and infer features of the nature of developmental potential by analyzing their repertoire and frequency of ESTs. Here we report the collection of approximately 250,000 ESTs, enriched for long-insert cDNAs, and signature genes associated with the potential of cells, various types of stem cells, and preimplantation embryos.

Results and Discussion

Novel Genes Derived from Early Mouse Embryos and Stem Cells

Twenty-one long-insert-enriched cDNA libraries with insert ranges from 2–8 kb (Piao et al. 2001) were generated from preimplantation embryos (unfertilized egg, fertilized egg, two-cell embryo, four-cell embryo, eight-cell embryo, morula, and blastocyst), ES cells (Anisimov et al. 2002) and EG cells (Matsui et al. 1992), trophoblast stem (TS) cells (Tanaka et al. 1998), adult stem cells (e.g., neural stem/progenitor [NS] cells) (Galli et al. 2002), mesenchymal stem (MS) cells (Makino et al. 1999), osteoblasts (Ochi et al. 2003), and hematopoietic stem/progenitor (HS) cells (Ortiz et al. 1999), their differentiated cells, and newborn organs (e.g., brain and heart). In total, 249,200 ESTs (170,059 cDNA clones: 114,437 5′ ESTs and 134,763 3′ ESTs) were generated and assembled together with public data into a gene index (see Materials and Methods).

Of 29,810 mouse genes identified in our gene index (Figure 1), 977 were not present as either known or predicted transcripts in other major transcriptome databases, such as RefSeq (Pruitt and Maglott 2001), Ensembl (Hubbard et al. 2002), and RIKEN (Okazaki et al. 2002). These genes represent possible novel mouse genes, as they either encode open reading frames (ORFs) greater than 100 amino acids or have multiple exons. In particular, 554 of the 977 genes remained novel with high confidence even after more thorough searches against GenBank and other databases. Comparisons of these 977 genes against all National Center for Biotechnology Information (NCBI) UniGene representative sequences showed that 377 genes did not match even fragmentary ESTs and are therefore unique to the National Institute on Aging (NIA) cDNA collection. A random subset of 19 cDNA clones representing these genes was sequenced completely to confirm their novelty (Figure 2). Protein domain searches using InterPro (Mulder et al. 2003) revealed that one of them, U004160, is an orthologue of human gene Midasin (MDN1), but the remaining 18 genes do not encode any known protein motifs. However, they were split into multiple exons in the alignment to the mouse genome sequences, and we therefore considered them genes. As these sequences are mainly derived from early embryos and stem cells, they most likely represent new candidates for genes specific to particular types of stem cells. RT–PCR analysis revealed that they are expressed in specific cell types (Figure 2). For example, the expression of gene U035352 was unique to ES cells, expression of U004912 unique to ES and TS cells, and expression of U001905 unique to ES and EG cells. In addition, one gene showed apparent specific expression in several stem cells and is thus a potential pan-stem cell marker (U029765). Taken together, these data suggest that most of the putative genes represented only in the NIA cDNA collection are bona fide genes that have not been previously identified.

Signature Genes That Characterize Preimplantation Embryos and Stem Cells

To identify genes that were consistently overrepresented in a given set of cDNA libraries when compared with other libraries, we performed the correlation analysis of log-transformed EST frequency combined with the false discovery rate (FDR) method (Benjamini and Hochberg 1995) (FDR = 0.1) (Figure 3).

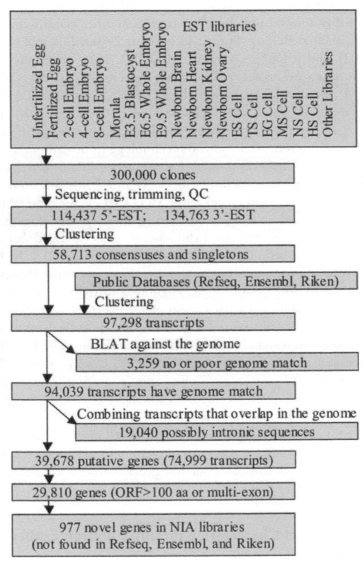

Figure 1. Flow Chart of Sequence Data Analysis. Using TIGR gene indices clustering tools (Pertea et al. 2003), 249,200 ESTs were clustered, generating 58,713 consensuses and singletons. NIA consensuses and singletons were further clustered with Ensembl transcripts (Hubbard et al. 2002), RIKEN transcripts (Okazaki et al. 2002), and RefSeq transcripts and transcript predictions (Pruitt and Maglott 2001). Alignments of these sequences to the mouse genome (UCSC February 2002 freeze data, available from ftp://genome.cse.ucsc.edu/goldenPath/mmFeb2002) (Waterston et al. 2002) using BLAT (Kent 2002) helped to avoid false clustering of similar sequences at nonmatching genome locations. Erroneous clusters were reassembled based on the analysis of genome alignment. A total 94,039 putative transcripts were thus generated and then grouped into 39,678 putative genes based on their overlap in the genome on the same chromosome strand and on clone-linking information. Using criteria of an ORF greater than 100 amino acids or of multiple exons (excluding sequences that are potentially located in a wrong strand), 29,810 mouse genes were identified. Finally, 977 genes unique to the NIA database were identified.

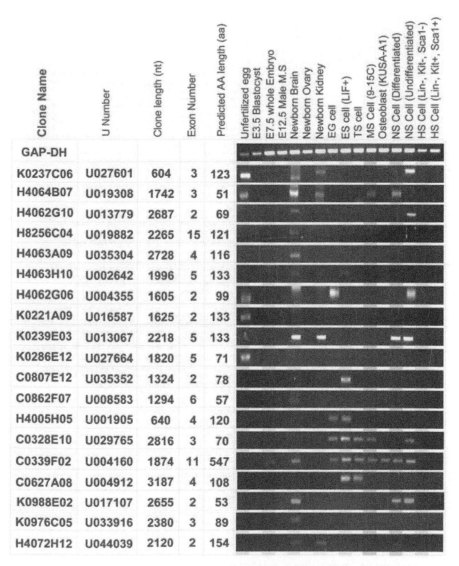

Clone Name	U Number	Clone length (nt)	Exon Number	Predicted AA length (aa)
GAP-DH				
K0237C06	U027601	604	3	123
H4064B07	U019308	1742	3	51
H4062G10	U013779	2687	2	69
H8256C04	U019882	2265	15	121
H4063A09	U035304	2728	4	116
H4063H10	U002642	1996	5	133
H4062G06	U004355	1605	2	99
K0221A09	U016587	1625	2	133
K0239E03	U013067	2218	5	133
K0286E12	U027664	1820	5	71
C0807E12	U035352	1324	2	78
C0862F07	U008583	1294	6	57
H4005H05	U001905	640	4	120
C0328E10	U029765	2816	3	70
C0339F02	U004160	1874	11	547
C0627A08	U004912	3187	4	108
K0988E02	U017107	2655	2	53
K0976C05	U033916	2380	3	89
H4072H12	U044039	2120	2	154

Figure 2. Examples of NIA-Only cDNA Clones and RT–PCR Results. Expression pattern of 19 novel cDNA clones in 16 different cell lines or tissues: unfertilized egg, E3.5 blastocyst, E7.5 whole embryo (embryo plus placenta), E12.5 male mesonephros (gonad plus mesonephros), newborn brain, newborn ovary, newborn kidney, embryonic germ (EG) cell, embryonic stem (ES) cell (maintained as undifferentiated in the presence of LIF), trophoblast stem (TS) cell, mesenchymal stem (MS) cell, osteoblast, neural stem/progenitor (NS) cell, NS differentiated (differentiated neural stem/progenitor cells), and hematopoietic stem/progenitor (HS) cells. Glyceraldegyde-3-phosphate dehydrogenase (GAP-DH) was used as a control. A U number is assigned to each gene in the gene index. The exon number was predicted from alignment with the mouse genome sequence, and the amino acid sequence was predicted with the ORF finder from NCBI.

Group	gene number	Gene symbols
A	196	Akap1; Alox12e; Apg5l; Arf6; Arf6ip2; Banp; Bcl2l10; Birc2; Bmp15; Bpgm; Btg4; Bub1b; Ccnb1rs1; Cdc25a; Cdc45l; Cryl1; Cyp11a; Dtx2; Eed; Epb4.1l2; Fbxw4; Fmn2; Folr4; Gdf9; Gtr2; H1fo; Hsd3b1; Ing1; Irf1; Itga9; Kcnh1; Kdel1; Krt2_16; Map2k6; Map4k5; Mapkbp1; Mater; Mdm4; Mitc1; Mmp23; Mrg1; Npc1; Oas1d; Oas1e; Obox3; Orc5l; Orc6l; P37nb; Pabpc4; Pcsk6; Phf1; Plat; Pld1; Pole3; Prkab1; Rbbp7; Rdx; Rfpl4; Rgs2; Rnf35; Rnpc1; Sh3d5; Sip1; Slc21a11; Smarcf1; Snrpb2; Tbn; Tcl1; Tcl1b1; Tcl1b3; Tes3; Tex14; Thrsp; Top1; Wbscr21; Xbp1; Zfp296; and 119 unknown genes.
D	81	Akp5; Bcl2l10; Bmp15; Bpgm; Btg4; Cdkap1; Ctsc; Dusp14; Fbxo15; Fxyd4; Gdf9; Hspa8; Klf5; Obox3; Ovgp1; Prkab1; Rfpl4; Rnf35; Rnpc1; Spin; Tcl1; Tcl1b3; Timd2; Ulk1; and 57 unknown genes.
E	143	Akp5; Arpc1b; Bcl2l10; Birc2; Bmp15; Bpgm; Btg4; Cd160; Cdkap1; Cry1; Daf1; Degs; Dusp14; E2f1; Fbxo15; Fkbp5; Fkbp6; Fmn2; Folr4; Fxyd4; Gdf9; Gja4; Gstm2; H1fo; Hspa8; Krt2_16; Lcn2; Magoh; Mater; Mbtd1; Mdm4; Mil1; Oas1d; Oas1e; Obox3; Ovgp1; P37nb; Prkab1; Rfpl4; Rgs2; Rnf33; Rnf35; Rnpc1; Rps8; Sat; Slc34a2; Spin; Tcl1; Tcl1b1; Tcl1b3; Timd2; Ybx3; and 91 unknown genes.
F	54	Bcl2l10; Bmp15; Btg4; Cdkap1; Degs; Dusp14; Fxyd4; Gdf9; Mitc1; Obox3; Ovgp1; Prkab1; Rfpl4; Rnf35; Rnpc1; Spin; Tcl1; Tcl1b3; Timd2; and 35 unknown genes.
I	140	Akap10; Akap12; Arnt; Ash2l; Atm; Birc6; Cask; Cbx5; Cdh11; CHD6; Crkol; Dnchc1; Ect2; Edr1; Enah; ENPP3; Fshprh1; Gabrg1; Galnt1; GPCR; Hells; Hmgcr; Hmmr; Impact; Kars_ps1; Kif10; Kif15; Knsl1; Lamc1; Lbr; Lox; Mad5; Mki67; Np95; Oazi; Odf2; Pald; Plrg1; Pola1; Ppp4r1; Ptch; Ptch2; Rasa2; Rb1; Rex2; Rpl31; Sh3bp3; Shcbp1; Ski; Slc27a1; Sms; Spna2; Synj2; Tnc; TRB_2; Trif; Trps1; Zfp148; Zfp191; and 80 unknown genes.
J	93	Abce1; Akap10; Ccnf; Cdh11; Col12a1; Crkol; Dda3; Dmd; Enh; Fanca; GPCR; Hmmr; Img; Lrig1; Iqgap1; Kars_ps1; Kif15; Lox; Morf; Nedd4; Opa1; Pald; Pola1; Ppp4r1; Ptprf; Rpl31; Sec23a; Sh3bp3; Slc27a1; Slc7a3; Snrp116; Tardbp; Thbs1; TRB-2; Trps1; Zfhx1a; Zfhx1b; Zfp191; and 55 unknown genes.
K	75	Abhd2; Ccnf; Cldn4; Dda3; Dnmt3b; ENPP3; F11r; Fkbp4; Foxh1; Grb7; Grcc8; Helb; Hmga1; Jmj; Jub; Lsm10; Map4k1; Mif; Morf; Mta3; Mybl2; Ncl; Nek4; Nfyb; Nol5; Pabpn1; Pcnxl3; Pdcd4; Ppp4r1; Prkar2a; Prss8; Rbbp6; Rfng; Rpl13; Rps2; Rps6ka1; Slc29a1; Slc7a3; Sntb2; Sox13; Tdh; Tsbp; Ubb; Unc13h1; Wee1; Xrcc2; Zfp42; and 28 unknown genes.
L	39	Akap10; Cldn4; Dnmt3b; ENPP3; Foxh1; Galk1; Grb7; Kars-ps1; Lcn7; Nfyb; Ngfrap1; Pcbp1; Rgds; Ubb; Unc13h1; Wee1; Xrcc2; Zfp278; Zfp42; and 20 unknown genes.
M	44	Atf2; Cbl; Ccne2; Cd44; Elf1; Fshprh1; Fyn; G7e; Herc3; IFI 203; Itga4; Jak1; MALT-1; Mbnl; Nab1; Nfat5; Phc3; SENP6; Stxbp4; Tde1l; Tex2; Tm6sf1; Wwp4; and 34 unknown genes.
N	108	Abce1; Abhd2; Akap10; Arl6ip2; Ccnf; Col12a1; Dda3; Dstn; Edr1; Enah; ENPP3; Fkbp4; Foxh1; Gfpt2; Helb; Impact; Ing5; Jmj; Jub; Mad5; Map4k1; Mkrn1; Morf; Mov10; Mta3; Mtf2; Mybl2; Pask; Pcnxl3; Pola1; Pola2; Ppp2r5e; Ppp4r1; Ptch2; Ranbp17; Rbbp6; Rest; Rex2; Rw1; Slc29a1; Slc7a3; Smarca4; Sntb2; Sox13; Tacc3; Tcof1; Tdgf1; Tdh; Zfp42; and 59 unknown genes.
O	113	Akap10; Akap12; Ash2l; Atm; Cipp; Col18a1; Dda3; Ect2; Edr1; Enah; ENPP3; Ermelin; Etl1; Gab1; Gfpt2; Hic2; Hmmr; Impact; Ing5; Itga3; Jmj; Lamc1; Lyar; Mad5; Mkrn1; Mtf2; Mybl2; Ncoa3; Np95; Opa1; Pald; Pola1; Ppp2r5e; Ptch2; Ranbp17; Rex2; Rnf17; Rw1; Slc29a1; Slc7a3; Smcx; Sms; Taf7; Tdh; Tex20; Trif; Wee1; Zfp110; and 65 unknown genes.

Column headers (left to right): Group; Unfertilized Egg; Fertilized Egg; 2-cell Embryo; 4-cell; 8-cell; Morula; E3.5 Blastocyst; E6.5 Whole Embryo; E7.5 Whole Embryo; E7.5 Embryonic Part; E7.5 Extraembryonic Part; E8.5 Whole embryo; E9.5 Whole Embryo; E12.5 Male Gonad/Mesonephros; E12.5 Female Gonad/Mesonephros; E13.5 VMB Dopamine Cell; Newborn Brain; Newborn Heart; Newborn Kidney; Newborn Ovary; ES Cell (LIF+); ES Cell (LIF-); TS Cell; EG Cell; MS Cell (9-15C); Osteoblast (KUSA-A1); NS Cell (Undifferentiated); NS Cell (Differentiated); HS Cell (Lin-, Kit+, Sca1+); HS Cell (Lin-, Kit+, Sca1-); HS Cell (Lin-, Kit-, Sca1+); HS Cell (Lin-, Kit-, Sca1-); gene number; Gene symbols.

Figure 3. Signature Genes for Specific Groups of Early Embryos and Stem Cells.

First, we analyzed various combinations of preimplantation stages and identified the following genes: (i) 196 genes specific to unfertilized eggs (oocytes) and fertilized eggs (Group A in Figure 3), (ii) 122 genes specific to two- to four-cell embryos (Group B in Figure 3), (iii) 119 genes specific to eight-cell embryos, morula, and blastocyst (Group C in Figure 3), (iv) 81 genes specific to all preimplantation embryos (Group D in Figure 3), and (v) 143 genes specific to all preimplantation embryos except for blastocysts (Group E in Figure 3). Blastocyst EST frequencies are unique even among preimplantation embryos, most likely reflecting the switch of the transcriptome from the maternal genetic program to the zygotic genetic program (Latham and Schultz 2001; Solter et al. 2002) or to the differentiation of the trophectoderm. At least 35 out of 196 genes in the egg signature gene list (Group A in Figure 3) have ATP-related protein domains. Genes in the following categories were also enriched in this gene list: the ubiquitin–proteasome pathway, the energy pathway, cell signaling (kinase and membrane) proteins, ribosomal proteins, and zinc finger proteins. Two SWI/SNF-related genes (5930405J04Rik, the homologue of human SMARCC2, and Smarcf1) and two Polycomb genes (Scmh1 and Sfmbt) overrepresented in eggs may be candidate genes for strong chromatin remodeling activity of eggs during nuclear transplantation of somatic cell nuclei.

Addition of ES and EG cells to preimplantation embryos (143 genes; Group E in Figure 3) yielded only 54 signature genes (Group F in Figure 3). Addition of adult stem cells, MS and NS, or MS, NS, and HS (Lin–, Kit+, Sca1+ and Lin–, Kit–, Sca1+) cells further reduced the number of signature genes to five and one, in Groups G and H, respectively. Taken together, these results seem to indicate that preimplantation embryos, particularly totipotent fertilized eggs and highly pluripotent cells (ES and EG cells), have quite distinct genetic programs, but that less pluripotent adult stem cells (MS, NS, and HS) have even more specialized genetic programs. This supports the notion of a gradual decrease of developmental potential from preimplantation embryos to stem cells to differentiated cells.

Additional analysis was done to determine genes that are enriched in stem cells, but not in preimplantation embryos and other tissues (see Figure 3). In this analysis, 140 genes were identified as signature genes for pluripotent stem cells (ES, EG, NS, and MS in Group I in Figure 3), whereas 93 genes were identified as signature genes for these stem cells and their differentiated forms (cultured cells in Group J in Figure 3). Similarly, 75 and 39 genes, respectively, were identified as ES- and TS-specific (Group K in Figure 3), whereas 44 genes were identified as signature genes for adult stem cells (NS, MS, and HS in Group M in Figure 3). Lists of these genes showed that distinctive sets of genes are responsible for cell specificity (Figure 3).

FDR analysis revealed that 113 genes were specifically expressed in ES and EG cells in Group O (the most pluripotent stem cells), but not in all other cell types examined (Figure 3). The most abundant group of these genes was transcription regulatory factors (about 30% of all specific genes), most of which were members of the zinc finger family, including Mtf2, Ing5, Mkrn1, Hic2, and the KRAB box zinc finger. Other abundant genes specifically expressed in ES and EG cells included matrix/cytoskeleton/membrane structural proteins such as Itga3, Dstn, Smtn, Dctn1, and Col18a1 and the DNA remodeling proteins such as Rcc1, Kars-ps1, Pola2, Mov10, and Rad54l . These two groups of genes may be associated with the unique feature of ES/EG cell cycle structure, where greater than 70% of the cell population are in S phase (Savatier et al. 1996).

Previous studies have identified genes specific to particular stem cells or genes common to a group of stem cells, although there was little agreement about which transcripts are commonly enriched in these studies (e.g., Anisimov et al. 2002; Ivanova et al. 2002; Ramalho-Santos et al. 2002; Tanaka et al. 2002). The difference in the method and platform used could be a major reason for the difficulty in identifying a common gene set. The analysis of limited number of cell types could also contribute to differences in the resulting gene lists, because genes that appeared specific to certain cell types may also be expressed in other cells that were not included in the analysis. In contrast, the current study has analyzed a large number of different stem cells, preimplantation embryos, and newborn organs from our own EST collections as well as all publicly available ESTs that were derived from a few hundred cell types. Combined with stringent FDR statistics (see Materials and Methods), the analysis of this large number of cell types may provide broader perspectives on this issue. Comparison between the gene lists of the present study and the gene lists from the previously published studies identified areas of agreement (common genes), but also revealed that many genes previously reported as specifically expressed in one cell type or group of cells are actually expressed in other cell types and thus are not specific. The signature genes identified in this study distinguish different stem cells, and this gene list may provide a way to recognize or purify specific stem cell types and provide insights into stem cell–specific functions.

Principal Component Analysis Identified Clusters of Cells/ Tissues with Similar EST Frequency

The global expression patterns of 2,812 relatively abundant genes (see Materials and Methods) were further analyzed by principal component analysis (PCA), which reduces high-dimensionality data into a limited number of principal components. The first principal component (PC1) captures the largest contributing factor of variation, which in this case corresponds to the average EST frequency in all tissues, and subsequent principal components correspond to other factors with smaller effects, which characterize the differential expression of genes. As we were interested in the differential gene expression component, we plotted the position of each cell type against the PC2, PC3, and PC4 axis in three-dimensional (3D) space by using virtual reality modeling language (VRML) (Figure 4A). Genes were also plotted in the same 3D space (a version of PCA called a biplot) (Chapman et al. 2002) to see their association with cell/tissue types. Close examination of the 3D model identified PC2 and PC3 as the most representative views of the 3D model (Figure 4B). A two-dimensional (2D) plot of PC2 and PC3 is therefore used for the following discussion, with references to the 3D model. It is important to keep in mind that the distance between cell types along principal components has a substantial error associated with randomness of clone counts in EST libraries. The estimated error range (2*SE) in the PC3 scale is about 7%–9% based on Poisson distribution (Figure 4B). Nonetheless, PCA identifies major trends and clusters in gene expression among these cell types.

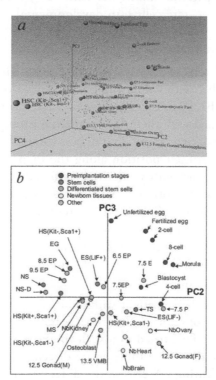

Figure 4. PCA Analysis of EST Frequency. The results were obtained by analyzing 2,812 genes that exceeded 0.1% in at least one library. (A) 3D biplot that shows both cell types (red spheres) and genes (yellow boxes). (B) 2D PCA of cell types. EST frequencies were log-transformed before the analysis. Names of some cells and tissues are abbreviated as follows: 6.5 EP, E6.5 whole embryo (embryo plus placenta); 7.5 EP, E7.5 whole embryo (embryo plus placenta); 8.5 EP, E8.5 whole embryo (embryo plus placenta); 9.5 EP, E9.5 whole embryo (embryo plus placenta); 7.5 E, E7.5 embryonic part only; 7.5 P, E7.5 extraembryonic part only; NbOvary, newborn ovary; NbBrain, newborn brain; NbHeart, newborn heart; NbKidney, newborn kidney; 13.5 VMB, E13.5 ventral midbrain dopamine cells; 12.5 Gonad (F), E12.5 female gonad/mesonephros; 12.5 Gonad (M), E12.5 male gonad/mesonephros; HS (Kit–, Sca1–), hematopoietic stem/progenitor cells (Lin–, Kit–, Sca1–); HS (Kit–, Sca1+), hematopoietic stem/progenitor cells (Lin–, Kit–, Sca1+); HS (Kit+, Sca1–), hematopoietic stem/progenitor cells (Lin–, Kit+, Sca1–); HS (Kit+, Sca1+), hematopoietic stem/progenitor cells (Lin–, Kit+, Sca1+); and NS-D, differentiated NS cells.

The most conspicuous trend was that cells that differ in their developmental potential appeared well separated along the PC3 axis. In Figure 4A and 4B, preimplantation embryos (unfertilized egg, fertilized egg, two-cell, four-cell, eight-cell, morula, and blastocyst) are positioned at the top of the PC3 axis; embryos and extraembryonic tissues from early- to mid-gestation stage, such as E6.5, E7.5, E8.5, and E9.5, are positioned at the middle; and cells and tissues mostly from terminally differentiated cells (newborn ovary, newborn heart, and newborn brain) are positioned at the bottom. PCA is unsupervised (performed without using knowledge of developmental stages of each cell types), and so this ordering along the PC3 axis seems to reflect the structures of global gene expression patterns among the cells. The PC2 axis provided an additional dimension to separate cells into developmental stages, functional groups, or both. The correlation of the PC2 axis to known biological stages, functions, or both, however, remains unclear.

Interestingly, both ES cells and adult stem cells are positioned at the middle of the PC3 axis together with whole-embryo libraries from early- to mid-gestation stages (Figure 4B). ES and EG cells were derived from embryos, and thus their positions matched with their developmental timing. Although NS, MS, and HS cells were all derived from adult organs (brain, bone marrow, and bone marrow, respectively), their position along the PC3 axis corresponded to early embryonic tissues and embryo-derived stem cells (ES and EG). The results are consistent with the notion that adult stem cells acquire or retain the pluripotency with characters of less-differentiated cell types. This also suggests that the PC3 axis does not represent just developmental timing, but also indicates the developmental potential of cells, with totipotent eggs at the top, pluripotent embryonic cells and stem cells at the middle, and terminally differentiated cells at the bottom.

This hypothesis seems to be consistent with another interesting observation that the differentiated forms of stem cells were always positioned lower than their stem cell counterparts (undifferentiated forms) in the PC3 axis (Figure 4A and 4B). For example, the position of NS (differentiated) cells, a mixture of neuron and glia obtained after culturing NS cells in the differentiation conditions, was lower and nearer to the terminally differentiated cells than were NS cells. Osteoblast cells, which are more differentiated than the MS cells from which they are derived, were again positioned lower than the MS cells. The same holds true for ES (LIF–) cells (lower PC3 position), which were obtained by culturing ES cells in the absence of leukemia inhibitory factor (LIF), allowing ES cells to differentiate into many different cell types, and ES (LIF+) cells (higher PC3 position), which were maintained as highly pluripotent by culturing them in the presence of LIF. For HS cells, all four cell types were selected first as lineage marker-negative cells, and thus they were all relatively undifferentiated cells. These cells were then sorted by c-Kit+ and Sca1+ into four separate fractions. The most pluripotent cells (Lin–, c-Kit+, Sca1+) were again positioned higher than other three cell types in the PC3 axis. Finally, TS cells were positioned at the least-potent place among stem cells, which seemed to fit to their known characteristics. It has previously been shown that TS cells are already committed to the extraembryonic lineage and are less pluripotent than ES and EG cells, because TS cells injected back to mouse blastocysts only differentiate into extraembryonic trophoblast lineages (Tanaka et al. 1998). The microarray analysis of TS cells also shows that they already express many placenta-specific genes, which is a sign of lineage-committed cells (Tanaka et al. 2002).

Finally, it is interesting to note that EG cells were positioned closely to E8.5 whole embryos and E9.5 whole embryos, whereas ES cells were positioned closely to blastocysts, E6.5, and E7.5 whole embryos (Figure 4). Because ES cells are derived from E3.5 blastocysts and EG cells are derived from primordial germ cells (PGCs) of E8.5 (in this particular line), these results indicate that the expression patterns of relatively abundant genes in ES and EG cells reflect their developmental stages of origin. Although ES and EG cells were established from different sources, EG cells are often considered to be ES cells and the distinction of their origin is ignored. However, the result here suggests potentially significant differences between the genetic programs of EG cells and ES cells.

Genes Correlated with the Developmental Potential of Cells

To identify a group of genes associated with the PC3 axis, we first fixed the coordinate of each cell type on PC3 and searched for genes whose log-transformed frequencies correlated with this coordinate in each cell type. Correlation analysis combined with the FDR method (FDR = 0.1) revealed 88 genes whose expression levels were significantly associated with PC3. To test how well these genes represent PC3, we plotted the sum of log-transformed EST frequencies for these 88 genes versus PC3 projections of the same cell types (Figure 5). Most cells were positioned diagonally relative to the original PC3 coordinates, indicating that the average expression levels of these 88 genes can roughly represent cell type position along the PC3 coordinate. Because the PC3 axis does not have a unit and cannot be directly translated to variables measured by molecular biological techniques, the possible use of 88 genes as a surrogate for the PC3 axis will help to test this working hypothesis in the future.

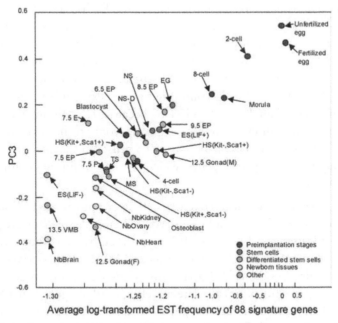

Figure 5. Relationship between PC3 and Average Expression Levels of 88 Signature Genes. A list of 88 genes associated with developmental potential: Birc2, Bmp15, Btg4, Cdc25a, Cyp11a, Dtx2, E2f1, Fmn2, Folr4, Gdf9, Krt2–16, Mitc1, Oas1d, Oas1e, Obox3, Prkab1, Rfpl4, Rgs2, Rnf35, Rnpc1, Slc21a11, Spin, Tcl1, Tcl1b1, Tcl1b3, 1810015H18Rik, 2210021E03Rik, 2410003C07Rik, 2610005B21Rik, 2610005H11Rik, 3230401D17Rik, 4833422F24Rik, 4921528E07Rik, 4933428G09Rik, 5730419I09Rik, A030007L17Rik, A930014I12Rik, E130301L11Rik, AA617276, Bcl2l10, MGC32471, MGC38133, MGC38960, D7Ertd784e, and 44 genes with only NIA U numbers.

What are the characteristics of these 88 potential correlating genes? Based on the available protein domain information, Gene Ontology (GO) annotation (Ashburner et al. 2000; http://www.geneontology.org/doc/GO.annotation. html), and literature, 58 genes can be classified into putative functional categories. For example, signature genes in the "transcriptional control" category include eight genes, such as MAD homologue 4 interacting transcription coactivator 1 (Mitc1), Drosophila Deltex 2 homologue (Dtx2), and oocyte-specific homeobox 5 (Obox5); the "RNA binding" category includes five genes such as RNA-binding region containing 1 (Rnpc1) and 2′-5′-oligoadenylate synthetase 1D (Oas1d); the "signal transduction" category includes ten genes, such as AMP-activated protein kinase (Prkab1) and regulator of G-protein signaling 2 (Rgs2); and the "proteolysis" category includes six genes, such as Ret finger protein-like 4 (Rfpl4) and ring finger protein 35 (Rnf35). These categories were diverse, and the domination of any specific categories was not observed.

Although all 88 genes shared the general trend of continuous decrease of expression levels from eggs to terminally differentiated tissues, these genes can be further subdivided by their expression patterns. First, 53 genes were those identified as preimplantation specific, particularly unfertilized and fertilized egg-specific genes, which include already well-known genes for their functions in oogenesis and zygotic gene activation, such as Gdf9, Bmp15, Rfpl4, Fmn2, Tcl1, Obox5, and Oosp1. Second, ten genes were represented as ESTs in both preimplantation embryos and postimplantation embryos, including Cyp11a and D7Ertd784e. Third, 25 genes were represented well as ESTs in preimplantation embryos, postimplantation embryos, and stem cells, including Mitc1, actin-binding Kelch family protein, Dtx2, Cdc25a, Spin, Rgs2, Prkab1, and Birc2. The seemingly continuous decrease of the expression of these genes is therefore not caused by passive dilution of transcripts that are abundant in oocytes, but is most likely caused by a specific mechanism that actively regulates the expression levels of these genes.

Concluding Remarks

The sequence information and cDNA clones collected in this work provide the most comprehensive database and resources for genes functioning in early mouse embryos and stem cells. All cDNA clones developed in this project have been made available through the American Type Culture Collection (ATCC). The subset of these cDNA clones have been rearrayed into the condensed clone sets, the NIA Mouse 15K cDNA Clone Set (Tanaka et al. 2000; Kargul et al. 2001) and the 7.4K cDNA Clone Set (VanBuren et al. 2002), which have been made available through designated academic distribution centers. Many genes that are uniquely or predominantly expressed in mouse early embryos and stem cells have been recently incorporated into a 60mer oligonucleotide microarray (Carter et al. 2003). Sequence information has been made available at public sequence databases (e.g., dbEST [Boguski et al. 1993]). Finally, all the information discussed here, as well as the graphical interfaces of the Mouse Gene Index, is available on our Web site at http://lgsun.grc.nia.nih.gov/cDNA/cDNA.html.

Although the full appreciation of these resources is yet to be realized, the initial assessment of the first comprehensive transcriptome of early mouse embryos and stem cells has already provided three major points presented in this report.

First, approximately 1,000 putative genes that were newly identified using our cDNA collection most likely represent mouse genes unidentified previously, as they either encode ORFs greater than 100 amino acids or have multiple exons. The RT–PCR analysis of 19 selected genes confirmed the notion that novel cDNAs from our libraries tend to be expressed specifically in cells and tissues that we used in this project. These gene candidates will be a rich source of genes that are expressed at low levels, but play major roles in ES cells and adult stem cells as well as in early embryos.

Second, the analysis provided lists of genes specific to particular embryonic stages or stem cells and not expressed in other cell types. For example, we have identified signature genes for the individual preimplantation stages, all preimplantation stages, ES cells, and adult stem cells.

Finally, the PCA of 2,812 genes with relatively abundant expression revealed 88 genes with average expression levels that correlate well to the developmental potentials of cells. These genes may provide the first scale to characterize the developmental potential of cells and tissues at the molecular level.

The developmental potential of cells is a fundamental concept in developmental biology, providing a conceptual framework of sequential transition from totipotent fertilized eggs to pluripotent embryonic cells and stem cells to terminally differentiated cells. It is worth noting that genes associated with developmental potential can be identified only by simultaneous analysis of preimplantation embryos and a variety of stem cells. The analyses of stem cells alone could not provide these broader perspectives (Ivanova et al. 2002; Ramalho-Santos et al. 2002; Tanaka et al. 2002). The 88 genes we have identified here may provide a set of marker genes for scaling the potential of cells. It is important to note that this scale is an operational construct. As such, further studies of the genes in the list will be required to test whether they provide critical clues to resolve the classic problem of the relation of stem cells to development. But the list could have immediate practical utility in assessing the effectiveness of treatments, gene manipulation, or both to convert differentiated cells such as fibroblasts into more potent cells such as ES—one of the most important goals required to achieve stem cell–based therapy.

Materials and Methods

Cdna Library Construction, Clone Handling, and Sequencing

Sources of tissue materials and RNA extraction methods are available as associated documents in the GenBank DNA sequence records (see also http://lgsun.grc.nia.nih.gov/cDNA/cDNA.html). cDNA libraries were constructed as described elsewhere (Piao et al. 2001).

Assembling of a Gene Index

See description in the legend to Figure 1.

Analysis of 19 cDNA Clones

Sequencing of full-length cDNA clones and RT–PCR analysis were done by the standard methods.

Identification of Differentially Expressed Genes

Most methods for selecting differentially expressed genes from EST frequencies are based on the assumption that each cDNA clone is a random sample from the mRNA pool in the cell and hence that EST frequencies correspond to the Poisson distribution (Audic and Claverie 1997). Real EST libraries, however, do not satisfy this assumption because even small changes in experimental conditions may affect the stability of particular species of mRNA, which in turn will cause a bias in EST frequency. Thus, a reliable detection of differentially expressed genes requires either library replications or comparison of classes of libraries. Because our EST libraries do not have true replications, we selected the latter approach, which yields genes that are specifically expressed in one class of tissues/stages and do not express in other tissues/stages. Some cDNA clones were represented by 5′ EST, some were by 3′ EST, and some were by both 5′ EST and 3′ EST. To avoid counting the same cDNA clone twice by 5′ EST and 3′ EST, all EST frequency analysis was done at the cDNA clone level.

To detect genes specific to a particular group of libraries, we first estimated the correlation between log-transformed clone frequencies, $\log(1000*n_i/N + 0.05)$, where n_i is the abundance of clone i in the library and N is the total number of clones, with membership indicated (0 or 1) in a particular group. The first three group classifications are targeted on oocytes. The next two classifications include all preimplantation stages with and without blastocysts. There are four classifications attempting to differentiate between pluripotent cells and other tissues. The final nine classifications capture various groups of stem cells. A subset of the data is shown in Figure 3. We analyzed only positive correlations because we were interested in genes that are overexpressed in tissues of interest, and P-values were estimated using a one-tailed t-test. Because P-values cannot be used for simultaneous assessment of multiple hypotheses, we determined significant genes using the FDR method (Benjamini and Hochberg 1995). The FDR was set to 0.1, which corresponds to the average proportion of false positives equal to 10%.

As this study is focused on embryo- and stem cell–specific genes, we analyzed EST frequencies in public databases (Boguski et al. 1993) to exclude those genes that are predominantly expressed in adult tissues. A total of 3,338,847 public ESTs have been grouped into the following categories: NIA Collection, Preimplantation, Embryo, Embryonic Stem Cells, Fetus, Neonate, Adult, Adult

Gonad, Adult Stem Cells, Adult Tumor, and Unclassified/Pooled Tissues. Of 29,810 mouse genes, 5,425 genes were not represented by ESTs, 11,574 genes were expressed predominantly in adult tissues (EST frequency in adult tissues exceeds one-third of the maximum EST frequencies in all tissues), and 12,811 were genes expressed in embryos or in gonads, tumors, and stem cells. By removing 2,055 gonad-specific and 56 tumor-specific genes (20 times more ESTs in gonad or tumors than in other tissues), we obtained 10,700 genes that are predominantly expressed in embryos and stem cells. Only ESTs matching to these genes were analyzed for differential expression.

PCA of Clone Frequencies

For the PCA shown in Figure 4, we selected 2,812 genes that had transcript frequencies of greater than or equal to 0.1% in at least one library. Clone/EST frequencies were log-transformed as $\log(1000*n_i/N + 0.05)$, where n_i is the number of clones in U-cluster i in the library, and N is the total number of all clones in this library.

Statistical significance of gene contribution to PC3 (see Figure 5) was evaluated using correlation between log-transformed clone frequencies in various libraries and library position on the PC3 axis. P-values, estimated using a one-tailed t-distribution, characterize the significance of correlation for a single clone. To control the proportion of false positives, we used FDR, which was set to 0.1.

Acknowledgements

We thank M. A. Espiritu, A. Ebrahimi, J. J. Evans, S. J. Olson, M. Roque-Briewer, and N. Caffo at Applied Biosystems for contract-based sequencing and S. Chacko for setting up the mouse genome database on Biowulf. This study utilized the high-performance computational capabilities of the Biowulf/LoBoS3 cluster at the National Institutes of Health (NIH), Bethesda, Maryland, United States of America. Sequencing of cDNA clones was solely supported by the research and development funds of the National Institute on Aging (NIA). The project was mainly supported by the Intramural Research Program of the NIA. The collection of HS cells has been funded in part with federal funds from the National Cancer Institute, under contract number NO1-CO-5600.

Authors' Contributions

MSHK conceived and designed the experiments. YP, RM, GF, PRM, CAS, UCB, YW, MGC, TH, KA, HA, LS, TST, WLK, TY, SAJ, SP, and MSHK performed the experiments. AAS, YP, RM, DBD, YQ, VV, GF, J. Kelso, WH, and MSHK analyzed the data. RN, KRB, DDT, RJH, DLL, DS, J. Keller, EK, GHK, AU, AV, JR, TK, BLMH, AC, MD, J. Kelso, and WH contributed reagents/materials/analysis tools. AAS, YP, RM, VV, GF, and MSHK wrote the paper.

References

1. Adams MD, Kelley JM, Gocayne JD, Dubnick M, Polymeropoulos MH, et al. (1991) Complementary DNA sequencing: Expressed seqeunce tags and human genome project. Science 252: 1651–1656.

2. Anisimov SV, Tarasov KV, Tweedie D, Stern MD, Wobus AM, et al. (2002) SAGE identification of gene transcripts with profiles unique to pluripotent mouse R1 embryonic stem cells. Genomics 79: 169–176.

3. Ashburner M, Ball CA, Blake JA, Botstein D, Butler H, et al. (2000) Gene ontology: Tool for the unification of biology—the Gene Ontology Consortium. Nat Genet 25: 25–29.

4. Audic S, Claverie JM (1997) The significance of digital gene expression profiles. Genome Res 7: 986–995.

5. Benjamini Y, Hochberg Y (1995) Controlling the false discovery rate: A practical and powerful approach to multiple testing. J R Stat Soc B Met 57: 289–300.

6. Boguski MS, Lowe TMJ, Tolstoshev CM (1993) dbEST: Database for "expressed sequence tags." Nat Genet 4: 332–333.

7. Carter MG, Hamatani T, Sharov AA, Carmack CE, Qian Y, et al. (2003) In situ-synthesized novel microarray optimized for mouse stem cell and early developmental expression profiling. Genome Res 13: 1011–1021.

8. Chapman S, Schenk P, Kazan K, Manners J (2002) Using biplots to interpret gene expression patterns in plants. Bioinformatics 18: 202–204.

9. Galli R, Fiocco R, De Filippis L, Muzio L, Gritti A, et al. (2002) Emx2 regulates the proliferation of stem cells of the adult mammalian central nervous system. Development 129: 1633–1644.

10. Hochedlinger K, Jaenisch R (2002) Nuclear transplantation: Lessons from frogs and mice. Curr Opin Cell Biol 14: 741–748.

11. Hubbard T, Barker D, Birney E, Cameron G, Chen Y, et al. (2002) The Ensembl genome database project. Nucleic Acids Res 30: 38–41.

12. Ivanova NB, Dimos JT, Schaniel C, Hackney JA, Moore KA, et al. (2002) A stem cell molecular signature. Science 298: 601–604.

13. Kargul GJ, Dudekula DB, Qian Y, Lim MK, Jaradat SA, et al. (2001) Verification and initial annotation of the NIA mouse 15K cDNA clone set. Nat Genet 28: 17–18.

14. Kent WJ (2002) BLAT: The BLAST-like alignment tool. Genome Res 12: 656–664.

15. Ko MSH (2001) Embryogenomics: Developmental biology meets genomics. Trends Biotechnol 19: 511–518.

16. Ko MSH, Kitchen JR, Wang X, Threat TA, Hasegawa A, et al. (2000) Large-scale cDNA analysis reveals phased gene expression patterns during preimplantation mouse development. Development 127: 1737–1749.

17. Latham KE, Schultz RM (2001) Embryonic genome activation. Front Biosci 6: D748–D759.

18. Makino S, Fukuda K, Miyoshi S, Konishi F, Kodama H, et al. (1999) Cardiomyocytes can be generated from marrow stromal cells in vitro. J Clin Invest 103: 697–705.

19. Marra M, Hillier L, Kucaba T, Allen M, Barstead R, et al. (1999) An encyclopedia of mouse genes. Nat Genet 21: 191–194.

20. Matsui Y, Zsebo K, Hogan BL (1992) Derivation of pluripotential embryonic stem cells from murine primordial germ cells in culture. Cell 70: 841–847.

21. Mulder NJ, Apweiler R, Attwood TK, Bairoch A, Barrell D, et al. (2003) The InterPro database 2003 brings increased coverage and new features. Nucleic Acids Res 31: 315–318.

22. Ochi K, Chen G, Ushida T, Gojo S, Segawa K, et al. (2003) Use of isolated mature osteoblasts in abundance acts as desired-shaped bone regeneration in combination with a modified poly-DL-lactic-co-glycolic acid (PLGA)–collagen sponge. J Cell Physiol 194: 45–53.

23. Okazaki Y, Furuno M, Kasukawa T, Adachi J, Bono H, et al. (2002) Analysis of the mouse transcriptome based on functional annotation of 60,770 full-length cDNAs. Nature 420: 563–573.

24. Ortiz M, Wine JW, Lohrey N, Ruscetti FW, Spence SE, et al. (1999) Functional characterization of a novel hematopoietic stem cell and its place in the c-Kit maturation pathway in bone marrow cell development. Immunity 10: 173–182.

25. Pertea G, Huang X, Liang F, Antonescu V, Sultana R, et al. (2003) TIGR gene indices clustering tools (TGICL): A software system for fast clustering of large EST datasets. Bioinformatics 19: 651–652.

26. Piao Y, Ko NT, Lim MK, Ko MSH (2001) Construction of long-transcript enriched cDNA libraries from submicrogram amounts of total RNAs by a universal PCR amplification method. Genome Res 11: 1553–1558.

27. Pruitt KD, Maglott DR (2001) RefSeq and LocusLink: NCBI gene-centered resources. Nucleic Acids Res 29: 137–140.

28. Ramalho-Santos M, Yoon S, Matsuzaki Y, Mulligan RC, Melton DA (2002) "Stemness": Transcriptional profiling of embryonic and adult stem cells. Science 298: 597–600.

29. Sasaki N, Nagaoka S, Itoh M, Izawa M, Konno H, et al. (1998) Characterization of gene expression in mouse blastocyst using single-pass sequencing of 3995 clones. Genomics 49: 167–179.

30. Savatier P, Lapillonne H, van Grunsven LA, Rudkin BB, Samarut J (1996) Withdrawal of differentiation inhibitory activity/leukemia inhibitory factor up-regulates D-type cyclins and cyclin-dependent kinase inhibitors in mouse embryonic stem cells. Oncogene 12: 309–322.

31. Shamblott MJ, Axelman J, Wang S, Bugg EM, Littlefield JW, et al. (1998) Derivation of pluripotent stem cells from cultured human primordial germ cells. Proc Natl Acad Sci USA 95: 13726–13731.

32. Slack JMW (1991) From egg to embryo. Regional specifications in early development. Cambridge, United Kingdom: Cambridge University Press. 348 p.

33. Solter D, de Vries WN, Evsikov AV, Peaston AE, Chen FH, et al. (2002) Fertilization and activation of the embryonic genome. In: Rossant J, Tam PPL, editors. Mouse development: Patterning, morphogenesis, and organogenesis. San Diego, California: Academic Press. pp. 5–19.

34. Tanaka S, Kunath T, Hadjantonakis AK, Nagy A, Rossant J (1998) Promotion of trophoblast stem cell proliferation by FGF4. Science 282: 2072–2075.

35. Tanaka TS, Jaradat SA, Lim MK, Kargul GJ, Wang X, et al. (2000) Genome-wide expression profiling of mid-gestation placenta and embryo using a 15,000 mouse developmental cDNA microarray. Proc Natl Acad Sci USA 97: 9127–9132.

36. Tanaka TS, Kunath T, Kimber WL, Jaradat SA, Stagg CA, et al. (2002) Gene expression profiling of embryo-derived stem cells reveals candidate genes associated with pluripotency and lineage specificity. Genome Res 12: 1921–1928.

37. Thomson JA, Itskovitz-Eldor J, Shapiro SS, Waknitz MA, Swiergiel JJ, et al. (1998) Embryonic stem cell lines derived from human blastocysts. Science 282: 1145–1147.

38. VanBuren V, Piao Y, Dudekula DB, Qian Y, Carter MG, et al. (2002) Assembly, verification, and initial annotation of the NIA mouse 7.4K cDNA clone set. Genome Res 12: 1999–2003.

39. Waddington CH (1957) The strategy of the genes: A discussion of some aspects of theortical biology. London: Allen and Unwin. 262 p.

40. Waterston RH, Lindblad-Toh K, Birney E, Rogers J, Abril JF, et al. (2002) Initial sequencing and comparative analysis of the mouse genome. Nature 420: 520–562.

41. Yanagimachi R (2002) Cloning: Experience from the mouse and other animals. Mol Cell Endocrinol 187: 241–248.

Transcriptional Profiling Reveals Barcode-Like Toxicogenomic Responses in the Zebrafish Embryo

Lixin Yang, Jules R. Kemadjou, Christian Zinsmeister,
Matthias Bauer, Jessica Legradi, Ferenc Müller,
Michael Pankratz, Jens Jäkel and Uwe Strähle

ABSTRACT

Background

Early life stages are generally most sensitive to toxic effects. Our knowledge on the action of manmade chemicals on the developing vertebrate embryo is, however, rather limited. We addressed the toxicogenomic response of the zebrafish embryo in a systematic manner by asking whether distinct chemicals would induce specific transcriptional profiles.

Results

We exposed zebrafish embryos to a range of environmental toxicants and measured the changes in gene-expression profiles by hybridizing cDNA to an oligonucleotide microarray. Several hundred genes responded significantly to at least one of the 11 toxicants tested. We obtained specific expression profiles for each of the chemicals and could predict the identity of the toxicant from the expression profiles with high probability. Changes in gene expression were observed at toxicant concentrations that did not cause morphological effects. The toxicogenomic profiles were highly stage specific and we detected tissue-specific gene responses, underscoring the sensitivity of the assay system.

Conclusion

Our results show that the genome of the zebrafish embryo responds to toxicant exposure in a highly sensitive and specific manner. Our work provides proof-of-principle for the use of the zebrafish embryo as a toxicogenomic model and highlights its potential for systematic, large-scale analysis of the effects of chemicals on the developing vertebrate embryo.

Background

Organisms are open systems that are in constant exchange with their environment. As a consequence, living systems have to adapt to environmental conditions by adjusting their physiology accordingly. Chemicals from natural sources or manmade pollution can represent rather adverse environmental conditions with a fatal outcome if the organism fails to adapt. It is a well-established fact that xenobiotics such as dioxin or cadmium can induce changes in gene expression [1-3]. The responsive genes include adaptive genes that are involved in detoxification or protection against oxidative or other cellular stresses and may also comprise genes that are directly responsible for the fatal effects of the toxicants. The early life stages of vertebrates are generally the most susceptible to adverse chemical impact [4]. Yet we do not have a detailed picture of the transcriptional response profiles of these early life stages.

There is a high demand by regulators and industry for reliable and ethically acceptable methods to evaluate the developmental toxicity of pharmaceuticals, industrial chemicals and waste products. For example, several tens of thousands of chemicals need to be assessed within the European Union REACH (Registration, Evaluation and Authorization of Chemicals) initiative for the safety testing and risk assessment of chemicals in the next years [5,6]. Cheap and reliable alternative methods are needed to cope with this enormous screening effort.

Toxicogenomics is a powerful tool for studies of toxicological mechanisms and for the detection of toxicity profiles [7] as it allows the simultaneous assessment of thousands of genes. To obtain the full potential of toxicogenomics for the evaluation of developmental toxicity, however, animal systems have to be used. The zebrafish embryo is a vertebrate system with great merits for this undertaking. The zebrafish was introduced more than two decades ago as a model to study development and neurobiology [8]. In parallel, the zebrafish embryo has evolved into a model for studies of chemical impact: it permits efficient compound screens [9] and is, for example, used in a standardized assay for sewage testing in Germany, replacing traditional toxicological tests with adult fish [10,11]. Given the experimental advantages such as small size of the embryo, cheap maintenance, availability of a genome sequence and many mutants, the zebrafish embryo is one of the most promising vertebrate systems for studies of toxicological mechanisms and toxicogenomics [12-14]. Most assays using zebrafish, however, rely on morphological endpoints, which display little discrimination between different toxicants.

Expression profiling has just recently entered zebrafish research [15-20] and only a few toxicogenomic studies exist [1,21,22]. Dioxin (TCDD) impairs fin regeneration in adult zebrafish, and expression profiling revealed TCDD-induced changes in the expression of genes involved in extracellular matrix formation [1,23]. Exposure of zebrafish to arsenic leads to changes in gene expression in adult zebrafish liver very similar to those reported for mammals, suggesting damage to protein and DNA and increased oxidative stress in the livers of arsenic-treated animals [22]. In another pilot study, zebrafish embryos were exposed to the reference compound 3,4-dichloroaniline and seven genes were significantly regulated [21].

Despite these advances, however, it is not known whether there are different responses to different toxicants and at different developmental stages. Would different toxic chemicals induce different genomic profiles, which might even be diagnostic for particular toxicants, or does the genome of the embryo respond in a general stress response. Would the sensitivity of whole-embryo exposure experiments be high enough to detect responses of genes that are restricted to small numbers of cells?

We established the toxicogenomic profiles of 11 toxicants. The gene-expression patterns induced by the 11 toxicants are related but sufficiently different to recognize toxicant-specific profiles and developmental stage-specific gene responses were also evident. Moreover, we could detect gene-expression changes at concentrations that do not have phenotypic consequences. We found synergistic effects when a mixture of compounds was applied at low doses, suggesting that the genomic response provides a more sensitive readout than morphological effects.

Results

Model Compounds Cause Similar Teratological and Toxic Effects in Zebrafish Embryos

We chose 11 model compounds, namely methylmercury chloride (MeHg), CdCl2 (Cd), PbCl2 (Pb), As2O3 (As), Aroclor 1254 (PCB), acrylamide (AA), tert-butylhydroquinone (tBHQ), 4-chloroaniline (4CA), 1,1-bis-(4-chlorophenyl)2,2,2-trichloroethane (DDT), 2,3,7,8-tetrachlorodibenzo-p-dioxin (TCDD) and valproic acid (VA). These are compounds known for their environmental toxicity [24] and VA is a teratogen and an anti-epileptic drug [25]. VA is known to inhibit histone deacetylases and Wnt signaling in mammals, thus adding an additional mode of toxic action [26].

We first established exposure protocols with which one can trigger toxicogenomic alterations with high likelihood and at the same time cause only a small amount of cell death or embryo mortality. We limited the exposure time to 20-24 hours in the expectation of focusing predominantly on primary responses rather than indirect, secondary effects. Finally, we decided to carry out these assays in embryos before they begin to feed, that is, before 120 hours post-fertilization (hpf). We tested a range of toxicant concentrations to determine the one that caused a morphologically visible toxic/teratological effect in the treated embryos after exposure at 96-120 hpf (Figure 1, Table 1). We were not able to discriminate unequivocally between toxicant-specific morphological effects (see Figure 1). Frequently the tails were bent, and the animals had difficulty swimming correctly; in some instances they developed pericardial edema (see Figure 1a). Vehicle-treated embryos did not show alterations (see Figure 1l-n) or did so only at very low frequency. Cell death as monitored by acridine orange staining was not, or only rarely, obvious immediately after treatment when animals were sacrificed for microarray analysis.

Between 96 and 120 hpf organogenesis has proceeded so far that the animals feed for the first time [8], marking the end of the embryonic stage. At this stage, gut, liver, pancreas, nervous system, musculature and the cardiovascular system are assumed to reflect adult physiology in many respects, including the response to toxicants. Younger embryonic stages are likely to have different responses to the toxicants. We therefore included two more stages in our initial experiments. The 4-24 hpf treatment covers late blastula, gastrula and segmentation stages, during which the overall body plan is laid down [8]. The treatment phase between 24 and 48 hpf coincides with the onset of organogenesis [8]. Early embryonic stages appear more sensitive to toxicant exposure than the older embryos (compare the 24-48 hpf and 96-120 hpf treatment groups in Table 1). The concentrations of the toxicants were adjusted accordingly (see Table 1).

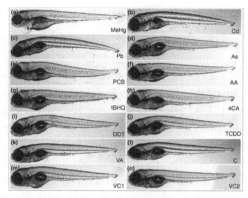

Figure 1. Toxicants induce similar morphological changes in 120 hpf zebrafish embryos. Embryos were treated with (a) methylmercury chloride (60 μg/l, MeHg); (b) CdCl2 (5 mg/l, Cd); (c) PbCl2 (2.8 mg/l, Pb); (d) As2O3 (79 mg/l, As); (e) Aroclor 1254 (33 mg/l, PCB); (f) acrylamide (71 mg/l, AA); (g) tert-butylhydroquinone (1.7 mg/l, tBHQ); (h) 4-chloroaniline (50 mg/l, 4CA); (i) 1,1-bis-(4-chlorphenyl)2,2,2-trichloroethane (15 mg/l, DDT); (j) 2,3,7,8-tetrachlorodibenzo-p-dioxin (500 ng/l, TCDD); (k) valproic acid (50 mg/l, VA); (l) vehicle 1 control (VC1): embryo water alone (for Cd, MeHg, Pb, As, VA, AA treatments); (m) vehicle 2 control (VC2): 0.2% ethanol control (for 4CA, DDT, tBHQ, PCB); (n) vehicle 3 control (VC3): 0.025% DMSO, 1.4 mg/l toluene (for TCDD). Embryos showed frequently a bent body axis and developed pericardial edema upon further cultivation.

Table 1. Summary of microarray experiments

Toxicants	Stage	Concentration	Arrays
4CA	24 hpf	15 ppm 15 mg/l 118 μM	8 (3)
	48 hpf	50 ppm 50 mg/l 390 μM	6 (3)
	120 hpf	50 ppm 50 mg/l 390 μM	8 (3)
	120 hpf	25 ppm 25 mg/l 195 μM	4 (1)
	120 hpf	5 ppm 5 mg/l 39 μM	4 (1)
	120 hpf	0.5 ppm 0.5 mg/l 3.9 μM	4 (1)
DDT	24 hpf	5 ppm 5 mg/l 14 μM	6 (3)
	48 hpf	15 ppm 15 mg/l 42 μM	6 (2)
	120 hpf	15 ppm 15 mg/l 42 μM	8 (3)
	120 hpf	1.5 ppm 1.5 mg/l 4.2 μM	4 (1)
	120 hpf	0.15 ppm 0.15 mg/l 0.42 μM	4 (1)
Cd	24 hpf	0.5 ppm 0.5 mg/l 2.7 μM	8 (4)
	48 hpf	5 ppm 5 mg/l 27 μM	8 (3)
	120 hpf	5 ppm 5 mg/l 27 μM	8 (3)
	120 hpf	2.5 ppm 2.5 mg/l 13.5 μM	4 (2)
	120 hpf	0.5 ppm 0.5 mg/l 2.7 μM	4 (2)
	120 hpf	50 ppb 50 μg/l 0.27 μM	4 (2)
TCDD	24 hpf	150 ppt 150 ng/l 0.47 nM	8 (3)
	48 hpf	500 ppt 500 ng/l 1.6 nM	4 (2)
	120 hpf	500 ppt 500 ng/l 1.6 nM	8 (3)
	120 hpf	250 ppt 250 ng/l 0.8 nM	4 (1)
	120 hpf	50 ppt 50 ng/l 0.16 nM	4 (2)
VA	24 hpf	15 ppm 15 mg/l 12.9 μM	8 (3)
	48 hpf	50 ppm 50 mg/l 43 μM	8 (3)
	120 hpf	50 ppm 50 mg/l 43 μM	8 (3)
	120 hpf	25 ppm 25 mg/l 21.5 μM	4 (1)
	120 hpf	5 ppm 5 mg/l 4.3 μM	4 (1)
	120 hpf	0.5 ppm 0.5 mg/l 0.43 μM	4 (1)
MeHg	24 hpf	50 ppb 50 μg/l 0.20 μM	8 (3)
	48 hpf	60 ppb 60 μg/l 0.24 μM	6 (2)
	120 hpf	60 ppb 60 μg/l 0.24 μM	10 (3)
	120 hpf	30 ppb 30 μg/l 0.12 μM	4 (2)
	120 hpf	6 ppb 6 μg/l 0.024 μM	4 (2)
As	120 hpf	79 ppm 79 mg/l 400 μM	8 (3)
	120 hpf	7.9 ppm 7.9 mg/l 40 μM	4 (1)
Pb	120 hpf	2.8 ppm 2.8 mg/l 10 μM	8 (3)
	120 hpf	0.28 ppm 0.28 mg/l 1 μM	4 (1)
PCB	120 hpf	33 ppm 33 mg/l 100 μM	8 (3)
AA	120 hpf	71 ppm 71 mg/l 1 mM	8 (3)
tBHQ	120 hpf	1.7 ppm 1.7 mg/l 10 μM	8 (3)
Mixture	120 hpf	Pb 1 μM. Cd 0.27 μM, As 40 μM, Hg 0.024 μM	6 (3)

Embryos were either treated from 4 to 24 (24 hpf) or from 24 to 48 hpf (48 hpf) or from 96 to 120 hpf (5 days). Arrays, total number of microarray hybridizations. Numbers in brackets indicate the number of independent biological repeats. 4CA, 4-chloroaniline; DDT, 1,1-bis-(4-chlorphenyl)-2,2,2-trichloroethane; Cd, cadmium chloride; TCDD, 2,3,7,8-tetrachlorodibenzo-p-dioxin; VA, valproic acid; MeHg, methylmercury chloride; As, arsenic (III) oxide; Pb, lead (II) chloride; AA, acrylamide; PCB, Aroclor 1254; tBHQ, tert-butylhydroquinone.

Stage-Specific Toxicogenomic Responses

To assess possible stage-specific differences, we analyzed and compared the toxicogenomic response to six compounds - MeHg, Cd, 4CA, DDT, TCDD, and VA - at the three different stages. We treated several hundred embryos with each of these compounds at each of the three stages (see Materials and methods). Principal component analysis (PCA) revealed distinct toxicogenomic responses to exposure with the six toxicants in the 24-48 and 96-120 hpf treatment groups (Figure 2a). Principal components were derived by singular value decomposition (SVD). SVD is based on the decomposition of the gene-expression matrix, whose entries are the log-transformed fold changes (M values) of gene expression, into unique orthonormal superpositions of genes and treatments. Expression changes of at least twofold and a padj < 0.025 were taken into account. The padj-value was adjusted for multiplicity testing by controlling the false discovery rate [27].

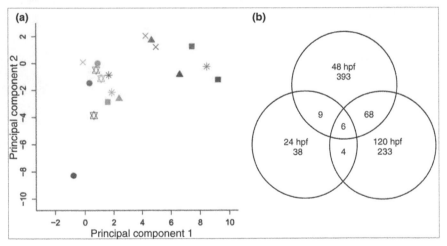

Figure 2. Distinct toxicogenomic expression profiles are induced by different toxicants. (a) Principal component analysis of the toxicogenomic profiles derived from three different embryonic stages. Embryos were exposed to vehicle controls or to one of six chemicals for the periods 4-24 hpf (green), 24-48 hpf (blue), or 96-120 hpf (red). Circles, TCDD: 150 ng/l (24 hpf), 500 ng/l (48 hpf), 500 ng/l (120 hpf). Squares, MeHg: 50 μg/l (24 hpf), 60 μg/l (48 hpf), 60 μg/l (120 hpf). Triangles, VA: 15 mg/l (24 hpf), 50 mg/l (48 hpf), 50 mg/l (120 hpf). Crosses, 4CA: 15 mg/l (24 hpf), 50 mg/l (48 hpf), 50 mg/l (120 hpf). Asterisks, Cd 500 μg/l (24 hpf), 5 mg/l (48 hpf), 5 mg/l (120 hpf). Stars, DDT: 5 mg/l (24 hpf), 15 mg/l (48 hpf), 15 mg/l (120 hpf). While the transcriptional profiles of the 4-24 hpf treatment group (green symbols) cluster closely, characteristic gene-expression profiles were induced by the 24-48 hpf (blue symbols) and the 96-120 hpf (red symbols) exposures to each of the different toxicants. (b) Venn diagram comparing the number of genes induced at the three stages by all six toxicants. Numbers indicate numbers of regulated or co-regulated genes at the different stages (more than 1.95-fold change and adjusted padj < 0.025).

The differences between the transcriptional profiles induced by the six toxicants were less prominent in the datasets from the 4-24 hpf treatment groups (see Figure 2a). This may be due to the fact that different toxicants caused similar gene effects at 24 hpf. For example, the expression of the gene for fast muscle troponin T (BE693169) was downregulated by Cd, MeHg, TCDD, and VA in embryos treated between 4 and 24 hpf but not at later stages (data not shown). Furthermore, many genes that are involved in organ physiology may not yet be responsive by 24 hpf, as organ development has not proceeded far enough. In agreement with this, the expression levels of only 57 genes were significantly altered by the 4-24 hpf treatment. In contrast, the expression levels of 476 and 311 genes were significantly affected by the 24-48 hpf and 96-120 hpf treatment regimens, respectively (see Figure 2b). Moreover, very few genes in the 4-24 hpf treatment set overlapped with the 24-48 and 96-120 hpf treatment groups (15 and 10 genes, respectively). The latter groups (24-48 h and 96-120 hpf) shared more gene responses (74 genes) but 393 and 233 gene responses were stage specific (see Figure 2b). The smaller number of affected genes in the 4-24 hpf regimen may also have been caused by the lower concentrations of toxicants that we had to apply to ensure sufficient survival at these younger stages. Irrespective of this, these data indicate a high stage specificity of the toxicogenomic effects in the three treatment windows.

The toxicogenomic responses triggered by different toxicants are highly specific

We focused further analysis on the 96-120 hpf stage and used the full set of 11 toxicants by including treatments with AA, PCB, As, tBHQ and Pb. Replicate hybridizations with mRNA from at least three independent toxicant treatments were performed (see Table 1). Toxicant effects were clustered based on their Euclidean distance to each other and the similarity of gene responses was determined by a Pearson correlation proximity measure. The expression profiles summarize clustering results for a subset of 199 genes across all 11 toxicant responses (Figure 3). The gene-selection criteria applied take into account the extent and significance of changes in gene expression (at least twofold, padj < 0.025) as well as differences and similarities in expression changes between toxicants (see Materials and methods). Distinct patterns of gene expression were noted for each of the 11 compounds. However, similarities in gene responses were also detected. One group of chemicals with related gene responses comprises Pb, As, Cd, tBHQ, MeHg and VA (see Figure 3, lanes 6 to 11). Another subgroup of related responses was induced by TCDD, 4CA, DDT and AA (see Figure 3, lanes 2-5), whereas the PCB triggered a more distinct expression profile (see Figure 3, lane 1).

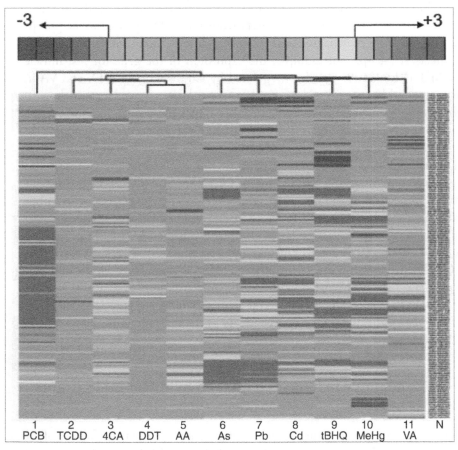

Figure 3. Toxicants induce highly specific toxicogenomic profiles. Hierarchical clustering of gene responses in embryos treated between 96 and 120 hpf with PCB (33 mg/l), TCDD (500 ng/l), 4CA (50 mg/l), DDT (15 mg/l), AA (71 mg/l), As (79 mg/l), Pb (2.8 mg/l), Cd (5 mg/l), tBHQ (1.7 mg/l), MeHg (60 μg/l), VA (50 mg/l). For each toxicant exposure, vehicle controls were carried out in parallel. The gene names are indicated (N) and are legible upon magnification of the PDF version of this figure. The key at the top indicates the color code for fold changes ranging from threefold upregulated (+3, red) to threefold downregulated (-3, blue). Fold changes greater than three are not indicated explicitly but are included. Only genes are listed whose mRNA levels changed by more than twofold (padj < 0.025) in at least one of the treatments. The data represent the average over all biological and technical repeats (see Table 1).

As verification, we carried out blind tests to identify the chemicals by their induced gene-expression profile. Fourteen out of the 15 chemicals were unambiguously identified (Table 2). In the case of 4CA, close matches were scored to the 4CA, the DDT and the AA response profiles (see Table 2). Thus, we identified the correct group of chemicals (see Figure 3, lanes 2-4). Taken together, the results from these blind trials underscore the reliability of the toxicogenomic profiles and furthermore suggest that it is possible to derive signatures of toxicogenomic responses predictive for specific chemicals or chemical groups from whole animal exposure experiments.

Table 2. Summary of results from blind experiments

Test	4CA	Cd	DDT	MeHg	TCDD	VA	AA	As	Pb	PCB	tBHQ
Pb	6.15	6.78	8.66	6.76	10.91	7.96	7.62	6.32	**5.01**	10.66	6.81
DDT	6.74	9.86	**4.51**	8.79	7.79	8.01	5.39	10.38	8.96	10.05	9.13
4CA	6.87*	10.96	6.69*	9.47	8.82	8.87	6.99*	11.21	8.42	10.75	9.42
TCDD	8.55	12.43	6.78	11.01	**5.42**	10.37	7.61	10.76	12.61	15.63	10.62
As	9.14	9.56	10.93	9.74	12.46	9.89	10.18	**6.88**	7.51	11.32	7.51
Pb	7.25	7.95	10.05	7.38	12.06	8.62	8.65	9.58	**4.22**	9.91	7.70
PCB	9.84	9.30	13.17	10.23	15.01	10.01	11.78	12.77	10.62	**3.93**	10.40
As	9.18	9.99	9.92	10.20	11.53	10.16	9.43	**5.48**	7.84	12.40	7.67
TBHQ	8.79	7.49	11.32	9.24	11.52	10.29	10.98	8.93	9.77	11.31	**6.33**
PCB	8.37	7.74	12.06	8.88	13.65	8.54	10.63	11.79	10.22	**5.71**	9.08
AA	7.18	10.79	4.29	9.28	7.21	8.63	**3.56**	7.67	8.12	13.04	8.86
TBHQ	7.49	7.26	11.22	8.30	11.66	9.00	10.40	9.57	9.58	10.47	**5.41**
Cd	8.32	**6.67**	10.76	8.39	12.79	8.90	10.16	10.89	6.96	8.05	8.15
VA	6.43	8.25	6.47	7.57	9.04	**4.57**	6.15	9.56	8.02	10.24	8.33
AA	10.26	14.38	7.09	12.23	9.61	11.59	**6.25**	11.26	10.54	13.89	11.84

The induced genes fell into different gene ontology groups such as genes involved in combating oxidative stress (Table 3) and genes encoding chaperones (Table 4). Another major class of genes that was significantly regulated by a number of toxicants comprised solute carriers (Table 5). We also carried out a computational analysis of the affected genes using the GoTreeMachine algorithm to identify more complex pathways and processes. An inflammatory response was induced by several compounds (As, 4CA, Cd, MeHg, Pb, PCB and tBHQ), whereas inductions characteristic of an immune response were evoked by MeHg and tBHQ. The latter compound also triggered genes involved in G-protein-coupled signaling and phototransduction. Induction of genes with a function in base-excision repair was noted in the case of exposure to As and PCB, suggesting that these compounds cause DNA damage in the embryo.

Table 3. Oxidative stress genes and their response to toxicants

Gene name	Gene ID	AA	As	4CA	Cd	MeHg	Pb	PCB	tBHQ	VA
Peroxiredoxin I	BI980610	3.9	13.6	4.1	3.1	7.7	9.9		7.5	
Thioredoxin	BI864190		14.2	3.5	4	4.4	8.1	2.5	6.1	
Glutathione S-transferase omega I	AW019036	3	6.3	2.1		2.6	3.5		2.2	
Glutathione S-transferase pi	AF285098	2	4.1			3.2	5.9		2.7	
Glutathione S-transferase omega 2	BI979918		5.6		2.7		5.2		2.1	
Thioredoxin interacting protein	BI892352				2					2
Glutathione peroxidase	AW232474					-4.2				

Table 4. Chaperone genes and their response to toxicants

Gene name	Gene ID	As	4CA	Cd	Pb	PCB	tBHQ	VA
Stress protein HSP70	AB062116	10.1	2.2	7	2.5	5.4	12.3	2.7
	AF210640	10.9		6.6	2.5	5	10.8	2.7
Hsp70 (2)	AF006007	7.6		6.3	2.1	4.6	10.1	2.4
Heat shock protein HSP 90-alpha	AF068773	2.6						
Heat shock cognate 70 kDa protein	BM024785	3						
DnaJ (Hsp40) homolog, family A, member 1	BI891737	3.6					3.6	
Ahsa1 protein	BM103957	2.2					2.1	

Table 5. Solute carrier family genes and their response to toxicants

Gene name	Gene ID	As	4CA	Cd	MeHg	Pb	PCB	VA
Solute carrier family 16 member 9 (1)	BE016639		2.5	2.9	2.4	3.6	4.1	2.3
Solute carrier family 16 member 9 (2)	BI474827		3.4	4.3	4.8	4.7	7.9	3
Solute carrier family 16 member 6	AW421040				2.9	4	4.6	2.9
Solute carrier family 2 member 5	AI477656	2.1				2.1	2.5	
Solute carrier family 6 member 8	BI980828			2		2.2	2.4	
Solute carrier family 43, member 2	BI887324						2.1	
Solute carrier family 3	BG985518						2.1	
Solute carrier family 20 (phosphate) member 1	BI890772						2.2	
Solute carrier family 6 (GABA) member 1	BF157011							-3.1
	BI563084							-2.1

Identification of Tissue-Specific Genes

We next verified the observed gene responses by methods other than microarray hybridization. First, the changes in gene expression were confirmed by re-evaluating a subset of gene responses by semi-quantitative reverse transcription PCR (RT-PCR). Out of 14 gene responses analyzed, all showed the up- or downregulation expected from the array data (Figure 4). This suggests that the changes in transcript levels measured by the microarray hybridizations reflect genuine responses to the toxicants.

We used in situ hybridization with selected probes to toxicant-treated and control embryos to assess the tissue-specific expression patterns of the response genes and whether these are altered in response to toxicant. Cytochrome P4501A1 mRNA (AF057713) was induced by 500 ng/l TCDD in endothelial cells (15/15 embryos, Figure 5a,b). The levels of glutathione peroxidase 1 (AW232474) mRNA in stomach and gut were repressed by 60 µg/l MeHg (11/15 embryos, Figure 5c,d), in agreement with microarray and RT-PCR data (see Figure 4).

Toxicants	Gene ID	Gene name	Fold change	cycle numbers	Cont	Treat
TCDD	AI397347	Similarity to keratin type 1 (human)	-2.6	30		
	AF057713	Danio rerio cytochrome p 4501A	37.7	25		
DDT	BI533854	Weakly similar to o-type lectin	-2.1	25		
Cd	BE201681	Danio rerio Oncomodulin A	-4.4	25		
	AW174507	Danio rerio materix metalloprotinase 9	8.8	25		
	AF210640	Danio rerio HSP 70	8.2	20		
	AW305943	Danio rerio materix metalloproteinase 13	7.5	30		
	BI864190	Similarity to thioredoxin	4.3	25		
Hg	AW232474	Danio rerio glutathione peroxidase 1	-4.7	25		
	BI980610	Similarity to natural killer cell enhacning factor	3.9	25		
	BI864190	Similarity to thioredoxin	3.8	25		
	BG727181	unknown	3.2	30		
VA	AY050500	Danio rerio cone transducin alpha subunit	-2.4	25		
	AW422298	Similarity to transcription factor ATF-3	4.2	25		
4 CA	BI843145	unknown	-2.8	30		
	BI980610	Similarity to natural killer cell enhancin g factor	3.9	30		
As	BI864190	Similarity to thioredoxin	11.4	30		
Embryo medium	β-actin		ND	30		
0.2% ethanol	β-actin		ND	30		
0.025%DMSO+ 1.4 mg/l toluene	β-actin		ND	30		

Figure 4. RT-PCR analysis confirms selected gene responses. Embryos were exposed to the indicated toxicants (500 ng/l TCDD; 15 mg/l DTT; 5 mg/l Cd; 60 μg/l MeHg; 50 mg/l VA; 50 mg/l 4CA; 79 mg/l As) or vehicle alone (embryo medium or 0.2% ethanol or 0.025% DMSO, 1.4 mg/l toluene) between 96 and 120 hpf. (a) cDNA was synthesized and subjected to PCR with primers specific for the selected genes indicated. Gene ID refers to the accession number in GenBank. The number of temperature cycles (cycle numbers) for every set of amplifications is indicated. The fold-change column summarizes the results from the microarray experiments for comparison with the RT-PCR results shown in (b). See legend of Figure 1 for details of treatments and controls. β-actin mRNA was used as a toxicant-insensitive reference. ND, not determined, as the actin gene response fell into the class of nonregulated genes in the microarray results.

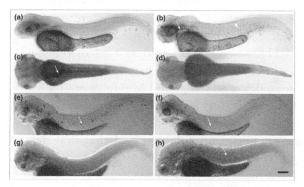

Figure 5. Examples of toxicant-responsive genes that are expressed in a highly tissue-restricted manner. (a) 48 hpf vehicle 3 control. Figure 1 indicated the exposure embryo from 96 to 120 hpf and (b) 500 ng/l TCDD-treated embryos hybridized to a cytochrome P450 1A1 antisense probe. TCDD-treated embryos showed increased levels of cytochrome P4501A1 mRNA in blood vessels Arrow, primary head sinus, arrowhead, intersegmental vessel. (c) 72 hpf vehicle control 1 and (d) 60 μg/l MeHg-exposed embryos hybridized to a glutathione peroxidase 1 probe. Embryos showed a reduction of mRNA levels in the gut (arrow). Embryos were treated from 4 to 72 hpf and were then fixed for in situ processing. (e) Control embryo and (f) 500 μg/l CdCl2-treated embryo hybridized to oncomodulin A antisense mRNA. Oncomodulin A mRNA levels are downregulated in the hair cells of the lateral-line organ (arrow) in response to Cd exposure. (g) Control and (h) Cd-treated embryos hybridized to a thioredoxin antisense probe. Thioredoxin is upregulated in the hair cells of the neuromasts (arrow). Embryos are oriented anterior to the left and dorsal up (a,b,e-h) or with dorsal side facing (c,d). Scale bar represents 220 μm.

The neuromasts of the zebrafish lateral line are very sensitive to a number of compounds including CdCl2 [28-30]. the mRNA for oncomodulin A (also called parvalbumin3a), which is expressed in the hair cells and supporting cells of neuromasts in untreated embryos, is barely detectable in the neuromasts of embryos treated with 500 μg/l CdCl2 (13/15 embryos, Figure 5e,f), in concordance with the Cd-induced, 4.4-fold decrease of oncomodulin A mRNA measured by microarray hybridization (see Figure 4). In contrast, thioredoxin-like mRNA (BI864190) is upregulated in hair cells (12/13 embryos, Figure 5g,h) in response to Cd. This suggests that Cd does not cause a complete loss of hair cells, even though staining with the dye DASPEI suggests that hair cells are strongly reduced (data not shown). The thioredoxin-like mRNA is also expressed in selected areas of the brain. These regions show also increased levels of expression in response to Cd (data not shown). In summary, these in situ expression studies show that the microarray procedure used permits detection of organ- and cell-specific gene responses with very high sensitivity. Moreover, these results also suggest that the gene responses occur in almost all of the embryos exposed to the toxicants.

The genome responds to very low toxicant concentrations

The concentrations of the toxicants were adjusted in the initial experiments so that they caused morphologically visible defects in exposed animals. We asked next whether one could measure changes in the expression profiles at lower concentrations that do not have apparent morphological effects. TCDD, DDT, Cd, 4CA, MeHg, and VA were used as a set of test compounds. We could detect significant changes in gene expression (at least twofold and padj < 0.025) in response to 0.5 mg/l Cd, 6 μg/l MeHg, 5 mg/l VA, 25 mg/l 4CA, 15 mg/l DDT, and 50 ng/l TCDD (Figure 6a-c, Table 6, and data not shown). With the exception of 6 μg/l MeHg and 25 mg/l 4CA, these low concentrations did not cause obvious morphological or behavioral defects (data not shown), suggesting that this assay can detect responses to toxicant concentrations that do not cause acute morphological effects. It is clear, however, that the number of genes with a significant response to the toxicants decreases (see Table 6). Cytochrome P4501a1 was fivefold upregulated by 50 ng/l TCDD, oncomodulin A was reduced 4.5-fold by 0.5 mg/l Cd and peroxiredoxin was still 3.5-fold induced by 6 μg/l MeHg. Thus, even though fewer genes respond to these lower concentrations, the measured changes in transcript levels are robust.

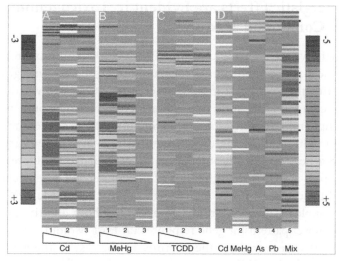

Figure 6. The concentration dependence of toxicogenomic responses and the synergistic effects of low doses. (a-c) Embryos were exposed to decreasing concentrations of Cd (a, lane 1, 5 mg/l: lane 2, 2.5 mg/l; lane 3, 0.5 mg/l), or MeHg (b, lane 1, 60 µg/l; lane 2, 30 µg/l; lane 3, 6 µg/l) or TCDD (c, lane 1, 500 ng/l; lane 2, 250 ng/l; lane 3, 50 ng/l). The low concentrations elicit significant changes in gene expression. (d) Embryos were exposed either to 50 µg/l Cd (lane 1) or 6 µg/l MeHg (lane 2) or 7.9 mg/l As (lane 3) or 280 µg/l Pb (lane 4) alone, or to a mixture (Mix, lane 5) of Cd (50 µg/l), Pb (280 µg/l), MeHg (6 µg/l) and As (7.9 mg/l). The mixture shows a strongly increased response with respect to the degree of changes of expression of individual genes (dark red and dark blue bars). Arrowheads point to examples of synergistic responses whereas the dots highlight genes whose response seems to be additive. The square indicates a gene that was downregulated by As and slightly upregulated by the mixture. All exposures were performed between 96 and 120 hpf. The color key for fold changes in gene expression in (a-c) is indicated on the left and ranges from threefold upregulated (red) to threefold downregulated (blue). The color key for (d) is on the right and ranges from fivefold upregulated (red) to fivefold downregulated (blue). White bars indicate missing data. Only genes were listed whose mRNA levels changed by at least twofold (padj < 0.025) in at least one of the treatments. The data represent the average over all biological and technical repeats (see Table 1).

Table 6. The number of regulated genes in response to different concentrations of toxicants

Toxicants	Stage	Concentration	Number of regulated genes
4CA	120 hpf	50 mg/l	201
		25 mg/l	2
		5 mg/l	0
Cd	120 hpf	5 mg/l	475
		2.5 mg/l	102
		0.5 mg/l	57
DDT	120 hpf	15 mg/l	25
		1.5 mg/l	0
		0.15 mg/l	0
TCDD	120 hpf	500 ng/l	34
		250 ng/l	34
		50 ng/l	4
VA	120 hpf	50 mg/l	335
		25 mg/l	1
		5 mg/l	4
MeHg	120 hpf	60 µg/l	417
		30 µg/l	20
		6 µg/l	9

Complex Synergistic Effects are Evident in Toxicogenomic Responses to Compound Mixtures

In the environment we are normally confronted with compound mixtures rather than pure substances. The components of these mixtures could act synergistically, thereby potentiating the toxic effect [31]. We therefore investigated whether synergistic effects of compound mixtures can be observed in toxicogenomic profiles. To this end, 96-hpf embryos were exposed to a mixture of low concentrations of Cd (50 µg/l), Pb (280 µg/l), MeHg (6 µg/l) and As (7.9 mg/l). About twice as many genes (158 genes) were significantly up- or downregulated (absolute change at least twofold, padj < 0.025) than the sum of the genes regulated by exposure to the individual toxicants (81 genes: Cd 48 genes; As 12 genes; MeHg 5 genes; Pb 16 genes). Complex expression profiles composed of both additive and synergistic as well as novel patterns of gene responses (at least twofold change, padj < 0.025; Figure 6d) were scored for the mixture. In the case of the genes with similarity to peroxiredoxin (BI980610) or the solute carrier family members 6 and 9 (BE016639, AW421040), the response to the mixture appears to be purely additive (see Figure 6d, dots; Table 6). In other instances, for example the Hsp70-related genes (AB062116, AF210640, AF006007) or the sequestosome1 gene (AW343560), the mixture induced a strong increase in transcript levels, whereas significant gene responses (more than twofold, padj < 0.025) were not induced by administration of the individual compounds (see Figure 6d, arrowheads, Table 6). These genes can, however, be induced by higher concentrations of the individual compounds (see Figures 3, 6a,b), suggesting that the observed synergy is the result of a lowered response threshold. Curiously, we also noted loss of gene responses on exposure to the compound mixture (see Figure 6d), suggesting suppressive effects of the combination. For example, the transcript levels of glutathione-S-transferase omega 1 (AW019036) are significantly altered by exposure to $PbCl2$, but not by the mixture (see Table 3). In a few instances we observed opposing effects, such as in the case of suppressor of cytokine signalling 3 (BI878700), which was 4.9-fold downregulated by As and 2.6-fold upregulated by the mixture. Taken together, these results show potentiated, additive, and nonadditive effects of the mixture in comparison to the individual compounds.

Discussion

We have shown that a diverse set of 11 chemicals induces highly specific gene responses in the zebrafish embryo. Moreover, synergy effects and responses to low-dose exposure were detectable in the genome-wide transcriptional response. Our work provides proof of principle that the zebrafish embryo can serve as a specific and highly sensitive whole-animal model to monitor the toxicogenomic impact of chemicals.

Although vertebrate cell lines and other in vitro test methods have great merits in assessing toxicological effects of drugs and pollutants, they cannot replace whole animal test systems entirely. The classical animal models such as mice, rats and rabbits are expensive and attract concerns from animal-rights groups. Zebrafish embryos before the feeding stage offer a cheap and ethically acceptable vertebrate model that will not only be useful in the toxicological assessment of the tens of thousands of compounds to be tested under the REACH program but can also help to evaluate the developmental toxicity of novel compounds at an early stage of drug development.

The requirement for adequate animal models for assessing developmental toxicology is further underscored by the remarkable stage dependence of the observed toxicogenomic profiles. These differences in gene responses are likely to be a reflection of the dynamics of cell differentiation and morphogenesis, which will be impossible to model in all their aspects in cell culture and other in vitro systems. The differences in gene responses were particularly striking at early stages, presumably reflecting the fact that many organs exist only as rudiments at these times and have not fully acquired their physiological function. It is also possible that the inter-embryonic variability of the gene responses is higher at this stage, blurring the gene-expression changes in the pooled cDNA.

Previous work showed that the sensitivity of the zebrafish embryo to toxicants equals that of the commonly used tests on adult freshwater fish, allowing a reliable prediction of the toxic potential of chemicals [10,11]. The embryonic DarT assay [10,11] uses an exposure paradigm from cleavage stages to 48 hpf and relies on a set of morphological endpoints and lethality. Morphological readouts provide little discrimination between the effects of different compounds, especially in the case of environmental toxicants with a broad spectrum of toxic effects on the embryo. In marked contrast to the morphological endpoints, we found highly specific patterns of transcriptional changes, resulting in barcode-like patterns of gene responses. With one exception, we were able to predict the chemical unequivocally by its pattern of induced gene-expression changes. In most cases, these patterns are related, forming distinct subgroups of profiles, but are still sufficiently different from one another to discriminate the individual compounds.

Strikingly, a general response to oxidative stress or protein damage does not seem to exist in the zebrafish embryo. A number of the chemicals (see Table 3) induced genes involved in the cellular systems that combat the effects of oxidative stress [32]. However, the induced oxidative-stress genes differed between chemicals, suggesting toxicant-specific effects (see Table 3). A similar observation was made with respect to chaperones (see Table 4). The tissue-specific expression of these genes as well as restricted tissue effects of the toxicant may be important in this context. For example, the expression of the thioredoxin-like gene is restricted to a small number of neurons in the brain. In in situ hybridization experiments, strong elevation of thioredoxin-like mRNA levels in response to Cd and MeHg was also noticed in the hair cells of the lateral line as well as in the brain. The differences in the type of induced defense genes and their tissue-restricted expression suggest tissue-specific effects of the different toxicants.

Another Gene Ontology (GO) group that is differentially regulated by exposure to a number of toxicants is represented by members of the solute carrier (SLC) family (Table 5). These transmembrane proteins have key roles in the transport of small molecules including neurotransmitters across vesicular and plasma membranes [33]. It is tempting to speculate that the specific downregulation of the GABA transporter SCL6 member 1 by VA (see Table 5) may be related to the therapeutic effect of VA as a suppressor of epileptic seizures.

The concentrations that elicited toxicogenomic responses are in the range of pollutant levels prevailing in the environment. We did not, however, exclude the possibility that compounds accumulate in the embryo, resulting in higher intra-embryonic concentrations than in the environment. Toxicogenomic responses were triggered by TCDD, Cd, DDT, and VA at concentrations that did not cause changes in morphology. Thus the genomic response appears to be more sensitive to toxic insult than is morphogenesis. A crucial question is whether the gene responses that are not obviously correlated with pathological alterations are indeed deleterious to the animal. For example, TCDD was shown to induce a battery of genes in the mouse paw (including homologs of genes we scored in our study) without obvious teratological consequences to paw development [34]. Future work will need to address whether the low-level effects on gene expression could be correlated with, and hence used to predict, chronic effects of long-term exposure.

The lowest concentration of MeHg (6 µg/l) triggered significant changes in gene expression. In addition, we also noted teratological effects on movement and tail development at these concentrations (L.Y. and J.R.K., unpublished work), indicating that low concentrations of MeHg are acutely toxic in the zebrafish embryo. Disturbingly, blood serum levels of MeHg in humans can be in the same concentration range [35]. The zebrafish embryo may be much more susceptible to MeHg, but defining blood serum levels that are regarded as safe in humans is an active area of research.

Application of a mixture of MeHg, Cd, As, and Pb at low concentrations resulted in synergistic effects with more than additive numbers of genes affected and also novel patterns of gene-expression changes. Clearly, some of the genes affected by exposure to the mixture would be induced or repressed by higher concentrations of the individual chemicals. Examples are the thioredoxin and Hsp70 genes. Thus, it appears that the threshold at which induction occurs is lowered. This agrees with previous studies of mixture effects that support the notion of 'concentration addition', in which a component of the mixture can be replaced by an equipotent concentration of another compound [31]. The patterns of gene-expression changes induced individually by the four chemicals differed, however, suggesting that other effects have to be taken into account that cannot be explained by an additive mechanism of action.

Expression levels of genes, and presumably also responses to environmental toxicants, can vary dramatically between individuals. In a systematic study of variation in gene expression in natural populations of fish of the genus Fundulus, significant differences in gene expression were noted in 18% of the 907 genes analyzed [36]. In this respect, zebrafish embryos have a big advantage over mammalian systems as one can easily obtain large numbers of embryos and can thus average the individual gene responses by using pooled cDNA prepared from many embryos. In the cases where we confirmed the gene responses by in situ hybridization, we found that most individuals showed the expected upregulation, suggesting that many of the observed responses have a high penetrance.

While the complete development outside of the mother and the transparency of the zebrafish embryos are certainly important advantages for observation, the small size of the embryos limits the possibility of dissecting particular organs for toxicogenomic analysis. To overcome these limitations, one can use transgenic animals expressing green fluorescent protein and fluorescence-activated cell sorting to enrich for particular cell types [37]. Moreover, even whole-embryo exposure protocols as we used here permit detection of highly tissue-restricted gene responses such as those seen, for example, in the lateral line, which comprises only a very small fraction of the whole embryo.

Conclusion

The induction of the Hsp70 gene was previously shown to be a sensitive biomarker in zebrafish for exposure to Cd and other heavy metals [38]. The work presented here adds a long list of other highly sensitive biomarkers to be developed as transgenic biosensors. We believe that the zebrafish could become a key model for molecular developmental toxicology. Functional studies of TCDD toxicity in the zebrafish embryo well illustrate this (reviewed in [39]). Forward genetics [40-43], targeting-induced local lesions in genomes (TILLING) [40], morpholino knockdown [44], transgenesis [38,45] and in situ expression studies [46,47] at cellular resolution represent a powerful technical repertoire for dissecting toxicological pathways. Moreover, a large number of developmental mutants have been isolated, some of which may serve as direct targets for drugs and toxicants [48,49]. We believe that the work on the toxicogenomics of zebrafish embryos reported here is a fundamental contribution to the use of the zebrafish embryo as a model system for molecular developmental toxicology.

Materials and Methods

Chemicals and Embryo Treatment

AA (acrylamide; $CH2 = CHCONH2$), PCB (Aroclor 1254), As (arsenic (III) oxide; $As2O3$), tBHQ (tert-butylhydroquinone; $(CH3)3CC6H3-1,4-(OH)2$), Cd (cadmium chloride; $CdCl22H2O$), 4CA (4-chloroaniline; $ClC6H4NH2$), DDT (1,1-bis-(4-chlorphenyl)-2,2,2-trichlorethane; $ClC6H4)2CHCCl3$), Pb (lead (II) chloride; $PbCl2$), MeHg (methylmercury chloride; $CH3ClHg$), TCDD (2,3,7,8-tetrachlorodibenzo-p-dioxin) and VA (valproic acid; $CH3(CH2)4CO2H$) were purchased from Sigma-Aldrich (St Louis, MO).

Wild-type zebrafish strains AB, ABO and Tübingen were kept and bred as described [50]. Embryos were grown in embryo medium (60 µg/ml Instant Ocean, Red Sea, Houston, TX). Different numbers of embryos were exposed to the chemicals: 4-24 hpf (600 embryos), 24-48 hpf (400 embryos) and 96-120 hpf (200 embryos). Vehicle controls used were embryo medium alone (for Cd, Hg, Pb, As, VA, AA treatments) or 0.2% ethanol in embryo medium (for tBHQ, 4CA, PCB, DDT treatments) or 0.025% DMSO, 1.4 mg/l toluene in embryo medium (500 ng/l TCDD) or 0.0075% DMSO, 420 µg/l toluene (150 ng/l TCDD) or 0.0025% DMSO, 140 µg/l toluene (50 ng/l TCDD). Toxicant concentrations were adjusted in such a way that embryo death was minimal. The few dead embryos were discarded before preparation of RNA. None of the vehicle controls had an apparent toxic effect on the embryos by itself. As the toluene-related chemical benzene can synergize with TCDD [51] we cannot completely exclude a synergistic effect between TCDD and toluene at the 1.4 mg/l toluene concentration.

Microarray analysis

A total of 16,399 gene-specific 65mers designed by Compugen (Jamesburg, NY) and produced by Sigma-Genosys (The Woodlands. TX) were purchased and the probes (40 mM) were spotted in duplicate in two separate subarrays using a Gene Machines Omnigrid 100 (San Carlos, CA) and TeleChem SMP3 pins (Sunnyvale, CA) on CodeLink activated slides (GE Healthcare, Chalfont St Giles, UK). Upon evaluation it turned out, however, that plates 29 to 43 had faulty amine linkers, impairing the retention of the oligonucleotides on the coat-link slides. As the companies were unable to replace the defective oligonucleotides, we used the reduced set of intact oligonucleotides (384-well plates 1 to 28).

Total RNA was isolated from toxicant- and vehicle-treated embryos in every experiment in parallel using the Nucleospin RNA L Kit (Macherey-Nagel, Düren, Germany) and mRNA was extracted with the Ambion Purist Kit (Austin, TX). Labelled cDNA was synthesized from 1-2 μg mRNA using the Amersham direct cDNA labeling kit (Amersham Europe, Freiburg, Germany). Upon removal of unincorporated nucleotides over Microcon 30 spin columns (Millipore, Bedford, MA), the concentrated probes were hybridized to the microarray in 1× DIG Easy-Hyb buffer (Hoffmann-La Roche, Basel, Switzerland) overnight at 42°C. Coverslips were removed from the slides by flushing with 4× SSC and slides were washed in prewarmed wash buffer 1 (2× SSC, 0.1% SDS) for 5 min at 42°C, then in buffer 2 (0.1× SSC, 0.1% SDS) for 10 min at room temperature, and finally in 0.1× SSC four times for 1 min at room temperature. The slides were briefly dipped into 0.01× SSC at room temperature before centrifugation for 7 min at 800 rpm in an Eppendorf 5810R centrifuge.

Arrays were scanned using the Axon model 4000B dual-laser scanner and the corresponding GenePix 6 software (Molecular Devices, Union City, CA). Both channels (532 nm for Cy3 and 635 nm for Cy5) were scanned in parallel and stored as 16-bit TIFF files. Each array was scanned three times (low, medium, and high scan) with different signal-amplification factors (voltage settings of the photomultiplier tubes), but with the same laser power. The channels for Cy3 and Cy5 were balanced in each scan for approximately the same intensity range. For the low scan no spot was saturated; in the high scan the signal amplification for Cy5 was set to approximately 80% of maximum and Cy3 amplification was adjusted to this. The settings used in the medium scan lie between the low and the high scan. The absolute intensity values span the range from 0 to 65536. The scans were performed with a resolution of 10 μm. From each spot with a mean diameter of 100 μm, 70-80 pixels were recorded. Individual local background areas around the spots were defined, which comprised approximately 400 pixels. For each channel, the spot signals were calculated as the median intensity of all foreground pixels minus the median intensity of all background pixels.

All Microarray Data from this Study have been Deposited in Ncbi's Gene Expression Omnibus under the Accession Number gpl4603.

Data Preprocessing, Quality Control, Transformation, and Normalization

Raw data was derived from the result files generated by the GenePix 6 suite and analyzed with the R software [52]. Preprocessing of data comprises mapping of scans, quality control, transformation, and normalization steps. Signal intensities from low, medium, and high scans are mapped onto the same scale by an affine transformation. Transformation parameters are estimated based on a least-squares optimization. Averaging the transformed intensities gives the consensus signals, which are independent of the voltage settings of the photomultiplier tube.

Quality control was performed on a spot and array level. Spots ideally have a diameter of 100 μm. Diameters less than 70 μm and greater than 140 μm are indicative of scratches and printing problems and the corresponding data was discarded. In addition, inconsistent spots with a coefficient of variance of pixels bigger than 0.7, and weak spots with a foreground signal less than 175% of the background signal were removed from further analyses. Strong but unreliable signals with at least 20% of pixels in saturation were discarded. Quality control on array level determined the overall quality of each single chip. Therefore, results from different arrays were compared with each other on the basis of correlation parameters, scatterplots and chi-plots for all combinations of arrays for a particular treatment [53,54]. Raw intensities were transformed with the natural logarithm. A locally weighted regression smoother (LOESS) was applied to correct intensity-dependent signal patterns [55]. The regression is a first-order polynomial that takes into account the subset of 25% of spots that yield a signal with similar intensities. Variance stabilization for weakly expressing genes was not performed as such effects were not apparent. All chips hybridized for a particular treatment were scaled to a common median absolute deviation from median (MAD) of the logarithmic fold change (M value) [56]. Statistical analysis was based on the assumption that the majority of genes are not changed in their expression and that the overall up- and downregulations compensate each other in sum.

Each individual gene was tested for difference in expression under toxic conditions with a t-test where an adjusted p value (padj) of less than 0.025 indicated significant differential expression. Statistical requirements of normal distribution and homoscedasticity are tenable. A robust variance estimation was derived by balancing gene-specific and pooled variance [57]. The number of false positives due to multiple testing was reduced by adjusting the resulting p values by controlling the Benjamini-Hochberg false discovery rate [27].

Multivariate analysis was based on a subset of genes of interest. Genes that remain unchanged under all conditions were ignored. Marker genes that are significantly changed by exposure to a particular toxicant were taken into account. In addition, the selected subset included genes that showed a global response across many chemicals. The selected subset included: the top 20 up- or downregulated genes based on fold change (minimum fold change > 2); the top seven genes with the highest correlation among at least two toxicants (minimum correlation > 0.7);

the top 100 genes with the highest MAD across all treatments; and the marker genes that are regulated at least threefold for just one treatment.

Most multivariate approaches require a complete dataset without missing values. Under the condition that more than 80% of the data for a particular gene is available, missing data for gene g are imputed by a k-nearest-neighbor algorithm [58]. Missing values are estimated as weighted average of the values for the k genes with the closest Euclidean distance to gene g.

The logarithmic fold changes (M values) of genes under toxic conditions are subjected to PCA and hierarchical clustering. The principal components of experimental data across all experiments were derived by SVD [59]. Gene-expression profiles summarize clustering information for toxicants and genes. Dissimilarity between toxicants is determined as Euclidean distance of their M values. In contrast, proximity between two genes is derived as the arc cosine transformed Pearson correlation coefficient [60].

GO analysis of toxicant-affected genes was carried out by extracting the human homologs from the Zebrafish Chip Annotation Database [61]. The GO trees and categories were established with the web-based GoTreeMachine [62]. The number of genes with significant alterations in expression levels in response to TCDD, DDT, and AA were too few to be analyzed by GoTreeMachine.

Expression Analysis

In situ hybridization and RT-PCR were carried out using standard procedures [46,63]. Embryos and RNA samples were derived from independent toxicant exposures. Cell death was monitored by acridine orange staining and examination by fluorescence microscopy [64].

Authors' Contributions

L.Y. and U.S. conceived the work and designed the experiments. U.S. supervised the project and wrote the manuscript. L.Y. performed all of the experimental work. J.R.K. optimized some of the concentrations of the toxicants. C.Z. and J.J performed statistical analysis of the microarray data. M.B. and M.P. provided technical assistance in microarray printing. F.M. gave technical advice at an early stage of the project. J.L. performed the gene ontological analysis.

Acknowledgements

We thank M. Rastegar and S. Rastegar for their help in lab organization, and J. Katzenberger, M. Bonnaus and N. Gretz for technical assistance. This work was supported through the Additional Funding Scheme of the Helmholtz-Gemein-schaft.

References

1. Andreasen EA, Mathew LK, Tanguay RL: Regenerative growth is impacted by TCDD: gene expression analysis reveals extracellular matrix modulation. Toxicol Sci 2006, 92:254–269.

2. Yoon CY, Park M, Kim BH, Park JY, Park MS, Jeong YK, Kwon H, Jung HK, Kang H, Lee YS, et al.: Gene expression profile by 2,3,7,8-tetrachlorodibenzo-p-dioxin in the liver of wild-type (AhR+/+) and aryl hydrocarbon receptor-deficient (AhR-/-) mice. J Vet Med Sci 2006, 68:663–668.

3. Koizumi S, Yamada H: DNA microarray analysis of altered gene expression in cadmium-exposed human cells. J Occup Health 2003, 45:331–334.

4. Hood RD: Handbook of Developmental Toxicology. Boca Raton, FL: CRC Press; 1997.

5. Combes R, Dandrea J, Balls M: A critical assessment of the European Commission's proposals for the risk assessment and registration of chemical substances in the European Union. Altern Lab Anim 2006, 34(Suppl 1):29–40.

6. Grindon C, Combes R: Introduction to the EU REACH legislation. Altern Lab Anim 2006, 34(Suppl 1):5–10.

7. Suter L, Babiss LE, Wheeldon EB: Toxicogenomics in predictive toxicology in drug development. Chem Biol 2004, 11:161–171.

8. Kimmel CB, Ballard WW, Kimmel SR, Ullmann B, Schilling TF: Stages of embryonic development of the zebrafish. Dev Dyn 1995, 203:253–310.

9. Zon LI, Peterson RT: In vivo drug discovery in the zebrafish. Nat Rev Drug Discov 2005, 4:35–44.

10. Nagel R: DarT: The embryo test with the Zebrafish Danio rerio - a general model in ecotoxicology and toxicology. Altex 2002, 19(Suppl 1):38–48.

11. Braunbeck T, Boettcher M, Hollert H, Kosmehl T, Lammer E, Leist E, Rudolf M, Seitz N: Towards an alternative for the acute fish LC(50) test in chemical assessment: the fish embryo toxicity test goes multi-species - an update. Altex 2005, 22:87–102.

12. Alestrom P, Holter JL, Nourizadeh-Lillabadi R: Zebrafish in functional genomics and aquatic biomedicine. Trends Biotechnol 2006, 24:15–21.

13. Langheinrich U: Zebrafish: a new model on the pharmaceutical catwalk. BioEssays 2003, 25:904–912.

14. Parng C: In vivo zebrafish assays for toxicity testing. Curr Opin Drug Discov Dev 2005, 8:100–106.

15. Mathavan S, Lee SG, Mak A, Miller LD, Murthy KR, Govindarajan KR, Tong Y, Wu YL, Lam SH, Yang H, et al.: Transcriptome analysis of zebrafish embryogenesis using microarrays. PLoS Genet 2005, 1:260–276.

16. Lam SH, Wu YL, Vega VB, Miller LD, Spitsbergen J, Tong Y, Zhan H, Govindarajan KR, Lee S, Mathavan S, et al.: Conservation of gene expression signatures between zebrafish and human liver tumors and tumor progression. Nat Biotechnol 2006, 24:73–75.

17. Linney E, Dobbs-McAuliffe B, Sajadi H, Malek RL: Microarray gene expression profiling during the segmentation phase of zebrafish development. Comp Biochem Physiol C Toxicol Pharmacol 2004, 138:351–362.

18. Lien CL, Schebesta M, Makino S, Weber GJ, Keating MT: Gene expression analysis of zebrafish heart regeneration. PLoS Biol 2006, 4:e260.

19. Xu J, Srinivas BP, Tay SY, Mak A, Yu X, Lee SG, Yang H, Govindarajan KR, Leong B, Bourque G, et al.: Genomewide expression profiling in the zebrafish embryo identifies target genes regulated by Hedgehog signaling during vertebrate development. Genetics 2006, 174:735–752.

20. Meijer AH, Verbeek FJ, Salas-Vidal E, Corredor-Adamez M, Bussman J, van der Sar AM, Otto GW, Geisler R, Spaink HP: Transcriptome profiling of adult zebrafish at the late stage of chronic tuberculosis due to Mycobacterium marinum infection. Mol Immunol 2005, 42:1185–1203.

21. Voelker D, Vess C, Tillmann M, Nagel R, Otto GW, Geisler R, Schirmer K, Scholz S: Differential gene expression as a toxicant-sensitive endpoint in zebrafish embryos and larvae. Aquat Toxicol 2007, 81:355–364.

22. Lam SH, Winata CL, Tong Y, Korzh S, Lim WS, Korzh V, Spitsbergen J, Mathavan S, Miller LD, Liu ET, et al.: Transcriptome kinetics of arsenic-induced adaptive response in zebrafish liver. Physiol Genomics 2006, 27:351–361.

23. Andreasen EA, Mathew LK, Lohr CV, Hasson R, Tanguay RL: Aryl hydrocarbon receptor activation impairs extracellular matrix remodeling during zebrafish fin regeneration. Toxicol Sci 2007, 95:215–226.

24. EPA: welcome to ECOTOX [http://cfpub.epa.gov/ecotox/quick_query.htm].

25. Nau H, Hauck RS, Ehlers K: Valproic acid-induced neural tube defects in mouse and human: aspects of chirality, alternative drug development, pharmacokinetics and possible mechanisms. Pharmacol Toxicol 1991, 69:310–321.

26. Wiltse J: Mode of action: inhibition of histone deacetylase, altering WNT-dependent gene expression, and regulation of beta-catenin--developmental effects of valproic acid. Crit Rev Toxicol 2005, 35:727–738.

27. Benjamini Y, Hochberg Y: Controlling the false discovery rate: A practical approach to multiple testing. J R Stat Soc 1995, 57:289–300.

28. Linbo TL, Stehr CM, Incardona JP, Scholz NL: Dissolved copper triggers cell death in the peripheral mechanosensory system of larval fish. Environ Toxicol Chem 2006, 25:597–603.

29. Santos F, MacDonald G, Rubel EW, Raible DW: Lateral line hair cell maturation is a determinant of aminoglycoside susceptibility in zebrafish (Danio rerio). Hear Res 2006, 213:25–33.

30. Chen WY, John JA, Lin CH, Chang CY: Expression pattern of metallothionein, MTF-1 nuclear translocation, and its DNA-binding activity in zebrafish (Danio rerio) induced by zinc and cadmium. Environ Toxicol Chem 2007, 26:110–117.

31. Escher BI, Hermens JL: Modes of action in ecotoxicology: their role in body burdens, species sensitivity, QSARs, and mixture effects. Environ Sci Technol 2002, 36:4201–4217.

32. Winyard PG, Moody CJ, Jacob C: Oxidative activation of antioxidant defence. Trends Biochem Sci 2005, 30:453–461.

33. Gether U, Andersen PH, Larsson OM, Schousboe A: Neurotransmitter transporters: molecular function of important drug targets. Trends Pharmacol Sci 2006, 27:375–383.

34. Bemis JC, Alejandro NF, Nazarenko DA, Brooks AI, Baggs RB, Gasiewicz TA: TCDD-induced alterations in gene expression profiles of the developing mouse paw do not influence morphological differentiation of this potential target tissue. Toxicol Sci 2007, 95:240–248.

35. Centers for Disease Control and Prevention [http://www.cdc.gov/exposurereport/pdf/factsheet_mercury.pdf].

36. Oleksiak MF, Churchill GA, Crawford DL: Variation in gene expression within and among natural populations. Nat Genet 2002, 32:261–266.

37. Dickmeis T, Plessy C, Rastegar S, Aanstad P, Herwig R, Chalmel F, Fischer N, Strahle U: Expression profiling and comparative genomics identify a conserved regulatory region controlling midline expression in the zebrafish embryo. Genome Res 2004, 14:228–238.

38. Blechinger SR, Warren JT Jr, Kuwada JY, Krone PH: Developmental toxicology of cadmium in living embryos of a stable transgenic zebrafish line. Environ Health Perspec 2002, 110:1041–1046.

39. Carney SA, Prasch AL, Heideman W, Peterson RE: Understanding dioxin developmental toxicity using the zebrafish model. Birth Defects Res A Clin Mol Teratol 2006, 76:7–18.

40. Sood R, English MA, Jones M, Mullikin J, Wang DM, Anderson M, Wu D, Chandrasekharappa SC, Yu J, Zhang J, et al.: Methods for reverse genetic screening in zebrafish by resequencing and TILLING. Methods 2006, 39:220–227.

41. Driever W, Solnica-Krezel L, Schier AF, Neuhauss SC, Malicki J, Stemple DL, Stainier DY, Zwartkruis F, Abdelilah S, Rangini Z, et al.: A genetic screen for mutations affecting embryogenesis in zebrafish. Development 1996, 123:37–46.

42. Mullins MC, Nüsslein-Volhard C: Mutational approaches to studying embryonic pattern formation in the zebrafish. Curr Opin Gen Dev 1993, 3:648–654.

43. Amsterdam A, Hopkins N: Mutagenesis strategies in zebrafish for identifying genes involved in development and disease. Trends Genet 2006, 22:473–478.

44. Nasevicius A, Ekker SC: Effective targeted gene 'knockdown' in zebrafish. Nat Genet 2000, 26:216–220.

45. Hill A, Howard CV, Strahle U, Cossins A: Neurodevelopmental defects in zebrafish (Danio rerio) at environmentally relevant dioxin (TCDD) concentrations. Toxicol Sci 2003, 76:392–399.

46. Strähle U, Blader P, Adams J, Ingham PW: Non-radioactive in situ hybridization procedure for tissue sections. Trends Genet 1994, 10:75–76.

47. Oxtoby E, Jowett T: Cloning of the zebrafish krox-20 gene (krx-20) and its expression during hindbrain development. Nucl Acids Res 1993, 21:1087–1095.

48. Behra M, Cousin X, Bertrand C, Vonesch JL, Biellmann D, Chatonnet A, Strahle U: Acetylcholinesterase is required for neuronal and muscular development in the zebrafish embryo. Nat Neurosci 2002, 5:111–118.

49. Behra M, Etard C, Cousin X, Strahle U: The use of zebrafish mutants to identify secondary target effects of acetylcholine esterase inhibitors. Toxicol Sci 2004, 77:325–333.

50. Westerfield M: The Zebrafish Book. Eugene: University of Oregon Press; 1993.

51. Mizell M, Romig ES: The aquatic vertebrate embryo as a sentinel for toxins: zebrafish embryo dechorionation and perivitelline space microinjection. Int J Dev Biol 1997, 42:411–423.

52. The R Project for Statstical Computing [http://www.R-project.org].

53. Fisher NL, Switzer P: Chi-plots for assessing dependence. Biometrica 1985, 72:253–265.

54. Fisher NL, Switzer P: Graphical assessment of independence: is a picture worth 100 tests? Am Stat 2001, 55:233–239.

55. Cui X, Kerr MK, Churchill GA: Transformations for cDNA microarray data. Stat Appl Genet Mol Biol 2003, 2:Article4.

56. Yang YH, Dudoit S, Luu P, Lin DM, Peng V, Ngai J, Speed TP: Normalization for cDNA microarray data: a robust composite method addressing single and multiple slide systematic variation. Nucleic Acids Res 2002, 30:e15.

57. Baldi P, Long AD: A Bayesian framework for the analysis of microarray expression data: regularized t-test and statistical inferences of gene changes. Bioinformatics 2001, 17:509–519.

58. Troyanskaya O, Cantor M, Sherlock G, Brown P, Hastie T, Tibshirani R, Botstein D, Altman RB: Missing value estimation methods for DNA microarrays. Bioinformatics 2001, 17:520–525.

59. Alter O, Brown PO, Botstein D: Singular value decomposition for genome-wide expression data processing and modeling. Proc Natl Acad Sci USA 2000, 97:10101–10106.

60. Eisen MB, Spellman PT, Brown PO, Botstein D: Cluster analysis and display of genome-wide expression patterns. Proc Natl Acad Sci USA 1998, 95:14863–14868.

61. Zebrafish Chip Annotation Database [http://giscompute.gis.a-star.edu.sg/~govind/zebrafish].

62. Gene Ontology Tree Machine [http://bioinfo.vanderbilt.edu/gotm].

63. Sambrook J, Russel DW: Molecular Cloning: A Laboratory Manual. 3rd edition. New York: Cold Spring Harbor Press; 2001.

64. Ton C, Lin Y, Willett C: Zebrafish as a model for developmental neurotoxicity testing. Birth Defects Res (Part A) 2006, 76:553–567.

Release of sICAM-1 in Oocytes and In Vitro Fertilized Human Embryos

Monica Borgatti, Roberta Rizzo, Maria Beatrice Dal Canto,
Daniela Fumagalli, Mario Mignini Renzini, Rubens Fadini,
Marina Stignani, Olavio Roberto Baricordi and
Roberto Gambari

ABSTRACT

Background

During the last years, several studies have reported the significant relationship between the production of soluble HLA-G molecules (sHLA-G) by 48–72 hours early embryos and an increased implantation rate in IVF protocols. As consequence, the detection of HLA-G modulation was suggested as a marker to identify the best embryos to be transferred. On the opposite, no suitable markers are available for the oocyte selection.

Methodology/Principal Findings

The major finding of the present paper is that the release of ICAM-1 might be predictive of oocyte maturation. The results obtained are confirmed using three independent methodologies, such as ELISA, Bio-Plex assay and Western blotting. The sICAM-1 release is very high in immature oocytes, decrease in mature oocytes and become even lower in in vitro fertilized embryos. No significant differences were observed in the levels of sICAM-1 release between immature oocytes with different morphological characteristics. On the contrary, when the mature oocytes were subdivided accordingly to morphological criteria, the mean sICAM-I levels in grade 1 oocytes were significantly decreased when compared to grade 2 and 3 oocytes.

Conclusions/Significance

The reduction of the number of fertilized oocytes and transferred embryos represents the main target of assisted reproductive medicine. We propose sICAM-1 as a biochemical marker for oocyte maturation and grading, with a possible interesting rebound in assisted reproduction techniques.

Introduction

Successful embryo formation and implantation are critical steps during in vitro fertilization procedure. Unfortunately, approximately 10% of retrieved oocytes and fewer than 20% of transferred embryos result in a successful delivery [1]. Analysis of the embryo morphology in still one of the most common approaches of selection in assisted reproduction, with the obvious drawback of being to some extent subjective.

Accordingly, there is urgent need of biochemical markers facilitating the prediction of successful oocyte fertilization and implantation of the in vitro fertilized (IVF) human embryos. In this respect, the only biochemical marker so far proposed for the selection of the most promising embryo obtained by IVF is represented by the release of in vitro cultured embryo (24-, 48- and 72-hours embryo) of soluble HLA-G (Histocompatibility Leukocyte Antigen-G) molecules. This has been consistently reported by several groups [2]–[7]. Using Enzyme-Linked Immunosorbent Assay (ELISA) and Bio-plex approaches, these groups reported that high expression of soluble HLA-G is associated with higher pregnancy and implantation rates.

On the other hand the analysis of oocyte maturation might be of great importance in predicting successful fertilization and embryo development. As far as oocyte morphological criteria, several have been claimed to correlate with outcome, including polar body morphology [8]; cytoplasm appearance [9], and more recently zona pellucida thickness, appearance and birefringence [10]–[12] and the position or shape of the spindle [13]. Also in this case biochemical markers helping in identifying oocytes completing in vitro maturation would be very interesting in IVF approaches. Markers of oocyte maturation are the presence of activated mitochondria and the ability to mobilize and release calcium for internal stores [14].

In this paper we analyze the release by oocytes and in vitro fertilized human embryos of proteins involved in inflammation, including several cytokines, chemokines and soluble Intercellular Adhesion Molecule 1 (sICAM-1). This study was carried on using three independent methodologies, such as ELISA, Bio-Plex assay [15], [16], and Western blotting.

Results

Release of Cytokines, Chemokines and Icam-1 By Human Embryos

We first performed a preliminary screening of 11 embryos using premixed multiplex beads of the Human 27-Plex Panel and the ICAM-1 Bio-Plex kit, obtaining the following results. IL-1β, IL-2, IL-4, IL-5, IL-10, IL-12 (P70), IL-15, IL-17, Basic FGF, G-CSF, GM-CSF, IFN-γ, MIP-1α, TNF-α were not present or undetectable in the analyzed supernatants. Presence of IL-1rα, IL-6, IL-7, IL-8, IL-9, IL-13, Eotaxin, IP-10, MCP-1 (MCAF), MIP-1β, PDGF-BB, RANTES, VEGF, ICAM-1 were detectable in 11, 1, 1, 10, 1, 1, 7, 1, 1, 1, 1, 1, 4, 11 embryos respectively. The only proteins present in the supernatant of all the screened embryos were ICAM-1 and IL-1rα. However, only ICAM-1 was expressed at high levels. In additional experiments on other IVS embryos (not included in this paper) we never found absence of ICAM-1 release, with the exception of few damaged embryos (data not shown).

Quantization of sICAM-1: Elisa and Bio-Plex Assay

In Figure 1 representative analysis is shown demonstrating that levels of ICAM-1 standards are detectable following both ELISA and the Bio-Plex assay. As expected, however, the Bio-Plex assay is more sensitive than ELISA. This is of course important for analysis of single cells, including oocytes. Accordingly, Bio-Plex analysis was chosen for studies involving human oocytes and fertilized embryos.

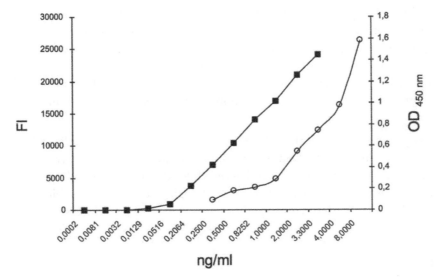

Figure 1. ELISA and Bio-Plex standard curves (white circles and black square respectively) have been obtained with 50 µl of ICAM-1 standard reagent at the concentrations of 0.25, 0.5, 1, 2, 4, 8 ng/ml or 0.0002, 0.00081, 0.0032, 0.0129, 0.0516, 0.20638, 0.82522, 3.3 ng/ml as indicated. FI: fluorescence intensity values. OD 450 nm: optic density at 450 nm wavelength.

Comparison of sICAM-1 Production in Mature and Immature Oocytes and In Vitro Fertilized Embryos

Figure 2 reports a sharp difference in sICAM-1 levels among immature and mature oocytes and fertilized embryos. The average sICAM-1 production by immature (n = 39) and mature (n = 73) oocytes was 6711.5±1502.4 and 2987±103.7 pg/ml/24 hours (mean±SD), respectively (Figure 2). This difference was very reproducible and statistically significant (Student t Test, p<0.0001). In addition, the levels of release of sICAM-1 levels by mature oocytes and in vitro fertilized embryos (n = 73), 1486.8±164.2 pg/ml/24 hours, were also found to be significantly different (Student t Test, p<0.0001) (Figure 2). Therefore, it appears that the release of sICAM-1 has a clear tendency to decrease from immature embryos, to mature embryos and to fertilized embryos.

The presence of sICAM-1 molecules in oocytes culture supernatants was also analyzed by western blotting. The results obtained are shown in Figure 3. Standard positive ICAM-1 controls are shown in lanes "a" and "b". As clearly evident, sICAM-1 is detectable both in mature (lane "d") and immature (lane "e") oocyte supernatants. In addition, sICAM-1 is present in mature oocytes in lower quantities in respect to immature oocytes, fully in agreement with the Bio-Plex data shown in Figure 3. These data were fully in agreement with ELISA assays (data not shown).

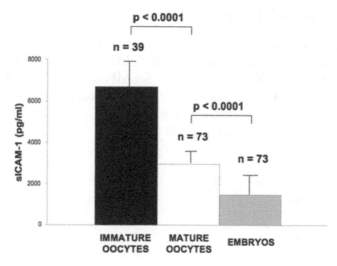

Figure 2. sICAM-1 release in immature oocytes (black box) compared to mature (white box) oocytes and to in vitro fertilized embryo (grey box). Oocytes were individually cultured in a 4-well culture dish as reported in the methods section. Following the maturation period 250 µl of supernatants were collected from each culture system and stored at –20°C until being tested for the presence released proteins. Mature and immature oocytes were identified, one by one, evaluating the presence or absence of the first polar body. In vitro fertilized embryos were individually cultured in 4-well culture dishes and 250 µl of supernatants collected from each embryo culture and stored at –20°C until being tested for the presence of released proteins. * Student t Test.

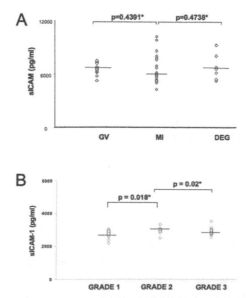

Figure 3. Western blotting analysis. The anti-ICAM-1 MoAb was used for the detection. a: standard positive control loaded at 8000 pg; b: plasma sample loaded at 10000 pg, accordingly to ELISA detection; c: medium negative control; d: mature oocyte supernatant loaded at 35 pg accordingly to ELISA detection; e: immature oocyte supernatant loaded at 100 pg accordingly to ELISA detection; M: protein ladder.

Production of sICAM-1 in Immature and Mature Oocytes

Figure 4a reports the levels of sICAM-1 in immature oocytes at different maturation stages (MI, Metaphase I; GV, germinal vesicle; DEG, degenerated). The average released sICAM-1 was 5900 pg/ml/24 hours for MI oocytes, 6600 pg/ml/24 hours for GV oocytes and 6600 pg/ml/24 hour for DEG oocytes. The difference between sICAM-1 production by MI, GV and DEG immature oocytes was not statistically significant (Student t Test, p = NS) (Figure 3a).

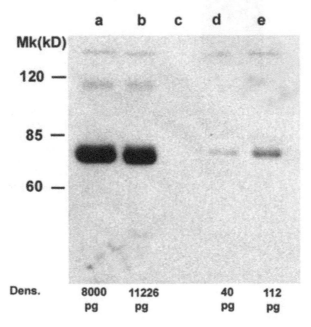

Figure 4. sICAM-1 levels in culture supernatants from immature (a) and mature oocytes (b). Immature oocytes were analysed individually for morphological characteristics to differentiate them as Metaphase I (MI), germinal vescicle (GV) and degenerated (DEG). (a). Mature oocytes were subdivided on the basis of the first polar body and cytoplasm characteristic in Grade 1, 2 and 3 (b). Preparation of oocyte supernatants was performed as described in the legend to Figure 2. * Student t Test.

Moreover, Figure 4b reports the analysis of ICAM-1 release in mature oocytes subdivided in grade 1, 2 and 3, as reported in the material and methods section. The average released sICAM-1 was 2804 pg/ml/24 hours for grade 1 oocytes, 2978 pg/ml/24 hours for grade 2 oocytes and 2923 pg/ml/24 hours for grade 1 oocytes (Figure 4b). Statistical analysis showed significant lower levels of sICAM-1 in grade 1 oocyte supernatants in comparison to grade 2 (Student t Test, p = 0.018) and grade 3 (p = 0.02) oocyte supernatants. Therefore, lower sICAM-1 levels in mature oocyte are predictive for the best grade oocytes (Grade 1).

sICAM-1 Levels and Embryo Grade

Figure 5 represents sICAM-1 levels in embryo culture supernatants graded as reported in the Methods section. The average levels of sICAM-1 were 1476.3±187 pg/ml/24 hours in the 29 Grade 1 embryos; 1522.4±206 pg/ml/24 hours in the 13 of Grade 2; 1481±116 pg/ml/24 hours in the 15 of Grade 3; 1461.9±143.9 pg/ml/24 hours in the 16 Grade 4 and 5 embryos.

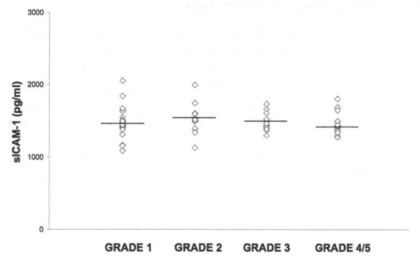

Figure 5. sICAM-1 levels in embryo culture supernatants subdivided into grades as reported in the Methods section. The differences between sICAM-1 production by Grade 1, 2, 3, 4 and 5 embryos were not statistically significant (Student t Test, p = NS).

sICAM-1 Levels in Oocyte Supernatants and Pregnancy Rate

Table 1 reports the sICAM-1 mean levels observed in mature oocyte and embryo culture supernatants subdivided for implantation and pregnancy outcome. There were no statistical differences (Student t Test, p = NS) in sICAM-1 levels observed in the supernatant of oocytes and embryo with a negative or positive implantation and pregnancy rate (Table 1).

Table 1. sICAM-1 release, implantation outcome and pregnancy outcome.

	Implantation outcome		Pregnancy outcome	
	positive	negative	positive	negative
frequency (n)	7	35	4	38
sICAM-1 oocytes (pg/ml)	2865.4±107.7	2856.9±520	2858.4±117.6	2860.7±403.1
sICAM-1 embryos (pg/ml)	1424.1±142.5	1493.8±170.7	1463.5±127.5	1483.9±171.2

The relationship between oocyte grade (Figure 4b) and implantation/pregnancy rate was not investigated, since we were not able to associate the pregnancy event to a specific embryo. In fact, our IVF protocol, in order to achieve the highest probability of pregnancy and meet law restrictions [17], allows the transfer of three embryos that could originate from different grade oocytes.

Comparison of sICAM-1 and sHLA-G Levels in Supernatants of Oocytes Using Bio-plex Technology

Since the release of soluble HLA-G (sHLA-G) molecules by in vitro fertilized embryos seems to help the morphological characterization in the selection of the most promising embryo obtained by IVF and has been proposed as a possible candidate marker for oocyte maturation [17] we compared the release of these two proteins in our samples. Representative analyses are shown in Figure 6, which clearly indicate that in human oocytes the release of sICAM-1 is far more efficient that release of sHLA-G molecules. In general, the release of sHLA-G molecules is very low in most of the oocytes employed. On the contrary, confident results are obtained studying sICAM-1, due to the high release of this protein by human oocytes.

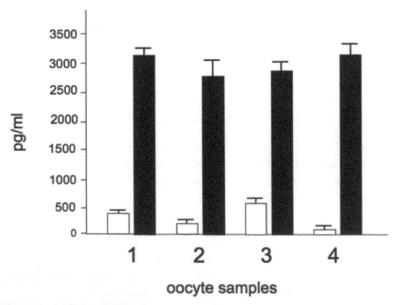

Figure 6. Comparison of levels of sICAM-1 (black boxes) and sHLA-G (white boxes) in supernatants of representative oocytes. For sHLA-G detection, covalent coupling of the anti-sHLAG antibodies to the carboxylated polystyrene microspheres (Bio-Rad, Hercules, CA, USA) was performed using the Bio-Plex amine coupling kit (Bio-Rad, Hercules, CA, USA). Bio-Plex assay was performed as elsewhere reported [2].

Discussion

The reduction of the number of fertilized oocytes and transferred embryos represents the main target of assisted reproductive medicine. During the last years, several studies have confirmed the significant relationship between the production of sHLA-G molecules by 48–72 hours early embryos and an increased implantation rate in IVF protocols [18]. As consequence, the detection of HLA-G modulation was suggested as a marker to identify the best embryos to be transferred. On the opposite, no suitable markers are available for the oocyte selection. The major finding of this paper is the detection, by a reliable technique, of soluble ICAM-1 molecules in the culture supernatants of human in vitro maturated oocytes and in vitro fertilized embryos. The data obtained showed a significant difference in sICAM-1 levels between immature and mature oocytes with significant higher amounts of sICAM-1 in the oocytes that failed to maturate. When the immature oocytes were morphologically classified in metaphase I, germinal vesicle and degenerated phenotypes we observed similar levels of sICAM-1 in the three groups. On the contrary, the mature oocytes subdivided into grade 1, 2 and 3 presented lower s-ICAM-1 levels in grade 1 group. Therefore, these results propose sICAM-1 levels as predictive for oocyte maturation and quality. Biochemical markers of the oocyte maturation are very important, due to the fact that (a) during in vitro oocyte maturation no ore than 50% of the oocytes isolated from a single woman reach grade 1 (Table 2); (b) only these oocytes are routinely considered for IVF. In addition, we like to point out that in some countries no embryo selection is allowed, only a limited number of oocytes are fertilized and all of the obtained embryos must be implanted [19].

Table 2. Details of the in vitro maturation procedure.

	Women (n = 42)
Age (years) (mean±SD)	35±3
Number of recovered oocytes per woman (mean±SD)	7±1
Mature oocytes per woman (%)	20–50

The culture supernatants of early embryos showed sICAM-1 levels lower in comparison to both mature and immature oocytes. Interestingly no significant differences were observed in sICAM-1 concentrations in the culture supernatants of early embryos subdivided into grades. These results underline the importance of sICAM-1 as a marker of the oocyte maturation process but not of the early embryos development.

This is the first report showing release of sICAM-1 in human oocytes and IVF human embryo. However the expression of ICAM-1 in human embryos is not surprising, when considering the implantation phase. In this context, ICAM-1 has been already presented as a protein involved in inflammation. In fact ICAM-1 knock-out mice do not develop inflammation and have less inflammatory cell infiltration [20], [21]. Mutations of ICAM-1 are associated with different diseases as infarct, biliary atresia, multiple sclerosis, obesity [21]–[24]. When the sICAM-1 levels are compared to sHLA-G, a soluble molecule involved in embryo implantation [3], sICAM-1 showed higher levels in oocyte supernatants than sHLA-G. These two molecules are both secreted by human oocytes but with a more efficient release of sICAM-1 than of sHLA-G molecules. In general, the release of sHLA-G molecules is very low in most of the oocytes employed. On the contrary, confident results are obtained studying sICAM-1, due to the high release of this protein by human oocytes.

The results obtained are confirmed using three independent methodologies, such as ELISA, Bio-Plex assay and Western blotting. Therefore, we propose this biochemical marker to be tightly linked to oocyte maturation. This finding is novel and, in our opinion, very important in the field of the selection of oocytes to be fertilized.

As known, the oocytes obtained under ovarian stimulation present a variable competence and although molecular approaches have been proposed [25], [26], the selection is still performed on morphological characteristics such as ploidy and chromosome/chromatin status. Since maturation of oocytes is so important for in vitro fertilization approaches, we suggest sICAM-1 to be a marker for testing different culture mediums under development by several laboratories to the aim to obtain optimal in vitro oocyte maturation.

In conclusion, our data encourage further studies from different laboratories/ networks using ICAM-1 as a marker for a positive oocyte maturation.

Materials and Methods

Patients

The oocytes employed in this study were obtained from regularly cycling patients attending the Biogenesi Reproductive Medicine Centre of Monza, Italy, for an Assisted Reproduction Technique with In Vitro Maturation Protocol (IVM). Couples included in the trial had an indication to IVF procedure because of infertility due to male factor, tubal factor, stage I/II endometriosis, polycystic ovarian syndrome (PCO) or unexplained cause. All the women included had regular cycles of 26–35 days. A written informed consensus was obtained from all participating couples. We considered just one cycle per couple, and after maturation process we used from one to three oocytes according to the Italian Law 40 on IVF. Following these criteria, 42 women were recruited for the study. Women characteristics are reported in Table 2.

Oocyte recovery was performed by means of transvaginal ultrasound–guided follicle aspiration, using a single lumen aspiration needle (Gynetics cod. 4551-E2 Ø17- gauce 35 cm) connected to a vacuum pump (Craft Pump pressure 80–90 mmHg). The retrieved oocytes were surrounded by granulosa cells forming a structure known as the cumulus ophorus complex (COC). The COCs were washed with prewarmed Flushing Medium with heparin (Medi-Cult product n. 10760125, Denmark).

The COCs, that for easiness we will define oocytes, were detected under a stereomicroscope, examined and classified on the basis of their morphology. Oocytes with signs of mechanical damage or atresia were discarded.

Immature oocytes were individually cultured in a 4-well culture dish with 0,5 ml of IVM Medium (vial 2 of IVM system medium; Medicult no. 82214010, Denmark) supplemented with rec–FSH 0,075 IU/ml (Serono, Italy), hCG 0,1 IU/ml (Serono, Italy) and 10% Serum Protein Substitute (SPS no. 3010– Sage Media- USA) for other 30 hrs.

Following the maturation period, 250 µl of supernatants were collected from each culture system containing a single oocyte and stored at –20°C until being tested for the presence released proteins.

The oocytes were then classified, one by one, evaluating the presence of the first polar body to confirm Metaphase II stage and their morphological characteristics.

Immature oocytes were classified as Metaphase I (MI), germinal vesicle (GV) and degenerated (DEG) whereas mature oocytes were classified on the first polar body and cytoplasm characteristics in Grade1: homogenous cytoplasm and round polar body; Grade 2: oocyte with variations in color or cytoplasm granularity and/or presence of inclusions, vacuoles or retractable bodies, but a round polar body: Grade 3: oocyte with variation in color or cytoplasm granularity and/or presence of inclusions, vacuoles or retractable bodies with a fragmented polar body.

Embryos were graded accordingly to cleavage (cell number) and cytoplasmic fragmentation. Embryos were graded as follows on Day 3: Grade 1, blastomeres have equal size and no cytoplasmic fragmentation; Grade 2, blastomeres have equal size and minor cytoplasmic fragmentation involving <10% of the embryo; Grade 3, blastomeres have unequal size and fragmentation involving 10–20% of the embryo; Grade 4, blastomeres have equal or unequal size, and moderate to significant cytoplasmic fragmentation covering 20–50% of the embryo; and

Grade 5, few blastomeres and severe fragmentation covering ≥50% of the embryo [17].

Measurement of sICAM-1 Levels by Enzyme Immunosorbent Assay (ELISA)

sICAM-1 concentrations were analyzed in triplicate on 1:2 diluted oocyte culture supernatants by the commercially available sICAM-1 kit (Diaclone, Besancon, FR) with a detection limit of 0.25 ng/ml.

Western Blotting

The presence of sICAM-1 molecules in oocyte culture supernatants was analyzed by Western Blot. Briefly, concentrated and albumin depleted (Enchant Life Science kit, Pall Corporation, MI, US) oocyte culture supernatants were loaded on 8% SDS-polyacrylamide gel, electrophoresed at 80 V for 2 hours and blotted onto PVDF membrane (Immobilon-P Millipore, Billerica, MA, US) by electrotransfer at 100 V for 45 minutes in 25 mM Tris Buffer, 190 mM Glycine, 2% SDS and 20% (V/V) Methanol. Blocking was carried out with 5% nonfat dry milk, Tris 100 mM pH 7.5, NaCl 150 mM over night at 4°C. After two washes, the membrane was incubated with monoclonal mouse-anti-human ICAM-1 (10 μg/ml) (Genzyme, MA, USA) for 3 hours at room temperature with gentle shaking. The sICAM-1 molecules were detected using Protein-G HRP (BioRad, Hercules, CA, US) at dilution of 1:5000 in 10 mM Tris pH 8.0, 150 mM NaCl, 0.1% Tween 20. Reactions were developed by chemiluminescence with SuperSignal enhanced chemiluminescence kit (Super Signal West Pico system, Pierce, Rockford, IL, US) and captured by Chemiluminescence Imaging Geliance 600 (PerkinElmer, CT, USA). The ELISA standard (sICAM-1 kit (Diaclone, Besancon, FR)) and a plasma sample were used as positive control, the culture medium alone as negative control. The molecular weights were determined with the BenchMark (Invitrogen, CA, US) pre-stained protein ladder (range 10–200 kD). Densitometric analysis was performed with the Gene Tools software (PerkinElmer, CT, USA).

Cyto/Chemokines and ICAM-1 Profiles

Cytokines and chemokines presence were measured in embryo culture supernatants by Bio-Plex cytokine assay (Bio-Rad Laboratories, Hercules, CA) [15], [16] described by the manufacturer. The Bio-Plex cytokine assay is designed for the multiplexed quantitative measurement of multiple cytokines in a single well using as little as 50 μl of sample. In our experiments, we used the premixed multiplex beads of the and Bio-Plex Human Cytokine singleplex Assay ICAM-1 (Bio-Rad, Cat. no. XF0-000003N) and Bio-Plex human cytokine Human 27-Plex Panel (Bio-Rad, Cat. no. 171-A11127) which included twenty-seven cytokines [IL-1β, IL-1rα, IL-2, IL-4, IL-5, IL-6, IL-7, IL-8, IL-9, IL-10, IL-12 (P70), IL-13, IL-15, IL-17, Basic FGF, Eotaxin, G-CSF, GM-CSF, IFN-γ, IP-10, MCP-1 (MCAF), MIP-1α, MIP-1β, PDGF-BB, RANTES, TNF-α, VEGF]. Briefly, 50 μl of cytokine/chemokine and ICAM-1 standards or samples (supernatants from IVF human embryos) were incubated with 50 μl of anti-cytokine/chemikine/ICAM-1 conjugated beads in 96-well filter plates for 30 min at room temperature with shaking. Plates were then washed by vacuum filtration three times with 100 μl of Bio-Plex wash buffer, 25 μl of diluted detection antibody were added, and plates were incubated for 30 min at room temperature with shaking. After three filter washes, 50 μl of streptavidin-phycoerythrin was added, and the plates were incubated for 10 min at room temperature with shaking. Finally, plates were washed by vacuum filtration three times, beads were suspended in Bio-Plex assay buffer, and samples were analyzed on a Bio-Rad 96-well plate reader using the Bio-Plex Suspension Array System and Bio-Plex Manager software (Bio-Rad Laboratories, Hercules, CA).

Statistical Analysis

Statistical analysis was conducted using the Stat View software package (SAS Institute Inc, Cary, NC, US). The data were analyzed by the Student t test for unpaired samples. Statistical significance was assumed for $p < 0.05$ (two tailed).

Authors' Contributions

Conceived and designed the experiments: RR OB RG. Performed the experiments: MB MS. Analyzed the data: MB RR MMR MS OB RG. Contributed reagents/materials/analysis tools: MBDC DF MMR RF. Wrote the paper: RF OB RG.

References

1. Deonandan R, Campbell MK, Østbye T, Tummon I (2000) Toward a more meaningful in vitro fertilization success rate. J Assist Reprod Genet 17: 498–503.

2. Rebmann V, Switala M, Eue I, Schwahn E, Merzenich M, Grosse-Wilde H (2007) Rapid evaluation of soluble HLA-G levels in supernatants of in vitro fertilized embryos. Human Immunol 68: 251–258.

3. Fuzzi B, Rizzo R, Criscuoli L, Noci I, Melchiorri L, et al. (2002) HLA-G expression in early embryos is a fundamental prerequisite for the obtainment of pregnancy. European J Immunol 32: 311–315.

4. Noci I, Fuzzi B, Rizzo R, Melchiorri L, Criscuoli L, et al. (2005) Embryonic soluble HLA-G as a marker of developmental potential in embryos. Human Reprod 20: 138–146.

5. Sher G, Keskintepe L, Batzofin J, Fisch J, Acacio B, et al. (2005) Influence of early ICSI-derived embryo sHLA-G expression on pregnancy and implantation rates: a prospective study. Human Reprod 20: 1359–1363.

6. Desai N, Filipovits J, Goldfarb J (2006) Secretion of soluble HLA-G by day 3 human embryos associated with higher pregnancy and implantation rates: assay of culture media using a new ELISA kit. 13: 272–277.

7. Yie SM, Balakier H, Motamedi G, Librach CL (2005) Secretion of human leukocyte antigen-G by human embryos is associated with a higher in vitro fertilization pregnancy rate. Fertil Steril 83: 30–36.

8. Ebner T, Yaman C, Moser M, Sommergruber M, Feichtinger O, Tews G (2000) Prognostic value of first polar body morphology on fertilization rate and embryo quality in intracytoplasmic sperm injection. Human Reproduction 15: 427–430.

9. Ebner T, Moser G, Tews G (2006) Is oocyte morphology prognostic of embryo developmental potential after ICSI? 12: 507–512.

10. Shen Y, Stalf T, Mehnert C, Eichenlaub-Ritter U, Tinneberg HR (2005) High magnitude of light retardation by the zona pellucida is associated with conception cycles. Human Reproduction 20: 1596–1606.

11. Montag M, Schimming T, Köster M, Zhou C, Dorn C, et al. (2008) Oocyte zona birefringence intensity is associated with embryonic implantation potential in ICSI cycles. 16: 239–244.

12. Pascale M-P, Chretien M-F, Malthiery Y, Reynier P (2007) Mitochondrial DNA in the oocyte and the developing embryo. Current Topics in Developmental Biology 77: 51–83.

13. Madaschi C, de Souza Bonetti TC, de Almeida Ferreira Braga DP, Pasqualotto FF, Iaconelli A Jr, Borges E Jr (2008) Spindle imaging: a marker for embryo development and implantation. Fertil Steril 90: 194–198.

14. Dumollard R, Duchen M, Carroll J (2007) The role of mitochondrial function in the oocyte and embryo. Curr Top Dev Biol 77: 21–49.

15. de Jager W, te Velthuis H, Prakken BJ, Kuis W, Rijkers GT (2003) Simultaneous detection of 15 human cytokines in a single sample of stimulated peripheral blood mononuclear cells. Clin Diagn Lab Immunol 10: 133–139.

16. Kerr JR, Cunniffe VS, Kelleher P, Smith J, Vallely PJ, Will AM, et al. (2004) Circulating cytokines and chemokines in acute symptomatic parvovirus B19 infection: negative association between levels of pro-inflammatory cytokines and development of B19-associated arthritis. J Med Virol 74: 147–155.

17. Veeck L (1999) An Atlas of Human Gametes and Conceptuses. New York: Parthenon Publishing. pp. 47–50.

18. Rizzo R, Fuzzi B, Stignani M, Criscuoli L, Melchiorri L, et al. (2007) Soluble HLA-G molecules in follicular fluid: a tool for oocyte selection in IVF? J Reprod Immunol 74: 133–142.

19. Fineschi V, Neri M, Turillazzi E (2005) The new Italian law on assisted reproduction technology. J Med Ethics 31: 536–539.

20. Hallahan DE, Virudachalam S (1997) Intercellular adhesion molecule 1 knockout abrogates radiation induced pulmonary inflammation. Proc Natl Acad Sci USA 94: 6432–6437.

21. Wang HW, Babic AM, Mitchell HA, Liu K, Wagner DD (2005) Elevated soluble ICAM-1 levels induce immune deficiency and increase adiposity in mice. FASEB J 19: 1018–1020.

22. Arikan C, Berdeli A, Kilic M, Tumgor G, Yagci RV, Aydogdu S (2008) Polymorphisms of the ICAM-1 gene are associated with biliary atresia. Dig Dis Sci 53: 2000–2004.

23. van den Borne SW, Narula J, Voncken JW, Lijnen PM, Vervoort-Peters HT, et al. (2008) Defective intercellular adhesioncomplex in myocardium predisposes to infarctrupture in humans. J Am Coll Cardiol 51: 2184–2192.

24. Mousavi SA, Nikseresht AR, Arandi N, Borhani Haghighi A, Ghaderi A (2007) Intercellular adhesion molecule-1 gene polymorphism in Iranian patients with multiple sclerosis. Eur J Neurol 14: 1397–1399.

25. Coticchio G, Sereni E, Serrao L, Mazzone S, Iadarola I, Borini A (2004) What criteria for the definition of oocyte quality? Annals New York Academy of Science 1034: 132–144.

26. Patrizio P, Fragouli E, Bianchi V, Borini A, Wells D (2007) Molecular methods for selection of the ideal oocyte. 15: 346–353.

Three-Dimensional Analysis of Vascular Development in the Mouse Embryo

Johnathon R. Walls, Leigh Coultas, Janet Rossant,
and R. Mark Henkelman

ABSTRACT

Key vasculogenic (de-novo vessel forming) and angiogenic (vessel remodelling) events occur in the mouse embryo between embryonic days (E) 8.0 and 10.0 of gestation, during which time the vasculature develops from a simple circulatory loop into a complex, fine structured, three-dimensional organ. Interpretation of vascular phenotypes exhibited by signalling pathway mutants has historically been hindered by an inability to comprehensively image the normal sequence of events that shape the basic architecture of the early mouse vascular system. We have employed Optical Projection Tomography (OPT) using frequency distance relationship (FDR)-based deconvolution to image embryos immunostained with the endothelial specific marker PECAM-1 to create a high resolution, three-dimensional atlas of mouse vascular development between E8.0 and E10.0 (5 to 30 somites). Analysis of the atlas has provided significant new information regarding normal development of intersomitic vessels, the perineural vascular plexus, the cephalic plexus and vessels connecting the embryonic and extraembryonic circulation. We describe examples of vascular remodelling that provide new insight into the mechanisms of sprouting angiogenesis, vascular guidance cues and artery/vein identity that directly relate to phenotypes observed in mouse mutants affecting vascular development between E8.0 and E10.0. This atlas is freely available at http://www.mouseimaging.ca/research/mouse_atlas.html and will serve as a platform to provide insight into normal and abnormal vascular development.

Introduction

The cardiovascular system is the first functional organ system to develop in the mammalian embryo. The blood vessels that initially comprise this organ originate by vasculogenesis, the aggregation of de novo-forming angioblasts (endothelial precursors) into simple endothelial tubes. Angioblasts in the mouse embryo first emerge from the mesoderm as Flk1+ cells around embryonic day (E) 7.0 and assemble a simple circulatory loop consisting of a heart, dorsal aorta, yolk sac plexus and sinus venosus by E8.0 [1], [2], [3]. Shortly after its formation, this early vascular circuit is remodelled by angiogenesis, the proliferation, sprouting and pruning of pre-existing vessels, transforming it into a complex network of branched endothelial tubes of varying diameter, length and identity. Such remodelling of pre-existing vessels is dependent on both genetically hardwired events and hemodynamic forces [4], [5].

Given the complex nature of the vascular system and the diversity of biological processes required for its assembly and refinement, it is hardly surprising that a large number of signalling pathways are employed in its development. Mutations in pathways required for vascular development frequently manifest phenotypes that result in embryonic lethality at mid gestation. In mice, mutations affecting Notch [6], [7], [8], TGFβ [9], [10], Hedgehog [11], [12], [13], VEGF [14], [15], [16], ephrin/Eph [17] and angiopoietin/Tie [18] signalling (among others) result in abnormal vascular development between E8.0 and E10.0 and ultimately embryonic lethality. The vascular activities of these pathways are not limited to this developmental time window, but extend to organogenesis [19], [20], maintenance of vascular homeostasis in adulthood [8], [9], [21], [22] and states of pathological angiogenesis [23], [24], [25], [26]. Correct interpretation of how these pathways regulate vascular development between E8.0 and E10.0 would therefore improve our understanding of how they contribute to later vascularization events. Such interpretation is often impeded however, by the complex nature of the vascular phenotypes, an inability to observe the vasculature of the mutants in its entirety and an incomplete understanding of the normal sequence of vascular remodelling events that occur during this period of development. Previous studies in zebrafish [27], [28] and chick [29], [30] have provided insight into normal vascular development, but have limited applicability to the sequence of vascular remodeling events in the mammalian embryo primarily due to differences in anatomy and the increased use of plexus bed intermediates in mammals compared to zebrafish. We have sought to address this issue by generating a high resolution, three-dimensional (3D) atlas of the developing mouse vasculature between E8.0 and E10.0 (5–30 somites).

The mouse embryo grows rapidly between E8.0 and E10.0 and undergoes complex morphological and conformational changes that present significant challenges to current imaging technologies. These challenges are further complicated by the inherent properties of the vascular system as a 3D network of branched, interconnected tubes of varying length and size. Accurate assessment of vascular development at this stage therefore requires a 3D imaging modality capable of visualizing the vasculature in its un-manipulated entirety in embryos of increasing size while retaining sufficient isotropic resolution (on the order of a few microns) to capture the details of the finest capillaries. Without these properties, significant positional information about the vasculature is lost and artefacts are introduced. While confocal microscopy has been used to generate an atlas of vascular development in zebrafish embryos [27] and study projections of the vasculature of dissected mouse embryos prior to E8.5 [2], it does not provide sufficient specimen coverage to create the 3D images necessary to visualize the complete embryonic vasculature in mouse embryos beyond this time point. Later embryonic stages have been studied using corrosion casting and electron microscopy [31]. While these investigations provided 3D representation, they did not retain information about vessels that cannot be perfused such as blind ended angiogenic sprouts, narrow vessels and vessels yet to form a complete lumen.

Recently, a new imaging modality named Optical Projection Tomography (OPT) [32] was developed to obtain molecular specificity and 3D cellular level resolution over complete specimens up to 1 cubic centimetre (cc) in size. OPT has been shown to support the use of multiple molecular markers [32], and has been previously used to visualize mouse embryos [32], [33], chick embryos [34], developing plant material [35], drosophila melanogaster [36] and whole adult mouse organs [37], [38]. We have further developed OPT using frequency-distance relationship [39] (FDR)-based deconvolution to obtain higher resolution images, on the order of a few microns, while still retaining the ability to image large specimen sizes [40], [41]. We set out to use this technique to study the normal sequence of mouse vascular development between E8.0 and E10.0.

Ideally one would be able to image a single living mouse embryo over an extended period of time in order to visualize the complete development of a single vasculature. 3D optical imaging of the live mouse embryo, however, is still a challenge due to the in utero development of mouse embryos and significant light scattering caused by living tissue. This in vivo time course can instead be approximated by static imaging of multiple genetically identical embryos collected at a range of ages throughout gestation. To this end, we performed FDR-based deconvolution OPT imaging on fixed PECAM-1-immunostained embryos ranging from 5 to 30 somites to study the normal development of the early embryonic mouse vasculature. PECAM-1 is major constituent of the endothelial cell intercellular junction and is widely used as a molecular marker for mature endothelial cells [2], [42], [43], [44]. With these images we are able to visualize vascular development across this age range, from vasculogenesis to capillary plexus to the development of larger vessels. The collection of all images comprises an atlas of normal development of the embryonic mouse vasculature, which is made freely available to other researchers. Analysis of this atlas has resulted in significant new information regarding the normal vascular development in the mouse embryo.

Results and Discussion

Acronyms for the vessels discussed in this paper are listed in Table 1.

Table 1. Acronyms Used. The list of vessel acronyms used in this paper.

Acronym	Vessel
AVS	Arteriovenous shunt
ACV	Anterior cardinal vein
CCV	Common cardinal vein
DA	Dorsal aorta
DLAV	Dorsal longitudinal anastomotical vessel
ICA	Internal carotid artery
ISA	Intersomitic artery
ISV	Intersomitic vein
PCV	Posterior cardinal vein
PVH	Primary head vein
PMA	Primitive maxillary artery
PNVP	Perineural vascular plexus
OA	Omphalomesenteric artery
OV	Omphalomesenteric vein
UA	Umbilical artery
UV	Umbilical vein
SV	Sinus venosus
VTA	Vertebral artery

3D Visualization of the Embryonic Mouse Vasculature

A sample OPT view of the Cy3-PECAM-1 signal from a 19 somite embryo is shown in Figure 1. At 0.9 degree angular increments of the specimen, 400 views were taken throughout a complete revolution. These views were then FDR-filtered as described (see Materials and Methods) and reconstructed using a standard parallel-ray filtered backprojection reconstruction algorithm [45]. The resulting 3D data can be digitally sliced and viewed along any angle. Example slices from the three orthogonal axes indicated by the blue, yellow, and green lines in Figure 1A and B are shown in Figure 1B–D. The pixel size and slice thickness of the digital slices is 2.0 μm. The technique has sufficient resolution to visualize the finest vessel structures, such as those of the perineural vascular plexus, with an estimated diameter of 4 μm (Figure 1E).

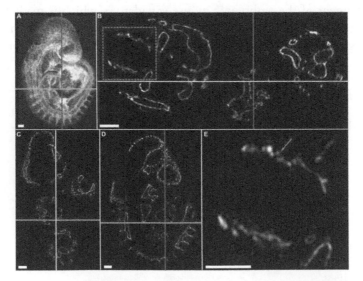

Figure 1. Example OPT view and slices from embryonic vascular mouse atlas data. (A) An example OPT view from a Cy3-PECAM-1 immunostained 19 somite mouse embryo demonstrating the complexity of the vascular pattern at this stage of development. The coloured lines (green, blue, yellow) in this and subsequent subfigures indicate the slice taken from the 3D reconstruction. (B,C,D) Slices from the reconstructed 3D FDR-deconvolution OPT data set can be viewed along any dimension. B is sliced along the green line in A, C is sliced along the blue line in A, and D is sliced along the yellow line in B and C. (E) Magnified view of the boxed region in B, depicting the perineural vascular plexus (pink arrow). All scale bars represent 100 microns.

Surfaces can be rendered from the 3D data set using an isosurface algorithm (see Materials and Methods) that creates a digital surface corresponding to constant intensity values in the reconstructed data. The resulting 3D surface contains those structures of the vascular network whose intensity values were greater than the chosen isovalue, which was selected to be approximately 1/4 the way between the background and maximum intensity value. The isosurface can be used to visualize the complete vasculature of the embryo as shown in Figure 2A. It can be zoomed in to give a magnified view of structures of interest, as in Figure 2B, and it can be arbitrarily rotated to view the surface from any angle, as illustrated in Figure 2C. The 3D data can also be manually segmented (see Materials and Methods), and rendered in different colours according to the segmentation label. This is demonstrated in Figure 2D which shows a surface rendering of the Cy3-PECAM-1 signal from a 19 somite embryo, with the dorsal aorta (DA), heart and internal carotid arteries (ICA) rendered in yellow, the umbilical vein (UV) in pink and the rest of the Cy3-PECAM-1 signal in blue. Any set of labels can be excluded from the rendering to simplify visualization of structures of interest. The removal of the vasculature in the head, for example, simplifies visualization of the ICA (Figure 2E). Much of the following data is represented in this way.

Volumes can be alternatively rendered from the 3D data set using a direct volume rendering algorithm (see Materials and Methods) that assigns to each voxel arbitrarily selected emissive and absorptive properties according to its reconstructed intensity value and projects an image of the volume onto a 2D image plane. Volume renderings are particularly useful when visualizing co-registered 3D data sets, or when a 3D data set has small features or features with weak intensity values. A rendered volume can be used to visualize the Cy3-PECAM-1 signal and the nonspecific tissue autofluorescence separately (Figure 3A,B respectively) or, as the data sets are co-registered, simultaneously (Figure 3C). This process was performed on all 24 embryos ranging from 5 to 30 somites (Table 2), from which six are shown in Figure 4, to provide a complete map of PECAM-1 expression throughout the entire embryo. Reconstructions from embryos older than 20 somites had insufficient resolution to resolve the finest vascular details in the complete embryo, but were sufficient to position larger vessels and overall structure (data not shown).

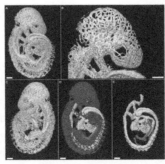

Figure 2. Surface renderings of embryonic vascular structures. (A) Reconstructed FDR-deconvolution OPT data of the 19 somite embryo is shown as a surface rendered object. (B) The surface rendered object can be zoomed in to any magnification, as in this magnified image of the vasculature in the mouse head. (C) The surface rendering can also be rotated so that it can be viewed from any angle. Viewing the rendering from the left side reveals structures in the heart (arrow) that are obscured by the tail in (A). (D) The 3D data can also be segmented as described in Materials and Methods. The DA, heart and ICAs are labelled yellow, the UV dark pink, and the unsegmented vasculature blue. (E) Segmentation of the data allows selective display of labelled structures. Exclusion of the unsegmented data provides better analysis of the ICAs and the pharyngeal arch arteries. All scale bars represent 100 microns.

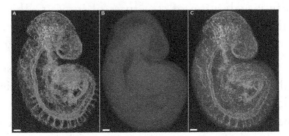

Figure 3. Volume renderings of embryonic vascular structures. (A) Reconstructed FDR-deconvolution OPT data can be visualized as a volume rendering. The vasculature of a 19 somite mouse embryo is visualized as a volume rendering using a hot metal colour map. (B) OPT data acquired for the mouse atlas include a co-registered 3D autofluorescence data set to visualize the vasculature in context of the rest of the embryo. The reconstruction from the OPT data of the autofluorescence channel of the 19 somite embryo can be visualized as a volume rendering using a blue colour map. (C) Since the data sets are co-registered, both volume renderings from (A) and (B) can be visualized in the same space. All scale bars represent 100 microns.

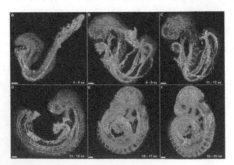

Figure 4. Stages of vascular development from 5 to 20 somites. (A) In the 5 somite embryo the vasculature, volume rendered with a hot metal colourmap, is confined mostly to a completed DA, a developing heart, the

allantois, the extraembryonic circulation, and clusters of PECAM-1 expressing cells dispersed throughout the cephalic mesenchyme. The autofluorescence is volume rendered with a transparent red colour map for overall positioning. (B) The 8 somite embryo has a rudimentary vascular plexus permeating the cephalic mesenchyme, the UV is elongating and the heart has initiated looping. (C) Occipital intersomitic vessels have begun to develop in the 11 somite embryo. (D) After turning, the cervical intersomitic vessels emerge in the 14 somite embryo. (E) The intersomitic vessels have begun to branch and connect together in the 16 somite embryo. (F) The vasculature of the 19 somite mouse embryo is a complicated but stereotypic structure. All scale bars represent 100 microns.

Table 2. Embryos Imaged.

Som #	5–6	7–8	8–9	9–10	10–11	12–13	13–14	15–16	16–17	18–19	19–20	20+	Total
n	3	1	2	2	1	2	1		2	1	2	1	6	24

Twenty four (24) embryos were imaged in total across an age range from 5 to 30 somites.
doi:10.1371/journal.pone.0002853.t002

These results demonstrate that the FDR-deconvolution OPT technique is capable of resolving the full range of vessels throughout whole embryos spanning the stages of 5–20 somites. Results presented below also illustrate one of the key benefits of the molecular specificity of FDR-deconvolution OPT as compared to imaging based on perfusion of the vascular lumen in that we were able to visualize endothelial cells and vasculature that is disconnected from the rest of the vascular tree and thus identify the origins of vessels at an earlier period of development. FDR-deconvolution OPT is the only imaging technique that has been shown to provide the combination of resolution, specimen coverage and molecular specificity necessary to image the complete vascular network in whole mouse embryos. Sufficient resolution is made possible in OPT imaging only by the inclusion of FDR-based deconvolution, which was shown to provide a factor of two resolution improvement over standard OPT imaging [41]. Our implementation of FDR-based deconvolution is performed entirely in software and can thus be used in conjunction with any OPT device without the need for hardware modification. FDR-deconvolution OPT is thus particularly well suited to the challenges presented by the developing vascular system of the mouse embryo and would be similarly suitable for imaging other detailed structures such as the developing nervous or lymphatic system.

We present here for the first time, a high-resolution three-dimensional atlas of the developing mouse vasculature in its native state between E8.0 and E10.0 of gestation (5–30 somites). The atlas comprises the collection of all 3D FDR-deconvolution OPT data sets of embryos ranging from 5 to 30 somites, as listed in Table 2. Videos of each of the data sets are made available at http://www.mouseimaging.ca/research/mouse_atlas.html. In the remainder of this paper, we present new information regarding the normal development of mouse vasculature that was obtained from analysis of the embryonic mouse vascular atlas.

Vascular Development Between the 5 and 8 Somite Stage

Continuous PECAM-1 expression in the 5 somite embryo was confined to the completed dorsal aorta (DA), the heart and the allantois as previously reported [2]. Disconnected clusters of PECAM-1 expression were evident throughout the cephalic mesenchyme and lateral mesoderm at this age (Figure 5). These discrete clusters of PECAM-1 expressing cells were not connected to established vessels, suggesting they were locations of vasculogenesis rather than angiogenesis, consistent with observations that cephalic mesoderm has intrinsic angiogenic potential and contributes to the vasculature of the head [46].

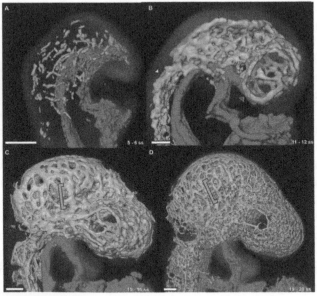

Figure 5. Development of the cephalic plexus between the 5 and 20 somite embryo. (A) The vasculature in the 5 somite mouse embryo is a series of disconnected clusters of PECAM-1-expressing cells. The DA and the heart are surface rendered red, PECAM-1 expression throughout the cephalic mesenchyme is surface rendered orange, and the autofluorescence of the mouse embryo is volume rendered with a hot metal colourmap. (B) By 11 somites, the cells have aggregated into a rudimentary vascular plexus. Larger vessels such as the PHV (blue arrowhead), the PMA (yellow arrowhead) and the ICA (green arrowhead) are visible. The PHV at this stage is a single large vessel that runs in an anterior-posterior direction starting from the cephalic flexure down to the first intersegmental vessel. (C) The cephalic plexus has remodelled into a more stereotypic pattern by 15 somites. The cephalic veins are easily distinguishable (green bracket). (D) At 19 somites the cephalic plexus has become more refined into recognizable structures. The cephalic veins are still visible at this stage (green bracket). All scale bars represent 100 microns.

By 7 somites, some PECAM-1 clusters in the cephalic mesoderm had begun to aggregate together forming a single larger vessel predominantly along the anterior-posterior axis and the future location of the primary head vein (PHV), while other cells remained as yet disconnected. By 11 somites virtually all PECAM-1 expression in the cephalic mesoderm was connected and formed a rudimentary vascular plexus that lined both the neural tube and the cephalic body wall. Recognizable structures such as the PHV, primitive maxillary artery (PMA) and the primitive internal carotid artery (ICA) were evident at this stage. Left-right communication between the two hemispheres of cephalic plexus was initiated at 7 somites by two vessels originating from the branch point of the PMA and the ICA, extending medially towards each other and forming a complete vessel by 13 somites (data not shown). By 14 somites the smaller vessels lateral to the plexus combined to form the anterior cardinal vein (ACV), as has been previously reported [44]. The plexus had also extended to surround the length of the neural tube and the otic vesicle, and the recognizable pattern of the mesencephalic artery and cephalic veins had begun to emerge in the cephalic body wall. This pattern continued to develop and refine up to the 20 somite stage.

Connection of Embryonic and Extraembryonic Circulation

The extraembryonic circulation of the mouse embryo is divided into two components: the omphalomesenteric (vitelline) vessels and the umbilical vessels. The former connect the embryo proper with the yolk sac, while the latter connect the embryo proper with the feto-maternal interface in the placenta.

The omphalomesenteric artery (OA) connects the yolk sac to the junction of the paired dorsal aorta at the most posterior tip of the embryo at E8.5. After turning, the OA leaves the embryo posterior to the developing heart, and eventually is folded back and fused to the dorsal aorta at E10.5 [47]. The OA has been previously reported to be present in the mouse as early as 7 somites [48]. We observed the OA as a single vessel in the 5 somite embryo that remained throughout the entire range of all embryonic stages studied and had folded back and fused to the DA by 30 somites (Figure 6).

The omphalomesenteric veins (OV) are paired vessels that connect the yolk sac to the sinus venosus (SV) of the embryo. The OV were located posterior to the developing heart throughout all stages imaged, initially draining directly into the SV (Figure 6A). The OV were exclusively extraembryonic until turning, at which point part of the OV traversed the future hepatic location in the embryo proper before connecting to the CCV. We observed the OV to be complete in the 5 somite embryo and present throughout the entire range of all embryonic stages studied (Figure 6).

The umbilical artery (UA) is initiated by de novo vasculogenesis in the allantois at E7.5 [49] and is of particular interest as it is implicated as a site of hematopoietic stem cell development [50], [51], [52], [53], [54]. We observed the UA in the 5 somite embryo as a vessel fully formed throughout the allantois but unconnected to the dorsal aorta (data not shown). The UA in the allantois fused to the paired dorsal aorta at the base of the allantois by 7 somites (Figure 6A), consistent with previous reports of 6 somites [49], and remained an extraembryonic vessel through the entire range of all embryonic stages studied (Figure 6).

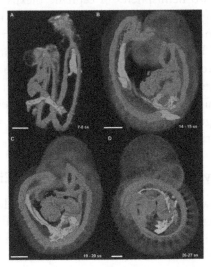

Figure 6. Connections between the embryonic and extraembryonic circulation in the early mouse embryo. (A) At 7 somites, the omphalomesenteric arteries and veins (green) are large structures. The UA (dark pink) connects to the DA and traverses the allantois, but is quite small. The DA and heart is surface rendered red, and the embryo autofluorescence volume rendered with a hot metal colourmap. The development of the UV is shown in Figure 6. (B) At 14 somites, the UA is accompanied by a series of smaller vessels connecting to nearby vasculature. At this point, it is still smaller than the OA and OV. After turning, the omphalomesenteric vessels branch off on one side of the embryo, and the umbilical vessels the other. (C) By 19 somites the UA has grown in size and is approximately equal in diameter to the OA. (D) By 26 somites, the OA is significantly smaller in diameter than the UA, suggesting that the balance of flow begins to favour the feto-maternal interface through the UA at some time between E9.0 and E9.5. All scale bars represent 250 microns.

The umbilical vein (UV), unlike the UA, has embryonic as well as extraembryonic components, leading us to question its origins. We were able to trace the origins of the UV back to the 5 somite stage embryo. At this stage it was observed bilaterally as a disconnected string of PECAM-1 expressing cells at the junction of the body wall (ectoderm) and the amnion extending from the sinus venosus (SV) to the posterior tip of the embryo (Figure 7A). The disconnected nature of the Cy3-PECAM-1 signal indicated that the UV is formed by vasculogenesis. These cells then aggregated in a primarily anterior-posterior fashion to extend the length of the embryo (Figure 7B). As the embryo turned, the posterior end of the nearly completed UV was brought into contact with the base of the allantois, thereby allowing connection between the extraembryonic (allantoic) and embryonic portions (Figure 7C). By the end of turning (~14 somites), the remaining cells had joined together to complete the UV as bilateral axial vessels running the length of the trunk from the SV through to the allanto is (Figure 7D). At the 7 somite stage, several small branches were observed to extend from the rudimentary UV, which, by the 14 somite stage, developed into a capillary plexus permeating the body wall surrounding the intraembryonic coelom dorsal to the UV (data not shown). This plexus continued to develop along the UV with increasing age, eventually becoming continuous with either the rudimentary PCV or the intersomitic veins (see below). To our knowledge, this is the first report of how the embryonic portion of the UV is established. The rudiments of this vessel, along with the cephalic plexus described above, were likely present in the embryos examined by Drake and Fleming [2] in their study of vasculogenesis including embryos from 5 to 8 somites, however they would not likely have been recognized as such, in part due to partial dissection of the embryos to facilitate imaging and the limited nature of the 2D confocal imaging used. These findings demonstrate the value of our technique of maintaining the original morphology of the embryo and 3D imaging of developing structures in their entirety.

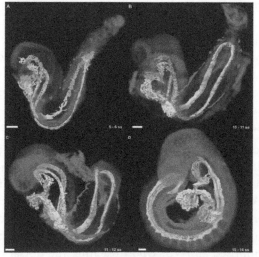

Figure 7. The development of the umbilical vein in the mouse embryo. (A) Discrete clusters of PECAM-1 expressing cells (dark pink) were evident along the length of the body wall immediately next to the junction of the body wall to amnion. The DA and heart is surface rendered yellow, and the embryo autofluorescence volume rendered with a hot metal colourmap. (B) The cells aggregated in a primarily anterior-posterior fashion beginning at the SV. (C) By 11 somites, the UV was almost complete, and had begun to develop a plexus which extended dorsolaterally. At the end of turning, the UV was complete and joined to the extraembryonic components of the vessel. (D) By 15 somites the UV is the second largest vessel in the embryo trunk. All scale bars represent 100 microns.

The 3D nature of FDR-deconvolution OPT data allows for measurements of vascular structures. We measured the relative diameter of the omphalomesenteric and umbilical vessels to serve as an indicator of relative blood flow volume. The diameter of the UA was less than that of the OA until the 19 somite stage embryo, at which point the two were approximately equal. By 28 somites, the diameter of the UA was approximately 1.5 times that of the OA. The UV, like the UA, were smaller in diameter than the OV until the 19 somite stage embryo, at which point the two were approximately equal. By the 26 somite stage, the UV was found to be approximately 1.5 times the diameter of the UV. Together, these results suggest that the volume of blood flow begins to favour the feto-maternal interface at approximately 20 somites or E9.0–E9.5, consistent with other data showing that the embryo becomes dependent on the chorioallantoic placenta by E10.0 [55].

Intersomitic Vessels of the Occipital Region

Intersomitic (intersegmental) vessels are a useful model for sprouting angiogenesis and vessel pathfinding. They are the first vessels in the embryo to form by sprouting angiogenesis and their navigation between somites is guided by the same cues that guide axon growth cones (reviewed in [56], [57], [58], [59]). Intersomitic arteries and veins (ISA and ISV) are branches of the DA and PCV respectively that extend dorsally between the borders of their adjacent somites. The development of intersomitic vessels over time can be followed in a single embryo at a single time point as they emerge in a temporally regulated fashion along the anterior posterior axis from oldest to youngest, similar to their adjacent somites. Comparison of embryos at different stages of development revealed that there was a distinction in the development of intersomitic vessels bounded by the first 5 (occipital) somites compared to those bounded by trunk somites (at least for somites 6–20). As such we refer to these intersomitic vessels as "occipital" and "trunk" respectively.

We observed the occipital intersomitic vessels to consist of three interconnected vessels: a transient ISA, a transient arteriovenous shunt (AVS), and a persistent ISV. The ISAs formed first, initiating bilaterally as dorsal sprouts from the DA as early as 5 somites. At the level of each dorsal ISA, a lateral branch originating from the DA was observed to extend towards the CCV, eventually connecting the two major vessels (Figure 8A). This created a direct communication between the DA and CCV and comprised an AVS. Upon reaching the dorsal margin of the bounding somites, the distal tips of the ISAs branched longitudinally, fusing with the neighboring ISAs to form the vertebral artery (VTA). Shortly after VTA formation, the ISV emerged as a branch between the CCV and the ISA. We cannot comment on the origin of this vessel as we did not observe any instances when the ISV was connected to solely either the CCV or the ISA. The fusion of the ISV to the ISA created a temporary triangular vascular structure involving the ISA, the AVS and the ISV. This was a short-lived structure, as regression of the ISA and AVS quickly followed (Figure 8B). All ISAs and AVS of the occipital region had regressed by the 18 and 28 somite stage respectively, leaving the CCV connected to the VTA via an ISV and the DA fully separated from the VTA and CCV. Overall, development of the occipital intersomitic vessels was a rapid process as evidenced by the rarity of capture of intermediate developmental stages.

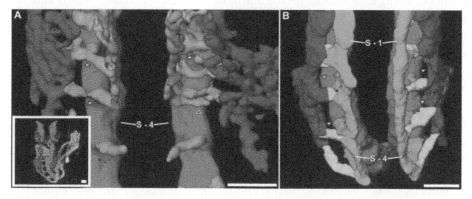

Figure 8. Development of the occipital intersomitic vessels. (A) The various stages of occipital intersomitic vessel development can be visualized in the 8 somite embryo (inset). The vessels surrounding somites 1 through 5 are segmented as DA (red), intersomitic vessels (pink) and the early cardinal vein (blue). The vessels initiate bilaterally, and an AVS originating from the DA connects to the cardinal vein (yellow arrowheads). An ISV then develops connecting the distal tip of the ISA to the cardinal vein (blue arrowhead). The ISA soon regresses (white arrowhead) leaving the vertebral artery connected to the cardinal vein. Somite number 4 is labelled as S-4. (B) The transient AVS have regressed (yellow arrowheads) by 15 somites, leaving only the expected dorsal ISA (pink), connected by the VTA (green), which connects via ISVs (light blue) to the ACV or CCV (blue). Somites numbers 1 and 4 are labelled as S-1 and S-4. All scale bars represent 100 microns.

The existence of transient DA-CCV arteriovenous shunts in the development of occipital intersegmental vessels has not been previously reported and it is unclear why they form at all. AVS similar to the DA-CCV shunts have been noted in the course of normal embryonic mouse vascular development connecting the primordial ACV directly to the DA, anterior to the first somite [44]. We were able to confirm the existence of these AVS in the same location in embryos with 8 somites (data not shown). Like the transient AVS between the DA and CCV, the AVS between the DA and the ACV were no longer present beyond the 18 somite stage, and were presumably pruned from the vasculature.

These transient connections, especially those between the DA and CCV, may be causative of the arteriovenous malformations (AVMs) that arise in mice defective for Notch [6], [60], [61] and TGFβ signalling [62], [63]. AVMs are miscommunications between arteries and veins that bypass the normal capillary plexus, resulting in the shunting of blood from the arterial circulation directly back to the venous circulation. Ink flow patterns and histological analysis suggested that AVMs in Notch and TGFβ mutants involved a shunting of blood from the DA into the ACV and CCV [6], [60], [61], [62], [63]. Importantly, these AVMs were observed at an age by which transient connections between the DA and the ACV and CCV should have fully regressed, suggesting that the AVMs could have arisen from failed regression of these naturally occurring connections. This would imply that Notch and TGFβ may be required for the coordinated regression of these naturally occurring connections between arterial and venous vessels. Notch and TGFβ may accomplish this through their known role in regulating artery/vein identity [6], [60], [61], [62], [63]. Artery/vein identity has been proposed to prevent AVMs either by keeping arterial and venous progenitors separate during vasculogenesis [64] or by preventing "promiscuous fusions between naïve endothelial sprouts" from pre-existing arteries and veins [62]. If we are correct, then artery/vein identity may have a third role: promoting the regression of pre-existing connections between arterial and venous territory. As hemodynamic forces are known to influence vascular remodelling [4], [5] it would be necessary to test whether Notch and TGFβ act downstream of blood flow to regulate such remodelling or are part of a true genetically pre-programmed event. Close examination of DA–CCV AVM shunt fate in Notch, TGFβ and blood flow mutant embryos would help shed light on this.

Intersomitic Arteries of the Trunk Region

Intersomitic vessels of the trunk differed from those of the occipital somites, both in their method of formation and final configuration. Trunk intersomitic vessels began to emerge bilaterally at the 8 somite stage as small dorsal protrusions in the paired DA, until the embryo began turning at the 11 somite stage (data not shown). Between this stage and the 12 somite stage (half turn), the protrusion at the level of the second trunk ISA (7th somite) had extended dorsally between the somites, and by 3/4 turning (13 somites), the protrusions down to the 11th somite had become extended branches. Subsequent ISA were observed to extend in more regulated intervals with advancing somite stage (Figure 9). Whereas zebrafish ISAs emerge from the aorta as narrow capillary-like projections composed of 3 or 4 linked endothelial cells [28], [65], each ISA in mouse appeared as a sheet-like evagination that was subsequently remodelled into a capillary-like ISA and DLAV (Figure 9). Unlike the occipital ISAs, transient branches directly connecting the DA and PCV were not observed, and ISAs were not seen to regress at any stage up to 30 somites. Distinct mechanisms of intersomitic vessel development in the occipital and trunk regions are fully consistent with differences in somitogenesis in these two regions. Unlike trunk somites, occipital somites do not form from presomitic mesoderm using the segmentation clock mechanism [66]. Furthermore, unlike trunk somites, occipital somites disperse and become undetectable as segments soon after their formation. As endothelial cells are highly responsive to their local microenvironment [67], differential patterning of occipital and trunk somites may explain why intersomitic vessels in the occipital region differ to those of the trunk.

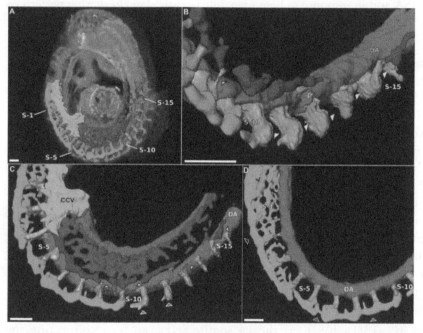

Figure 9. Development of the cervical intersomitic vessels. (A) The various stages of cervical intersomitic vessel development can be segmented as visualized as surface renderings in the 16 somite mouse embryo. The vessels along the right side of the embryo and surrounding somites 1 through 16 are labelled as: DA (red), ISA (pink), ISV (blue), VTA, DLAV and PNVP (green), ACV and CCV (cyan), UV (dark pink), UV plexus (purple), and PCV (blue). Somites 1, 5, 10 and 15 are numbered as S-1, S-5, S-10 and S-15. (B) Branches of PECAM-1 expression originating from the tips of the ISAs (yellow arrowheads) were observed to turn towards the location of the future PCV. (C) A second branch from the ISA was also observed to extend in a predominantly anterior direction (pink arrowheads) to connect up with other ISAs, eventually forming the DLAV. PECAM-1 expression along the location of the expected PCV was observed to lag development of the ISAs and is discontinuous (yellow arrowheads). (D) The PNVP develops through remodelling of the VTA and DLAV. Branches initiate medially from the DLAV (pink arrowhead), begin to remodel into simple mesh (blue arrowhead), and eventually remodel into a fine structured capillary plexus surrounding the neural tube. Note at this stage that the first ISA has regressed. Scale bars represent 100 microns.

Intersomitic Veins and the Posterior Cardinal Vein

As the ISA approached the dorsal boundary of the somites, a branch was observed to extend laterally from the ISA toward the location of the future PCV, which is lateral to the DA (Figure 9B). This lateral branch became progressively more defined with age and formed the rudiment of the ISV, eventually connecting to either a rudimentary PCV or directly to the capillary plexus from the UV. A small number of branches of PECAM-1 expression were also observed to extend from the UV capillary plexus or the rudimentary PCV towards the rudimentary ISV. As for occipital ISA, each trunk ISA was paired with an accompanying ISV. This arrangement is in contrast to zebrafish, where the ISV sprouts from the PCV to join the ISA and in response to hemodynamic pressure, an alternating pattern of ISA/ISV is established [28].

PECAM-1 expression in the region of the expected PCV was first noted at 15 somites extending posteriorly from the CCV and connecting to the ISVs in a predominantly anterior-posterior fashion (Figure 9). Discontinuous PECAM-1 expression was observed along the path of the future PCV strongly suggesting its formation was a vasculogenic process, as occurs in the chick and zebrafish [29], [68]. Branches extended ventrally from the PCV and became continuous with the UV plexus as described above. By 16 somites PECAM-1 expression was continuous to the 8th ISV, extending discontinuously down to the 15th ISV, and by 19 somites the expression was continuous through all ISV. The vasculogenic activity in the vicinity of the PCV and its juxtaposition to the newly emerging ISVs makes it difficult to determine whether the ISV truly sprouts from its accompanying or is generated by (PECAM-1 negative) angioblasts recently added to the PCV by vasculogenesis. Detailed analysis of Flk1 expression as a marker of immature angioblasts combined with live imaging studies will be needed to resolve this issue.

PCV formation was disrupted in the vicinity of the developing limb bud (somites 8/9–13/14 [69]). At the 14 somite stage, an AVS connecting the DA to the rudimentary PCV was observed at the location of the 13th ISA, immediately posterior to the developing limb bud. This vessel was observed to develop bilaterally, but not necessarily at the same somite level. At the 20 somite stage, this vessel was still present in the same location, and at the 25 somite stage had developed into an artery feeding the developing limb. Additional branches were noted at this stage, originating from the DA and the neighbouring ISAs and connecting to the capillary plexus of the developing limb bud. The forelimb field, as defined by Tbx5 expression in lateral plate mesoderm, is first evident at the 8 somite stage [70]. We noted the limb bud itself to begin its formation at approximately 12 somites, and spanned the 8th to the 12th ISA. Together, these results suggest that the appearance of these limb arteries lags the development of the limb bud, and thus may be in response to a changing environment rather than a programmatic or preemptive occurrence.

Anterior Branching of Intersomitic Arteries Establishes the DLAV

Upon reaching the dorsal margin of the somite and after formation of an ISV bud, trunk ISA tips branched longitudinally and fused with their adjacent ISAs forming the DLAV. This longitudinal branching was strongly biased in an anterior direction suggesting that it was guided by an attractive or repulsive mechanism (Figure 9C). The anterior bias of dorsal ISA branchings was not absolute, as some dorsal branches were seen to extend in a posterior direction, and in some cases bi-directionally. From 42 trunk ISAs for which branching directionality could be demonstrated, the number of anterior:posterior:bi-directional branching was 36:4:2, suggesting a strong bias for anterior branching of trunk ISAs. In the occipital region, ISAs observed were either completely connected to the VTA or had not yet begun to branch, thus we were unable to determine whether bias exists in occipital ISA branching. We observed instances of some trunk ISAs reaching the dorsal somite margin and initiating longitudinal branching prior to ISAs located anteriorly, indicating that timing of ISA remodelling is not absolute and is a highly dynamic process.

Anterior bias of trunk ISA branching is in contrast to the zebrafish, where ISAs branch in both anterior and posterior directions after reaching the dorsal boundary of the somite [28]. As described above, intersomitic vessel branching and pathfinding are regulated by the same guidance cues used by the axon growth cone. Somites are divided into rostral and caudal halves and many genes associated with axon guidance show polarized expression in these halves [71]. An anterior bias of ISA branching could be explained by attractive cues in the caudal half-somite anterior of the sprout or repulsive cues in the rostral half-somite posterior to the sprout.

EphrinB2 is one possible ISA attractant. In addition to being expressed in the ISA itself, ephrinB2 is expressed in the caudal half-somite during the time that longitudinal ISA branching is occurring [44], [72]. In vitro, ephrinB2 can induce endothelial sprouting [72], [73]. In vivo, ephrinB2-/- mice show defective vascular sprouting into the CNS [74] and reduced lymphatic sprouting [75], consistent with an attractive role for ephrinB2. While ephrinB2-/- mice display defective intersomitic vessel patterning that can be attributed to vascular specific ephrinB2 [44], these mutants were assessed after the DLAV had formed. Analysis of earlier time points in ephrinB2-/- embryos or ideally a somite specific ephrinB2 knockout would need to be performed to determine whether ephrinB2 is required for the anterior branching bias.

Our observations of preferential anterior branching toward the caudal half-somite would seem inconsistent with the expression of a known repulsive vascular cue, Sema3E, in this part of the somite [76]. One possibility is that Sema3E is not expressed early enough in the caudal half-somite to repel initial dorsal branching of the ISA, which we observed to occur before somites dispersed into sclerotome and dermamyotome. Sema3E expression in the caudal half-somite and defects in intersomitic vessel branching were reported at E10.5 and E11.5, after somites have begun to disperse and form sclerotome and dermamyotome compartments [76], [77]. Sema3E may therefore affect intersomitic branching that occurs subsequent to the primary branching we describe here. Similarly, the restriction of neural crest migration and peripheral nervous system axon pathfinding to the rostral half-somite occurs after formation of the sclerotome [71], [78], [79]. Guidance cues active in somites at this time such as the anti-angiogenic thrombospondin-1 in the rostral half-somite [80] are therefore not likely to be relevant to the vascular guidance we observed. Whatever the mechanisms are that govern the guidance of this branch, they are likely to involve a complex interplay of multiple attractive and repulsive factors, which together provide an anterior bias.

Formation of the Perineural Vascular Plexus

The perineural vascular plexus (PNVP) is the precursor to the blood brain barrier and is recruited to surround the neural tube in response to VEGF between E8.5 and E9.5 [81], [82]. In mouse, vascular sprouts from the PNVP invade the neurepithelium around E10.0 in a stereotypic fashion [83], [84], [85]. PNVP sprouting into the neurepithelium is mediated by VEGF [86], ephrinB2 [74] and Tie1 [87], while subsequent branching and remodelling of the sprouts in the neuroepithelium is regulated by Np1 [85], heparin-binding VEGF isoforms [86], [88], Dll4/Notch [43] and netrin1/Unc5b [89] signalling. Failure of the PNVP to invade the neurepithelium results in neurodegeneration and neonatal lethality demonstrating the importance of this plexus to organogenesis [86].

The plexus was first evident around the 8 somite stage in the occipital somite region as rudimentary branches extending ventrally from the VTA, which by the 10 somite stage had extended as a capillary plexus around the neural tube at the level of the first somite. The PNVP in the cervical somite region (somites 6–12) was first observed at the 12 somite stage as rudimentary branches extending ventrally from the VTA, and then at the 16 somite stage as a capillary plexus extending from the VTA and DLAV and surrounding the neural tube at the level of the fourth somite (Figure 9D). The appearance of the rudimentary branches occurred soon after fusion of the VTA between two adjacent ISA. The PNVP extended to the 8th somite at the 20 somite stage, and down to the 20th somite by the 30 somite stage. Consistent with previous reports, the PNVP was first observed to invade the neural tube at the 27 somite stage (data not shown) [83], [84].

Quail-chick and mouse-quail chimera studies have shown that somites ([81] and references therein) and lateral mesoderm ([81] and references therein) are major sources of PNVP endothelial cells in the trunk. The fine chimerism between host and graft derived cells in the PNVP led the authors of one study to conclude that somite derived angioblasts migrated to and incorporated into the PNVP by a vasculogenic process [90]. While our study does not lend itself to fate mapping the cells comprising the PNVP, our results strongly suggest that during the stages we examined, the PNVP in the trunk remodels directly from the VTA and DLAV by angiogenesis, while in the cervical region it remodels from the cephalic plexus.

As the VTA and DLAV originate from ISAs, which in turn arise from the dorsal wall of the DA, we would argue that the DA is the initial source of PNVP endothelial cells. In addition to the PNVP, somites [90], [91], [92], [93] and lateral mesoderm [92] both contribute to the dorsal wall of the DA and to ISAs in quail-chick chimeras. It is plausible to suggest that somite and lateral mesoderm contribution to the PNVP initially comes from a contribution to the DA, which subsequently donates its cells to the ISA and VTA/DLAV by angiogenesis. This would be consistent with zebrafish, where individual lateral mesoderm cells were found to migrate to the DLAV after incorporating into the DA and ISA [28], [65]. Somites may also make a second contribution to the PNVP, after the stages we imaged. Somite derived angioblasts may incorporate into or replace cells of the preformed PNVP as they have been demonstrated to do in the chick DA [92], [93], or, alternatively, somite derived vascular beds in the body wall may simply fuse to the pre-existing PNVP and contribute to it in that way.

Conclusion

We have employed FDR-deconvolution OPT to generate a high-resolution three-dimensional atlas of the developing mouse vasculature in its native state between E8.0 and E10.0 of gestation (5–30 somites). Analysis of the 3D reference atlas we have constructed has revealed significant new information regarding normal development of the embryonic mouse vasculature. The need for an atlas such as this is critical, as numerous pathways required for vascular development exhibit severe vascular phenotypes during this time period when disrupted. This atlas can thus be used as a tool for better interpretation of these vascular phenotypes and as a platform to provide insight into normal mammalian vascular development. The observations in this paper represent only a portion of the information available in this atlas, which is provided for further study at http://www.mouseimaging.ca/research/mouse_atlas.html.

Materials and Methods

Embryo Collection and Staining

Wild type ICR embryos were collected between the ages of embryonic day (E) 8.0 (5 somites) and E10.0 (30 somites). Noon of the plug day was considered to be E0.5. Embryos were dissected from their deciduas and Reichert's membranes, then, to maintain natural shape, were fixed for 1 h in 4% paraformaldehyde before remaining extraembryonic tissues were removed. For incompletely turned embryos, the amnion and the portion of yolk sac contiguous with the embryo were left attached to prevent disruption of embryonic-extraembryonic circulation. Embryos were then dehydrated through a graded series of methanol (25%, 50%, 75%, 100%) and stored at –20°C. Before staining, embryos were rehydrated and endogenous peroxidase activity was quenched with 3% H2O2. Non-specific antibody binding was blocked by pre-incubating embryos in 1% heat inactivated FCS (Hyclone, Logan UT) and 1% normal goat serum (Cedarlane, Burlington ON). Embryos were then stained overnight with 5 µg/mL anti-PECAM-1 antibody (Mec13.3) (BD Pharmingen). Primary antibody was detected by staining overnight with anti-rat HRP secondary antibody (Biosource, Camarillo CA) followed by incubation with tyramide-Cy3 reagent (1:50) for 1 h (PerkinElmer, Boston MA). Experiments were approved by the Animal Care Committee of Mount Sinai Hospital (Toronto, ON, Canada) and were conducted in accordance with guidelines established by the Canadian Council on Animal Care.

Although all embryos presented in this study were processed according to the above protocol, we have since determined that methanol fixation slightly decreased the signal to noise ratio. While this effect did not significantly affect our imaging or findings, we would recommend replacing methanol fixation with a longer (4 h) paraformaldehyde fixation time followed by treatment with 50 mM sodium azide prior to quenching endogenous peroxidase by H_2O_2 treatment. A detailed description of the protocol is available at http://www.sickkids.ca/rossant/custom/protocols.asp.

Optical Projection Tomography (OPT) of Embryos

Optical projection tomography was performed as described previously [41]. Specimens were embedded in 1% low melting point (LMP) agarose and subsequently cleared using a 1:2 mixture of benzyl alcohol and benzyl benzoate (BABB). The index-matched specimen was suspended from a stepper motor and immersed in a BABB bath with optically flat parallel glass windows. Images of the specimen were formed using a Leica MZFLIII stereozoom microscope equipped with a 0.5× objective lens and a 1.0× camera lens. Typical zoom settings used for image formation were between 4.0×–6.3×, resulting in numerical apertures from 0.0465 to 0.0620. Images (termed views) were acquired with a Retiga Exi CCD camera with pixel size 6.45×6.45 microns. Light from a mercury lamp was directed onto the specimen and filter sets were used to create fluorescent images of the specimen. An autofluorescence view was captured with the GFP1 filter set in the illumination and detection light path, and a view of the Cy3 fluorescence from the specimen was captured using the Cy3 filter set in the illumination and detection light path. The sample was rotated stepwise with a 0.9° step size through a complete revolution and views were acquired at each step.

Each OPT view approximates a parallel ray projection through the specimen. The temporal sequence from a row of detectors on the CCD forms a sinogram that is used to reconstruct the corresponding slice through the specimen using the standard convolution filtered back-projection algorithm [45]. The reconstruction of all slices yielded a 3D volumetric representation of the specimen. The stack of Cy3 views from a single specimen were subjected to Frequency Distance Relationship (FDR)-based filtering as described below, and both the filtered and unfiltered views reconstructed separately. The resulting 3D reconstruction of autofluorescence views and its corresponding 3D reconstruction of either filtered or unfiltered Cy3 views were co-registered.

Point Spread Function Acquisition

The point spread function of the optical system was required for the FDR-deconvolution process described below. A solution of silica beads (micromod sicastar-greenF 40-02-403) was mixed into 1% LMP agarose. A plug was cut out of the agarose and subjected to the same clearing process as the specimens. The plug was hung from the stepper motor, and images were acquired using the GFP1 filter set and the 4× zoom setting. The motorized focus moved the focal plane through the specimen, and an image of the bead plug was acquired at each step. An isolated bead was found in the stack. The data was resampled to approximate the PSF of the system at a wavelength of 600 nm rather than 535 nm, and at zoom settings of 5× and 6.3×.

Frequency Distance Relationship (FDR)-Based Deconvolution and Filtering

The stack of Cy3 views acquired over a complete revolution were subjected to Frequency Distance Relationship (FDR)-based deconvolution as described previously [41] and according to the equation

$$P(R_x, R_z, \Phi) = H(R_x, R_z, l = -\frac{\Phi}{R_x})^{-1} P_b(R_x, R_z, \Phi) \tag{1}$$

where (Rx, Rz, Φ) is the Fourier equivalent of the sinogram space (rx, rz, ϕ). Specifically, (rx, rz) are the axes of detector element (perpendicular to the rotational axis) and detector row (parallel to the rotational axis) respectively, l is slope of the line in the (Rx, Φ) plane and also the distance of the object from the lens, Pb(Rx, Rz, Φ) is the 3D Fourier Transform (FT) of the blurred sinogram, and P(Rx, Rz, Φ) is the 3D FT of the unblurred sinogram. $H(R_x, R_z, l = -\frac{\Phi}{R_x})$ is the FT of the distance dependent PSF, and is evaluated at each sample (Rx, Rz, Φ) using the FDR.

The filter H^{-1} is constructed from four distinct components, as described in the equation

$$H_{final}^{-1} = H_{\lim}^{-1} \cdot W_r \cdot W_w \cdot W_b \tag{2}$$

where $H_{\lim}^{-1}$ is a max-limited recovery filter designed according to the FDR using the experimentally acquired PSF, Wr is a slope-based roll-off filter to exclude out of focus data, W_w is a Wiener filter to deemphasize noise, and Wb is a bandlimiting roll-off filter for high frequencies. The individual components are described in the equations:

$$\left| H_{\lim}^{-1} \right| = \begin{cases} \left| H^{-1} \right| : \left| H^{-1} \right| \leq C_t \\ \\ C_t + C_r \left(1 - \exp\left[-\frac{\left| H^{-1} \right| - C_t}{C_r} \right] \right) : \left| H^{-1} \right| > C_t \end{cases} \tag{3}$$

$$W_r(R_X, R_Z, \Phi) = \begin{cases} 1.0 : l = -\frac{\Phi}{R_X} \leq 0 \\ \\ \cos^2\left(\frac{\pi}{2} \frac{|l|}{w} \right) : w > l > 0 \\ \\ 0.0 : l > w \end{cases} \tag{4}$$

$$W_W(R_X, R_Z, \Phi) = \frac{P_S}{P_S + P_n} \tag{5}$$

and

$$W_{bx}(R_X, R_Z, \Phi) = \begin{cases} 1.0 : R_X < 0.90b \\ \cos^2\left(\dfrac{\pi}{2}\dfrac{R_X - 0.90b}{0.1b}\right) : b > R_X > 0.90b \\ 0.0 : R_X > b \end{cases} \tag{6}$$

with the parameters $Ct = 10$, $Cr = 10$, $w = 0.3$, and b set according to the NA of the system by the equation

$$b = \frac{4\pi \cdot NA}{\lambda} \tag{7}$$

assuming a wavelength $\lambda = 630$ nm.

Reconstructions resulting from the FDR based filtered projections were intensity normalized according to [41]. Both the filtered and unfiltered reconstructions were blurred by a 3D Gaussian with a full width at half-maximum of 40 pixels. The inverse of the ratio of the two blurred reconstructions was used as the ratio for intensity normalization of the filtered reconstruction.

The resolution achieved was estimated to be 5 microns at the 6.3× zoom setting, 6.5 microns at the 5× zoom setting, and 8 microns at the 4× zoom setting.

Data Visualization and Segmentation

3D OPT reconstructions were loaded into Amira 3.1 (TGS, Inc.) for visualization. Surface renderings were created using the Amira "Isosurface" module with a threshold chosen just above the noise floor. Volume renderings were created using the Amira "Voltex" module, with the low threshold being chosen just above the noise floor and the high threshold chosen to maximize vessel visibility. Volume renderings of the autofluorescence reconstruction overlapping the Cy3-PECAM-1 reconstruction were created using a red colourmap for the autofluorescence and a hot metal colourmap for the Cy3-PECAM-1.

Reconstructions were segmented using the Amira module "LabelField" to maximize vessel visibility and aid in image interpretation. Observations were confirmed in both the unfiltered and filtered Cy3-PECAM-1 reconstructions. Surface renderings of the segmented vessels were performed using the same threshold but different colourmaps.

Acknowledgements

We thank Jorge Cabezas for expert animal husbandry and care and all members of the Rossant laboratory for insightful discussions and James Sharpe and the British Medical Research Council for providing us with an OPT device.

Authors' Contributions

Conceived and designed the experiments: JRW RMH. Performed the experiments: JRW LC. Analyzed the data: JRW LC. Wrote the paper: JRW LC JR RMH.

References

1. Ema M, Takahashi S, Rossant J (2006) Deletion of the selection cassette, but not cis-acting elements, in targeted Flk1-lacZ allele reveals Flk1 expression in multipotent mesodermal progenitors. Blood 107: 111–117.

2. Drake CJ, Fleming PA (2000) Vasculogenesis in the day 6.5 to 9.5 mouse embryo. Blood 95: 1671–1679.

3. Huber TL, Kouskoff V, Fehling HJ, Palis J, Keller G (2004) Haemangioblast commitment is initiated in the primitive streak of the mouse embryo. Nature 432: 625–630.

4. le Noble F, Moyon D, Pardanaud L, Yuan L, Djonov V, et al. (2004) Flow regulates arterial-venous differentiation in the chick embryo yolk sac. Development 131: 361–375.

5. Lucitti JL, Jones EAV, Huang CQ, Chen J, Fraser SE, et al. (2007) Vascular remodeling of the mouse yolk sac requires hemodynamic force. Development 134: 3317–3326.

6. Gridley T (2007) Notch signaling in vascular development and physiology. Development 134: 2709–2718.

7. Hofmann JJ, Iruela-Arispe ML (2007) Notch signaling in blood vessels - Who is talking to whom about what? Circ Res 100: 1556–1568.

8. Niessen K, Karsan A (2007) Notch signaling in the developing cardiovascular system. Am J Physiol Cell Physiol 293: 1–11.

9. Lebrin F, Deckers M, Bertolino P, Ten Dijke P (2005) Tgf-β receptor function in the endothelium. Cardiovasc Res 65: 599–608.

10. Rossant J, Howard L (2002) Signaling pathways in vascular development. Annu Rev Cell Dev Bio 18: 541–573.

11. Astorga J, Carlsson P (2007) Hedgehog induction of murine vasculogenesis is mediated by Foxf1 and Bmp4. Development 134: 3753–3761.

12. Byrd N, Becker S, Maye P, Narasimhaiah R, St-Jacques B, et al. (2002) Hedgehog is required for murine yolk sac angiogenesis. Development 129: 361–372.

13. Vokes SA, Yatskievych TA, Heimark RL, McMahon J, McMahon AP, et al. (2004) Hedgehog signaling is essential for endothelial tube formation during vasculogenesis. Development 131: 4371–4380.

14. Coultas L, Chawengsaksophak K, Rossant J (2005) Endothelial cells and VEGF in vascular development. Nature 438: 937–945.

15. Carmeliet P, Ferreira V, Breier G, Pollefeyt S, Kieckens L, et al. (1996) Abnormal blood vessel development and lethality in embryos lacking a single VEGF allele. Nature 380: 435–439.

16. Ferrara N, Carver-Moore K, Chen H, Dowd M, Lu L, et al. (1996) Heterozygous embryonic lethality induced by targeted inactivation of the VEGF gene. Nature 380: 439–442.

17. Kuijper S, Turner CJ, Adams RH (2007) Regulation of angiogenesis by Eph-ephrin interactions. Trends Cardiovasc Med 17: 145–151.

18. Thurston G (2003) Role of angiopoietins and tie receptor tyrosine kinases in angiogenesis and lymphangiogenesis. Cell Tissue Res 314: 61–68.

19. Pola R, Ling LE, Silver M, Corbley MJ, Kearney M, et al. (2001) The morphogen Sonic hedgehog is an indirect angiogenic agent upregulating two families of angiogenic growth factors. Nat Med 7: 706–711.

20. White AC, Lavine KJ, Ornitz DM (2007) FGF9 and SHH regulate mesenchymal Vegfa expression and development of the pulmonary capillary network. Development 134: 3743–3752.

21. Lee S, Chen TT, Barber CL, Jordan MC, Murdock J, et al. (2007) Autocrine VEGF signaling is required for vascular homeostasis. Cell 130: 691–703.

22. Eremina V, Baelde HJ, Quaggin SE (2007) Role of the VEGF—a signaling pathway in the glomerulus: evidence for crosstalk between components of the glomerular filtration barrier. Nephron Physiol 106: 32–37.

23. Ferrara N, Kerbel RS (2005) Angiogenesis as a therapeutic target. Nature 438: 967–974.

24. Thurston G, Noguera-Troise I, Yancopoulos G (2007) The Delta paradox: DLL4 blockade leads to more tumour vessels but less tumour growth. Nat Rev Cancer 7: 327–331.

25. Hellström M, Phng LK, Hofmann J, Wallgard E, Coultas L, et al. (2007) Dll4 signalling through Notch1 regulates formation of tip cells during angiogenesis. Nature 445: 776–780.

26. Lobov IB, Renard RA, Papadopoulos N, Gale NW, Thurston G, et al. (2007) Delta-like ligand 4 (Dll4) is induced by VEGF as a negative regulator of angiogenic sprouting. Proc Natl Acad Sci USA 104: 3219–3224.

27. Isogai S, Horiguchi M, Weinstein BM (2001) The vascular anatomy of the developing zebrafish: an atlas of embryonic and early larval development. Dev Biol 230: 278–301.

28. Isogai S, Lawson ND, Torrealday S, Horiguchi M, Weinstein BM (2003) Angiogenic network formation in the developing vertebrate trunk. Development 130: 5281–5290.

29. Coffin JD, Poole TJ (1988) Embryonic vascular development: immunohistochemical identification of the origin and subsequent morphogenesis of the major vessel primordia in quail embryos. Development 102: 735–748.

30. Poole TJ, Coffin JD (2005) Vasculogenesis and angiogenesis: Two distinct morphogenetic mechanisms establish embryonic vascular pattern. J Exp Zool 251: 224–231.

31. Hiruma T, Nakajima Y, Nakamura H (2002) Development of pharyngeal arch arteries in early mouse embryo. J Anat 201: 15–29.

32. Sharpe J, Ahlgren U, Perry P, Hill B, Ross A, et al. (2002) Optical projection tomography as a tool for 3D microscopy and gene expression studies. Science 296: 541–545.

33. Lickert H, Takeuchi JK, Von Both I, Walls JR, McAuliffe F, et al. (2004) Baf60c is essential for function of BAF chromatin remodelling complexes in heart development. Nature 342: 107–112.

34. Fisher ME, Clelland AK, Bain A, Baldock RA, Murphy P, et al. (2008) Integrating technologies for comparing 3D gene expression domains in the developing chick limb. Dev Biol 317: 13–23.

35. Lee K, Avondo J, Morrison H, Blot L, Stark M, et al. (2006) Visualizing plant development and gene expression in three dimensions using optical projection tomography. Plant Cell 8: 2145–2156.

36. McGurk L, Morrison H, Keegan LP, Sharpe J, O'Connel MA (2007) Three-dimensional imaging of Drosophila melanogaster. PLoS ONE 2: e834.

37. Alanentalo T, Asayesh A, Morrison H, Lorén CE, Holmberg D, et al. (2007) Tomographic molecular imaging and 3D quantification within adult mouse organs. Nat Methods 3: 31–33.

38. Hajihosseini MK, De Langhe S, Lana-Elola E, Morrison H, Sparshott N, et al. (2008) Localization and fate of Fgf10-expressing cells in the adult mouse brain implicate Fgf10 in control of neurogenesis. Mol Cell Neurosci 37: 857–868.

39. Xia W, Lewitt RM, Edholm PR (1995) Fourier correction for spatially variant collimator blurring in SPECT. IEEE Trans Med Im 14: 100–115.

40. Walls JR, Sled JG, Sharpe J, Henkelman RM (2005) Correction of artefacts in optical projection tomography. Phys Med Biol 50: 4645–4665.

41. Walls JR, Sled JG, Sharpe J, Henkelman RM (2007) Resolution improvement in optical projection tomography. Phys Med Biol 52: 2775–2790.

42. Chaturvedi K, Sarkar DK (2006) Isolation and characterization of rat pituitary endothelial cells. Neuroendocrinology 83:

43. Suchting S, Freitas C, le Noble F, Benedito R, Bréant C, et al. (2007) The Notch ligand Delta-like 4 negatively regulates endothelial tip cell formation and vessel branching. Proc Natl Acad Sci USA 104: 3225–3230.

44. Gerety SS, Anderson DJ (2002) Cardiovascular ephrinb2 function is essential for embryonic angiogenesis. Development 129: 1397–1410.

45. Slaney M, Kak AC (1988) Principles of Computerized Tomographic Imaging. IEEE Press.

46. Couly G, Coltey P, Eichmann A, Ledouarin NM (1995) The angiogenic potentials of the cephalic mesoderm and the origin of brain and head blood-vessels. Mech Dev 53: 97–112.

47. Garciaporrero JA, Godin IE, Dieterlen-Lièvre F (1995) Potential intraembryonic hemogenic sites at pre-liver states in the mouse anatomy and embryology. Anat and Embryol 192: 427–437.

48. Wood HB, May G, Healy L, Enver T, Morriss-Kay GM (1997) CD34 expression patterns during early mouse development are related to modes of blood vessel formation and reveal additional sites of hematopoiesis. Blood 90: 2300–2311.

49. Inman KE, Downs KM (2007) The murine allantois: emerging paradigms in development of the mammalian umbilical cord and its relation to the fetus. Genesis 45: 237–258.

50. de Bruijn MF, Ma X, Robin C, Ottersbach K, Sanchez MJ, et al. (2002) Hematopoietic stem cells localize to the endothelial cell layer in the midgestation mouse aorta. Immunity 16: 673–683.

51. Li Z, Chen MJ, Stacy T, Speck NA (2006) Runx1 function in hematopoiesis is required in cells that express Tek. Blood 107: 106–110.

52. de Bruijn MFTR, Speck NA, Peeters MCE, Dzierzak E (2000) Definitive hematopoietic stem cells first develop within the major arterial regions of the mouse embryo. EMBO Journal 19: 2465–2474.

53. Gekas C, Dieterlen-Lièvre F, Orkin SH, Mikkola H (2005) The placenta is a niche for hematopoietic stem cells. Dev Cell 8: 365–375.

54. Ottersbach K, Dzierzak E (2005) The murine placenta contains hematopoietic stem cells within the vascular labyrinth region. Dev Cell 8: 377–387.

55. Mu J, Adamson SL (2006) Developmental changes in hemodynamics of uterine artery, utero- and umbilicoplacental, and vitelline circulations in mouse throughout gestation. Am J Physiol Heart Circ Physiol 291: H1421–1428.

56. Eichmann A, Makinen T, Alitalo K (2005) Neural guidance molecules regulate vascular remodeling and vessel navigation. Genes Dev 19: 1013–1021.

57. Carmeliet P, Tessier-Lavigne M (2005) Common mechanisms of nerve and blood vessel wiring. Nature 436: 193–200.

58. Suchting S, Bicknell R, Eichmann A (2006) Neuronal clues to vascular guidance. Exp Cell Res 312: 668–675.

59. Jones C, Li DY (2007) Common cues regulate neural and vascular patterning. Curr Opin Genet Dev 17: 332–326.

60. Krebs LT, Shutter JR, Tanigaki K, Honjo T, Stark KL, et al. (2004) Haploinsufficient lethality and formation of arteriovenous malformations in notch pathway mutants. Genes Dev 18: 2469–2473.

61. Duarte A, Hirashima M, Benedito R, Trindade A, Diniz P, et al. (2004) Dosage-sensitive requirement for mouse dll4 in artery development. Genes Dev 18: 2474–2478.

62. Sorensen LK, Brooke BS, Li DY, Urness LD (2003) Loss of distinct arterial and venous boundaries in mice lacking endoglin and a vascular-specific TGF-β coreceptor. Dev Biol 261: 235–250.

63. Urness LD, Sorensen LK, Li DY (2000) Arteriovenous malformations in mice lacking activin receptor-like kinase-1. Nat Genet 26: 328–331.

64. Lawson ND, Weinstein BM (2002) Arteries and veins: making a difference with zebrafish. Nat Rev Genet 3: 674–682.

65. Childs S, Chen JN, Garrity DM, Fishman MC (2002) Patterning of angiogenesis in the zebrafish embryo. Development 129: 973–982.

66. Dale K, Pourquié O (2000) A clock-work somite. Bioessays 22: 72–83.

67. Nikolova G, Lammert E (2003) Interdependent development of blood vessels and organs. Cell Tissue Res 314: 33–42.

68. Zhong TP, Childs S, Leu JP, Fishman MC (2001) Gridlock signalling pathway fashions the first embryonic artery. Nature 414: 216–220.

69. Burke AC, Nelson CE, Morgan BA, Tabin C (1995) Hox genes and the evolution of vertebrate axial morphology. Development 121: 333–346.

70. Agarwal P, Wylie JN, Galceran J, Arkhitko O, Li C, et al. (2003) Tbx5 is essential for forelimb bud initiation following patterning of the limb field in the mouse embryo. Development 130: 623–633.

71. Kuan CY, Tannahill D, Cook GM, Keynes RJ (2004) Somite polarity and segmental patterning of the peripheral nervous system. Mech Dev 121: 1055–1068.

72. Adams RH, Wilkinson GA, Weiss C, Diella F, Gale N, et al. (1999) Roles of ephrinB ligands and EphB receptors in cardiovascular development: demarcation of arterial/venous domains, vascular morphogenesis, and sprouting angiogenesis. Genes Dev 13: 295–306.

73. Zhang XQ, Takakura N, Oike Y, Inada T, Gale N, et al. (2001) Stromal cells expressing ephrin-b2 promote the growth and sprouting of ephrin-b2(+) endothelial cells. Blood 98: 1028–1037.

74. Wang HU, Chen ZF, Anderson DJ (1998) Molecular distinction and angiogenic interaction between embryonic arteries and veins revealed by ephrin-b2 and its receptor eph-b4. Cell 93: 741–753.

75. Makinen T, Adams RH, Bailey J, Lu Q, Ziemiecki A, et al. (2005) Pdz interaction site in ephrinb2 is required for the remodeling of lymphatic vasculature. Genes Dev 19: 397–410.

76. Gu C, Yoshida Y, Livet J, Reimert DV, Mann F, et al. (2005) Semaphorin 3e and plexin-d1 control vascular pattern independently of neuropilins. Science 307: 265–268.

77. Kaufman MH, Bard JBL (1999) The Anatomical Basis of Mouse Development. Academic Press.

78. Krull CE (2001) Segmental organization of neural crest migration. Mech Dev 105: 37–45.

79. Sauka-Spengler T, Bronner-Fraser M (2006) Development and evolution of the migratory neural crest: a gene regulatory perspective. Curr Opin Genet Dev 16: 360–366.

80. Tucker RP, Hagios C, Chiquet-Ehrismann R, Lawler J, Hall RJ, et al. (1999) Thrombospondin-1 and neural crest cell migration. Dev Dyn 214: 312–322.

81. Hogan KA, Ambler CA, Chapman DL, Bautch VL (2004) The neural tube patterns vessels developmentally using the vegf signaling pathway. Development 131: 1503–1513.

82. Evans HM (1909) On the development of the aortae and cardinal and umbilical veins and and the other blood vessels of vertebrate embryos from capillaries. Anat Rec 3: 498–519.

83. Nagase T, Nagase M, Yoshimura K, Fujita T, Koshima I (2005) Angiogenesis within the developing mouse neural tube is dependent on sonic hedgehog signaling: possible roles of motor neurons. Genes Cells 10: 595–604.

84. Nakao T, Ishizawa A, Ogawa R (1988) Observations of vascularization in the spinal-cord of mouse embryos, with special reference to development of boundary membranes and perivascular spaces. Anat Rec 221: 663–677.

85. Gerhardt H, Ruhrberg C, Abramsson A, Fujisawa H, Shima D, et al. (2004) Neuropilin-1 is required for endothelial tip cell guidance in the developing central nervous system. Dev Dyn 231: 503–509.

86. Haigh JJ, Morelli PI, Gerhardt H, Haigh K, Tsien J, et al. (2003) Cortical and retinal defects caused by dosage-dependent reductions in vegf-a paracrine signaling. Dev Biol 262: 225–241.

87. Sato TN, Tozawa Y, Deutsch U, Wolburg-Buchholz K, Fujiwara Y, et al. (1995) Distinct roles of the receptor tyrosine kinases tie-1 and tie-2 in blood vessel formation. Nature 376: 70–74.

88. Ruhrberg C, Gerhardt H, Golding M, Watson R, Ioannidou S, et al. (2002) Spatially restricted patterning cues provided by heparin-binding VEGF-A control blood vessel branching morphogenesis. Genes Dev 16: 2684–2698.

89. Lu X, Le Noble F, Yuan L, Jiang Q, De Lafarge B, et al. (2004) The netrin receptor unc5b mediates guidance events controlling morphogenesis of the vascular system. Nature 432: 179–186.

90. Ambler CA, Nowicki JL, Burke AC, Bautch VL (2001) Assembly of trunk and limb blood vessels involves extensive migration and vasculogenesis of somite-derived angioblasts. Dev Biol 234: 352–364.

91. Wilting J, Brand-Saberi B, Huang R, Zhi Q, Kontges G, et al. (1995) Angiogenic potential of the avian somite. Dev Dyn 202: 165–171.

92. Pardanaud L, Luton D, Prigent M, Bourcheix L, Catala M, et al. (1996) Two distinct endothelial lineages in ontogeny, one of them related to hemopoiesis. Development 122: 1363–1371.

93. Pouget C, Gautier R, Teillet MA, Jaffredo T (2006) Somite-derived cells replace ventral aortic hemangioblasts and provide aortic smooth muscle cells of the trunk. Development 133: 1013–1022.

Nucleologenesis and Embryonic Genome Activation are Defective in Interspecies Cloned Embryos Between Bovine Ooplasm and Rhesus Monkey Somatic Cells

Bong-Seok Song, Sang-Hee Lee, Sun-Uk Kim, Ji-Su Kim,
Jung Sun Park, Cheol-Hee Kim, Kyu-Tae Chang, Yong-Mahn Han,
Kyung-Kwang Lee, Dong-Seok Lee and Deog-Bon Koo

ABSTRACT

Background

Interspecies somatic cell nuclear transfer (iSCNT) has been proposed as a tool to address basic developmental questions and to improve the feasibility of cell therapy. However, the low efficiency of iSCNT embryonic development is a crucial problem when compared to in vitro fertilization (IVF) and intraspecies SCNT. Thus, we examined the effect of donor cell species on the early development of SCNT embryos after reconstruction with bovine ooplasm.

Results

No apparent difference in cleavage rate was found among IVF, monkey-bovine (MB)-iSCNT, and bovine-bovine (BB)-SCNT embryos. However, MB-iSCNT embryos failed to develop beyond the 8- or 16-cell stages and lacked expression of the genes involved in embryonic genome activation (EGA) at the 8-cell stage. From ultrastructural observations made during the peri-EGA period using transmission electron microscopy (TEM), we found that the nucleoli of MB-iSCNT embryos were morphologically abnormal or arrested at the primary stage of nucleologenesis. Consistent with the TEM analysis, nucleolar component proteins, such as upstream binding transcription factor, fibrillarin, nucleolin, and nucleophosmin, showed decreased expression and were structurally disorganized in MB-iSCNT embryos compared to IVF and BB-SCNT embryos, as revealed by real-time PCR and immunofluorescence confocal laser scanning microscopy, respectively.

Conclusion

The down-regulation of housekeeping and imprinting genes, abnormal nucleolar morphology, and aberrant patterns of nucleolar proteins during EGA resulted in developmental failure in MB-iSCNT embryos. These results provide insight into the unresolved problems of early embryonic development in iSCNT embryos.

Background

The derivation of human embryonic stem cells (hESCs) from somatic cell nuclear transfer (SCNT) blastocysts represents an innovative strategy for overcoming immune rejection during transplantation. However, autologous human therapeutic cloning using human donor cells and oocytes has been continuously faced with legal and moral quandaries. Thus, monkey primary cells and bovine oocytes have been used as alternative donor and recipient cells for SCNT, respectively. In addition, interspecies SCNT (iSCNT) shows promise as a technique for examining nucleocytoplasmic interactions [1], stem cells [2], and the cloning of endangered animals whose oocytes are difficult to obtain [3,4]. The most important application of iSCNT lies in its potential to facilitate the reprogramming of human somatic cells into embryonic stem cells, thus avoiding ethical issues associated with using human oocytes. Therefore, iSCNT may increase the feasibility of human therapeutic cloning by providing comprehensive information about a variety of developmental events.

Many iSCNT embryonic studies have used bovine oocytes or oocytes from a variety of other species, such as pigs, rats, sheep, and monkeys [1,5-8]. The bovine oocyte is one of the most popular recipient cytoplasts for iSCNT because of the large number of oocytes that can be retrieved and because the in vitro culture system is well established. Although bovine oocytes support development beyond the morula stage in dogs [9], humans [10] and monkeys [6], the poor developmental efficiency of iSCNT embryos remains a crucial problem when compared to in vitro fertilization (IVF) and intraspecies SCNT techniques. Some studies have reported that high rates of abnormal iSCNT development may result from aberrant gene expression [5,11,12] or epigenetic modification by DNA methylation [2].

Among mammals, embryonic genome activation (EGA) is the most critical event for viability during early development [13]. EGA occurs at the 2-cell stage in mice [14], at the 8- to 16-cell stage in humans [15] and bovines [16], and at the 6- to 8-cell stage in monkeys [17]. It requires the expression of the housekeeping genes HSP70 (cell cycle regulation), PGK1, and PDHA1 (glucose metabolism) [18], as well as imprinting genes such as NDN (a transcription activator) and XIST (X chromosome × inactivator) [19]. In addition, the transcription of ribosomal RNA (rRNA) serves as a marker for EGA and coincides with a dramatic increase in nucleolar gene activation in mice [20], bovines [21], and pigs [22], resulting from the formation of functional nucleoli. When the inactive nucleolus, or nucleolar precursor body (NPB), is transformed into an active nucleolus, it consists of the innermost fibrillar centers (FCs) surrounded by dense fibrillar components (DFCs), which are bordered by granular components (GCs) [23]. The FCs contains rDNA transcriptional enzymes, such as RNA polymerase I and upstream binding transcription factor (UBTF). The DFC, which delivers pre-mature rRNA to the GC, contains fibrillarin. The GC includes nucleophosmin and nucleolin, which are associated with the processing of premature rRNA [23]. The various nucleolar proteins must be localized in a specific nucleolar region for the formation of a functional nucleolus [24].

Impaired nucleologenesis often coincides with failed early development in SCNT embryos. The bovine ooplasm successfully supports initial nucleolar assembly in embryos cloned from bovine and porcine cells [25], whereas delayed nucleolar assembly and decreased fibrillarin content were found in mouse [26] and monkey embryos [27]. However, little or no data have been gathered regarding the reprogramming of a monkey donor genome within ooplasms of different species, especially that of the bovine, although the iSCNT strategy facilitates monkey embryogenesis studies and their subsequent applications.

The aim of the present investigation was to understand why monkey-bovine (MB)-iSCNT embryos have poor developmental rates to the blastocyst stage compared to IVF and bovine-bovine (BB)-SCNT) embryos. The objectives were to: 1) compare the developmental competence of IVF, BB-SCNT, and MB-iSCNT embryos to the blastocyst stage, 2) investigate the expression of housekeeping and imprinting genes, 3) compare nucleolar ultrastructure by transmission electron microscopy (TEM), and 4) compare the expression of nucleolar component proteins among IVF, BB-SCNT, and MB-iSCNT embryos by immunofluorescence confocal laser scanning microscopy and real-time polymerase chain reaction (PCR).

Results

Development of IVF, BB-SCNT, and MB-iSCNT Embryos

Rhesus monkey ear fibroblast cells (Figure 1A) were fused with enucleated bovine oocytes (MB-iSCNT) using the same fusion parameters used for bovine fibroblasts (BB-SCNT). IVF bovine embryos were also cultured to control for the effect of the nuclear transfer (NT) procedure. Eight-cell stage IVF, BB-SCNT, and MB-iSCNT embryos were observed (Figure 1B) after 3 days of culture. PCR-based mitochondrial DNA (mtDNA) analysis was used to confirm that 8-cell-stage MB-iSCNT embryos had fused. PCR amplification of D-loop mtDNA was performed using species-specific primers. Monkey and bovine mtDNA were detected in 8-cell iSCNTs (Figure 1C). No difference in cleavage rate was observed among MB-iSCNT ($89.3 \pm 2.7\%$, 99/110), IVF (86.3 ± 1.3, 101/117), or BB-SCNT embryos ($85.3 \pm 2.5\%$, 83/99; Table 1). The developmental rates of IVF and BB-SCNT embryos were $33.5 \pm 2.8\%$ and $28.2 \pm 2.7\%$, respectively. However, the MB-iSCNT embryos did not develop into blastocysts (Table 1).

Table 1. Developmental capacity of embryos derived from IVF, BB-SCNT, and MB-iSCNT

Group	No. of embryos cultured	No. (%) of embryos developed to				
		Day 3				Day 7
		2-cell	4-cell	8- to 16-cell	Cleavage	Blastocyst
IVF	117	21 (17.9 ± 2.8)	37 (31.5 ± 2.7)	43 (36.9 ± 3.4)	101/117 (86.3 ± 1.3)	38 (33.5 ± 2.8)[a]
BB-SCNT	99	13 (14.1 ± 2.8)	13 (13.7 ± 2.1)	58 (57.4 ± 3.4)	83/99 (85.3 ± 2.5)	29 (28.2 ± 2.7)[a]
MB-iSCNT	110	12 (16.1 ± 4.1)	20 (22.3 ± 3.7)	67 (51.0 ± 5.1)	99/110 (89.3 ± 2.7)	0 (0)[b]

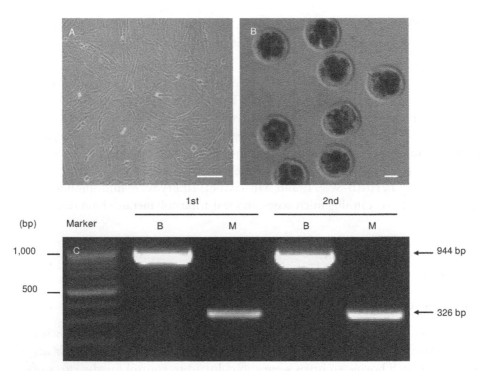

Figure 1. Analysis of mitochondrial DNA (mtDNA) in monkey-bovine MB-iSCNT embryos on day 3. Rhesus monkey ear skin fibroblasts (A), image of MB-iSCNT embryos at day 3 (B), and expression of mtDNA in MB-iSCNT embryos. Bar, 50 μm. B, bovine; M, rhesus monkey.

Down-Expression of Housekeeping and Imprinting Genes During EGA in MB-iSCNT Embryos

We used the terminal transferase dUTP nick-end labeling (TUNEL) assay to examine the frequency of apoptotic cells in 8- to 16-cell-stage MB-iSCNT embryos on day 3.5; however, no difference was observed compared to the IVF and BB-SCNT embryos (data not shown). Next, we determined whether pre-EGA (4-cell stage) and post-EGA (8-cell stage) MB-iSCNT, IVF, and BB-SCNT embryos expressed the HSP70, PDHA, and PGK1 housekeeping genes and the NDN and XIST imprinting genes. Interestingly, there was an increase in the expression of the three housekeeping and imprinting genes in the 8-cell-stage IVF and BB-SCNT embryos compared to their respective 4-cell stages, but the MB-iSCNT embryos only expressed the HSP70 housekeeping gene during the same period (Figure 2). These results suggest that the failure of MB-iSCNT embryos to develop into blastocysts was due to the down-regulation of housekeeping and imprinting gene transcription during EGA.

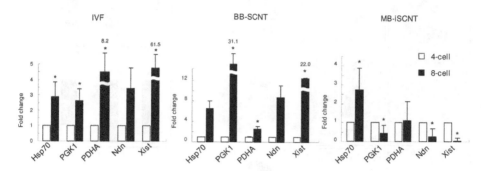

Figure 2. Real-time PCR quantification of mRNA expression in IVF, BB-SCNT, and MB-iSCNT embryos. Relative abundance of housekeeping and imprinting genes in IVF, BB-SCNT, and MB-iSCNT embryos between the 4- (before EGA) and 8-cell stages (after EGA). *Values are significantly different from the 4-cell control (P < 0.05).

Abnormal Nucleolar Morphology During EGA in MB-iSCNT Embryos

Given the abnormal housekeeping and imprinting gene expression during EGA in the MB-iSCNT embryos, we suspected abnormal transcription of the nucleus. Thus, we investigated the nucleolar ultrastructure of MB-iSCNT embryos during EGA and compared it to IVF and BB-SCNT embryos using TEM. We described the nucleologenic stages of the embryos according to Kopecny et al. [28] and evaluated the proportion of IVF, BB-SCNT, and MB-iSCNT embryos showing a given nucleolar stage during EGA. Control nuclei from IVF and BB-SCNT embryos contained NPBs with large and small vacuoles. Approximately 48% of the IVF and 60% of the BB-SCNT EGA-stage embryos were in stage 3 of nucleolar development (Figure 3B, top and middle, respectively), whereas few stage 3 nucleoli were observed among EGA-stage MB-iSCNT embryos (Figure 3A, a2, and 3A, b1). Interestingly, abnormal nucleolar structures, which were scattered near the nuclear membrane, were prevalent in MB-iSCNT embryos (68.2% of embryos; Figure 3A, c2, and 3B, bottom), indicating that they failed to develop normal nucleoli. Otherwise, the nuclear membrane, mitochondria, the Golgi apparatus, and other ultrastructural features were similar among all three types of embryos (data not shown).

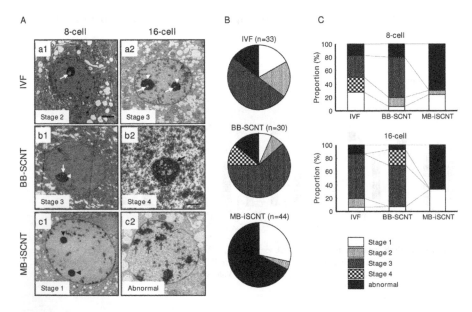

Figure 3. TEM evaluation of nucleolar developmental stages in IVF, BB-SCNT, and MB-iSCNT embryos. Primary vacuole (a1, white arrow) and both primary and secondary vacuoles (a2, white arrow head) in an IVF embryo. Both primary and secondary vacuoles (b1) and reticular nucleolus (b2, black arrow) in a BB-SCNT embryo. NPBs (c1, black arrow head) and abnormally structured nucleolus (c2, gray arrows) in an MB-iSCNT embryo. (B) Proportion of each nucleolar developmental stage in IVF, BB-SCNT, and MB-iSCNT embryos. (C) Proportion of each nucleolar development stage in both 8-cell and 16-cell stage embryos. Bar, 2 μm (A1, A2, B1, and C1); 1 μm (B2 and C2).

Given that nucleologenesis is related to the embryonic developmental stage [29], we investigated the difference in the proportions of nucleolar-stage IVF, BB-SCNT, and MB-iSCNT embryos from early (8–11 cells) to late (12–16 cells) EGA. As shown in Figure 3C, there was a greater proportion of stage 3 nucleoli (IVF, 33.3%; BB-SCNT, 60.0%) than stage 2 (IVF, 22.2%: BB-SCNT 13.3%) or stage 1 (IVF, 27.7%; BB-SCNT, 6.6%) nucleoli among early-EGA IVF and BB-SCNT embryos. Abnormal nucleoli were found in 16.6% of IVF and 20.0% of BB-SCNT embryos. Similar results were observed for late-EGA IVF and BB-SCNT embryos. Interestingly, stage 4 (26.6%) nucleoli were only observed in late-EGA BB-SCNT embryos. Conversely, the MB-iSCNT embryos exhibited stage 1 (early, 24.1%; late 33.3%) and stage 2 (early, 6.8%; late 0%) nucleoli, but showed very large numbers of abnormal nucleoli (early, 68.9%; late, 66.6%). These results suggest that bovine enucleated oocytes have a limited capacity for supporting nucleologenesis in embryos produced via nuclear transfer from rhesus monkey somatic cells.

Abnormal Expression and Disorganization of Nucleolar Component Proteins in MB-iSCNT Embryos

A variety of nucleolar component proteins must be expressed and localized in the appropriate nucleolar regions to form a functional nucleolus. However, our results indicate that failed early development in MB-iSCNT embryos is closely associated with abnormal nucleologenesis, as well as defects in EGA.

We investigated the expression and localization of nucleolar component proteins, including UBTF, fibrillarin, nucleophosmin, and nucleolin, in IVF, BB-SCNT, and MB-iSCNT embryos during EGA. UBTF is typically found in focal clusters in putative nucleoli [30]. Although UBTF was found in small focal clusters in interphase IVF and BB-SCNT cells (Figure 4C1 and 4C2), very few clusters were observed in MB-iSCNT embryos (Figure 4C3) and some nuclei did not express UBTF (data not shown). In previous studies, fibrillarin was localized in clusters of intensely labeled foci, and nucleolin and nucleophosmin were found in a shell-like structure that appeared as a ring-like image [29]. Unlike IVF and BB-SCNT embryos, fibrillarin was detected in some blastomeres derived from MB-iSCNT embryos (Figure 5B1, B2, and 5B3). However, fibrillarin expression was reduced in MB-iSCNT embryos and it was spot-shaped (Figure 5C3). Nucleophosmin and nucleolin were co-localized in the nucleolus in almost all IVF and BB-SCNT embryos and appeared as shell-like structures (Figure 6C1 and 6C2), whereas nucleophosmin and nucleolin expression levels were lower in MB-iSCNT embryos and were spot-shaped (Figure 6C3).

We quantified the expression of these proteins in individual blastomeres from IVF, BB-SCNT, and MB-iSCNT embryos in early (8–11 cells) and late EGA (12–16 cells). Although we observed significantly lower expression in early- and late-EGA stage MB-iSCNT blastomeres, we found similar protein expression rates in early and late IVF and BB-SCNT blastomeres (Figure 7A and 7B). Immunofluorescence confocal laser scanning microscopy results indicated that UBTF, fibrillarin, nucleophosmin, and nucleolin expression levels were significantly reduced in MB-iSCNT embryos compared to IVF and BB-SCNT embryos (Figures 4, 5 and 6). This may suggest that the abnormal expression and disorganization of nucleolar component proteins resulted in the formation of abnormal nucleoli in MB-iSCNT embryos.

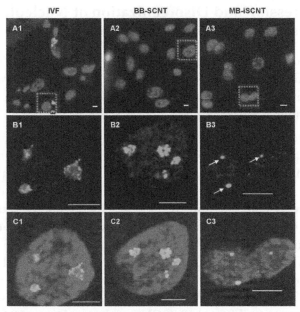

Figure 4. Expression and localization of upstream binding transcription factor (UBTF) during embryonic genome activation (EGA). IVF (A1, B1, and C1) and BB-SCNT (A2, B2, and C2) embryos showed clusters of foci, whereas MB-iSCNT embryos showed only small foci (A3, B3, and C3). DNA was stained with DAPI. Arrow indicates small UBTF foci in an MB-iSCNT embryo. Bar, 10 μm.

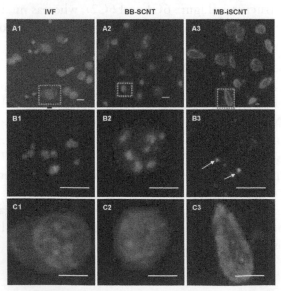

Figure 5. Expression and localization of fibrillarin proteins during embryonic genome activation (EGA). IVF (A1, B1, and C1), BB-SCNT (A2, B2, and C2), and MB-iSCNT (A3, B3, and C3) embryos were immunostained with antibodies specific to fibrillarin. DNA was stained with DAPI. Arrow indicates small fibrillarin foci in an MB-iSCNT embryo. Bar, 10 μm.

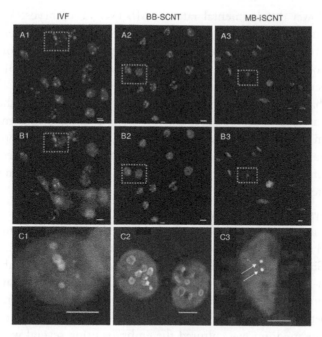

Figure 6. Expression and localization of the nucleophosmin and nucleolin proteins during embryonic genome activation (EGA). IVF (A1, B1, and C1), BB-SCNT (A2, B2, and C2), and MB-iSCNT (A3, B3, and C3) embryos were immunostained with antibodies specific to nucleophosmin and nucleolin. DNA was stained with DAPI. Arrow indicates small foci of nucleophosmin and nucleolin staining in an MB-iSCNT embryo. Bar, 10 μm.

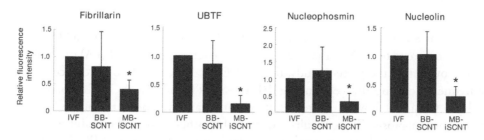

Figure 7. Evaluation of nucleolar protein expression in embryos derived from IVF, BB-SCNT, and MB-iSCNT. The expression of UBTF, fibrillarin, nucleophosmin, and nucleolin were relative to the mean expressed cells to the total nuclei in early (A) and late (B) EGA-stage embryos. Early EGA, 8- to 11-cell stage; late EGA, 12- to 16-cell stage. *Values are significantly different from the other groups (P < 0.05).

Discussion

Our results demonstrate that the abnormal expression and disorganization of nucleolar component proteins during EGA results in abnormal nucleolar structure, which leads to lower developmental rates in MB-iSCNT embryos. When the developmental rates of IVF, BB-SCNT, and MB-iSCNT embryos were examined, the cleavage rates of MB-iSCNT embryos, which received monkey fibroblast nuclei and bovine oocyte cytoplasm, were similar to those of IVF and BB-SCNT embryos; however, they failed to develop to the blastocyst stage (Table 1). These results suggest that MB-iSCNT embryos could not proceed through the transition from maternal to embryonic development. These findings also demonstrate that the ooplasm at the second metaphase, and not the nucleus, is principally responsible for early embryonic development prior to EGA, a finding consistent with results reported in iSCNT mice [12] and cats [31] generated using bovine enucleated oocytes.

The culture system is a very important factor affecting the efficiency of nuclear transfer. Non-human primate iSCNT embryos derived from donor human adult somatic cells and recipient bovine oocytes have been successfully cultured in IVC medium [1,3]. Similarly, Park et al. [12] reported that MB-iSCNT embryos developed to the 8-cell stage in CR1-aa medium, even though EGA occurs at the 2-cell stage in mouse embryos. Therefore, to identify whether the limited developmental competency of the MB-iSCNT embryos described here is related to the type of culture medium, we cultured the embryos using several well-known culture media, including CR1-aa, IVC-1, IVC-3, G-1, G-2, complete early cleavage, and complete blastocyst medium. However, no culture medium was capable of supporting the early embryonic development of MB-iSCNT embryos. These results suggest that the developmental competency of iSCNT embryos may be more dependent upon the donor species than the recipient species or the culture medium used.

A recent study showed that 20% of MB-iSCNT embryos exhibited DNA damage compared to SCNTs that were parthenogenetically activated (PA) on day 7 [8]. However, in the present study, the TUNEL assay revealed no difference in the frequency of apoptotic cells among IVF, BB-SCNT, and MB-iSCNT embryos on day 3 (data not shown). Therefore, it appears that developmental failure was not due to apoptosis. These results led us to hypothesize that the inability to progress through EGA was responsible for developmental failure in MB-iSCNT embryos. EGA is associated with transcriptional activation [32] and its timing is species-specific [14-16]. A failure to regulate the timing of activation or the types of genes expressed can lead to developmental arrest during preimplantation [13], and previous studies have shown that MB-iSCNT embryos undergo developmental arrest due to failed gene expression at the appropriate time [12]. In agreement with previous studies, we found an increase in the expression of HSP70, PGK1, PDHA, Ndn, and Xist in 8-cell-stage IVF and BB-SCNT embryos (Figure 2) [18,33]; in contrast, PDHA, PGK1, Ndn, and Xist expression decreased in

MB-iSCNT embryos, indicating that blastocyst development did not occur in the 8-cell stage MB-iSCNT embryos due to a failure to complete EGA.

It is generally accepted that the transcriptional activity of the nucleus is associated with a change from a compact nucleolus (stages 1 and 2) to a functional nucleolus (stages 3 and 4), and that greatly increased transcription is indicative of EGA [34]. We observed stage 1 and 2 nucleoli during early EGA in MB-iSCNT embryos, but stage 3 nucleoli were sparsely observed in early or late EGA; furthermore, these embryos showed irregularly shaped NPBs (Figure 3). This finding coincided with those observed when transcription is inhibited by alpha-amanitin, which causes the disintegration of NPBs and embryonic developmental arrest [35]. Conversely, a greater proportion of stage 3 nucleoli were also observed in early-EGA BB-SCNT embryos than in early-EGA IVF embryos, and stage 4 nucleoli were only observed in late-EGA BB-SCNTs. We assumed that BB-SCNTs developed to the nucleolar stage at a faster rate than the IVF embryos, as in previous studies [36]. In addition, previous studies have reported that the nucleoli in late-EGA, 8-cell-stage IVF-derived embryos have nucleolar vacuoles, whereas 4-cell stage BB-SCNT embryos display both small and large vacuoles [21,37]. Interestingly, we previously observed that normal blastocyst development in goat-bovine iSCNT embryos was associated with functional changes in nucleoli (unpublished data), and development was similar to that of goat-goat SCNT embryos [34]. From an ultrastructural viewpoint, it is clear that MB-iSCNT embryos were unable to form a functional nucleolus during EGA. These results suggest that functional nucleoli are critical for the normal development of iSCNT embryos and that successful development may be donor species-dependant.

We demonstrated that developmental arrest in MB-iSCNT embryos may be caused by aberrant expression of nucleolar proteins during early embryonic development. Numerous additional studies have shown that abnormal nucleologenesis is observed during the development of iSCNT embryos [34]. However, until now, no direct evidence was available as to whether this nucleolar defect is a cause or a result of developmental failure in iSCNT embryos.

It was recently suggested that the aberrant expression of nucleolar proteins causes abnormal nucleologenesis in SCNT embryos [37], which leads to failed EGA and developmental arrest [35]. In particular, the expression of nucleolar proteins is frequently dependent upon epigenetic modifications such as genomic methylation. We attempted to determine whether the reduced nucleolar protein expression observed in MB-iSCNT embryos is due to epigenetic modification. However, no difference was observed in the methyl profiles of CpG islands in fibrillarin promoters between monkey donor cells and MB-iSCNT embryos (data not shown). Given these findings, we propose two possible hypotheses or mechanisms regarding the reduced expression of nucleolar proteins such as fibrillarin.

One possible explanation is differences in histone modification, which were not examined in this study. Alternatively, it is possible that the bovine oocyte-derived transcriptional machinery was unable to express the monkey fibrillarin gene. For example, successful embryonic or full-term development of iSCNT embryos may depend on the accurate spatio-temporal action of donor cell species-specific transcription factors. To investigate these hypotheses, we are currently developing rescue strategies, including a lentiviral vector system expressing same-species transcription factors.

A functioning nucleolus requires numerous components [24,38], but the location of these components is not completely understood [23]. It is, however, well known that various nucleolar component proteins such as UBTF, fibrillarin, nucleophosmin, and nucleolin must be expressed and localized in the appropriate nucleolar region to form a functional nucleolus. UBTF is detected in first-cell cycle zygotic embryos, it is localized to small spherical bodies, and it is one of several transcription factors required for the binding of RNA polymerase I to rDNA in IVF embryos [39]. Fibrillarin is localized to the FCs and dense DFCs of nucleoli [29] and is associated with rRNA modification [40,41], ribosome assembly [42], nucleolar assembly [43], and early embryonic development [44], an indicator of EGA [45]. Nucleolin is a phosphorylated protein present in large amounts in nucleoli during active ribosomal biogenesis [46]; it is localized to DFCs and GCs of nucleoli [47] and plays essential roles in rDNA transcription, rRNA maturation, ribosome assembly, nucleocytoplasmic transport, and nucleologenesis [24,48]. Nucleophosmin is involved in the shuttling of proteins into the nucleolus [49]. As shown in Figures 3, 4, 5 and 6, the initial targeting of nucleolar proteins appears to be normal; however, the reduced expression of nucleolar protein seems to be associated with impairment in further nucleologenesis during EGA. Although we could not address the relationship between nucleolar protein expression and nucleologenesis, our findings suggest that impaired nucleologenesis is a major hurdle in iSCNT technology.

We examined the expression patterns of nucleolar component proteins and the localization of functional nucleoli in MB-iSCNT embryos. As in previous studies [21,37], UBTF and fibrillarin were displayed as clusters of small foci in the putative nucleolus of 8-cell-stage IVF and BB-SCNT embryos (Figures 4 and 5), and nucleolin and nucleophosmin appeared as ring-shaped structures at EGA (Figure 6); these results are consistent with a previous study [29]. However, in MB-iSCNT embryos, UBTF, fibrillarin, nucleolin, and nucleophosmin were sporadically detected as small foci (Figures 4, 5 and 6). These results indicate that the dysregulation of these nucleolar proteins may have led to nonfunctional nucleoli, which might be separated into DFCs and GCs, and the down-regulation of housekeeping and imprinting genes during EGA. As the result, MB-iSCNT embryos exhibited developmental arrest.

Finally, we observed significantly lower expression of these nucleolar proteins in individual MB-iSCNT blastomeres than in IVF or BB-SCNT blastomeres (Figure 7). These results suggest that it is important for MB-iSCNT embryos to continuously maintain the expression of nucleolar proteins during EGA. Our future studies will use an adenoviral vector to examine whether an increase in nucleolar component proteins during the first three cleavage cycles overcomes nucleolar dysfunction and allows development to reach the morula or blastocyst stage.

Conclusion

MB-iSCNT embryos derived from donor monkey fibroblasts and bovine recipient oocytes did not develop to the blastocyst stage. We determined that this failure was caused by the down-regulation of EGA, and that it was not related to apoptosis. Instead, impaired nucleologenesis and aberrant nucleolar formation in MB-iSCNT embryos lead to developmental arrest. Abnormal expression and disorganization of the nucleolar component proteins also resulted in the down-regulation of EGA in MB-iSCNT embryos. These results provide insight into the early stages of development in iSCNT embryos, useful model for understanding nucleoar biology and will assist in the development of techniques to resolve these issues in iSCNT technology and spur the development of further applications.

Methods

Chemicals

Chemicals were purchased from Sigma Chemical Co. (St. Louis, MO, USA) unless otherwise indicated.

In Vitro Maturation (IVM) and IVF

Bovine ovaries were collected from a local slaughterhouse and transported to the laboratory in 0.9% saline at 25–30°C. Cumulus-oocyte complexes (COCs) were aspirated from follicles (2–6 mm in diameter) using a disposable 10-ml syringe with an 18-gauge needle. Aspirated COCs with at least three layers of compact cumulus cells and homogeneous cytoplasm were washed three times in TL-HEP-ES (1 mg/ml bovine serum albumin [BSA] and low-carbonate TALP [50]). Ten oocytes were matured in 50 µl of the in vitro maturation medium in a 60-mm dish (Nunc, Roskilde, Denmark) under mineral oil for 20–22 h at 38.5°C under an atmosphere of 5% CO_2 in air. The medium used for oocyte maturation was TCM-199 (Gibco-BRL, Grand Island, NY, USA) supplemented with 10% (v/v) fetal bovine serum (FBS; Gibco-BRL), 10 IU/ml pregnant mare's serum gonadotropin (PMSG), 0.6 mM cysteine, 0.2 mM sodium pyruvate, and 1 µg/ml 17β-estradiol. Following IVM, 15 oocytes were fertilized in 50 µl of fertilization medium with frozen-thawed sperm at a concentration of 2 × 106 cells/ml. When sperm were added to the fertilization drops, 2 µg/ml heparin, 20 µM penicillamine, 10 µM hypotaurine, and 1 µM epinephrine (PHE) were also added. After 18 h, cumulus-enclosed oocytes were stripped using gentle pipetting and transferred to CR1-aa medium containing 0.3% BSA for in vitro culture [51].

Culture of Donor Cells

All of the experimental and animal care protocols were conducted in accordance with the Korea Research Institute of Bioscience and Biotechnology (KRIBB) Guidelines for the Care and Use of Laboratory Animals. Cell culture and assessment procedures have been previously described [33]. We used bovine ear skin fibroblast (bESF) cells and monkey ear skin fibroblast (mESF) cells as donor cells for intra- and interspecies nuclear transfer, respectively. To prepare primary bESF and mESF cells, bovine and monkey ear skins were manually cut into small pieces of approximately 1 cm2 and chopped with a surgical blade on a 100-mm culture dish. The chopped tissue was incubated in 10 ml of 0.25% (w/v) trypsin/3.65 mM EDTA solution (Gibco-BRL) at 37°C for 30 min. The trypsin was inactivated by adding an equal volume of growth medium (Dulbecco's modified Eagle's medium [DMEM, Gibco-BRL] supplemented with 10% FBS). After removal of cellular debris and undigested cell masses, the cells were re-suspended in growth medium, seeded into 100-mm cell culture dishes, and cultured at 38.5°C under 5% CO_2 in air for approximately 2 weeks until confluent. The fibroblasts were passaged three times before use as a source of donor nuclei for intra- or interspecies nuclear transfer. Cells were frozen in DMEM with 10% FBS and 10% dimethylsulfoxide (DMSO) and stored in liquid nitrogen until use.

Somatic Cell Nuclear Transfer and In Vitro Culture (IVC)

SCNT was performed using bESF and mESF cells as donor cells. Donor cells were plated in six-well plates and cultured in DMEM (Gibco-BRL) with 10% FBS until confluent. Donor cells were washed with phosphate-buffered saline (PBS, Gibco-BRL), digested with 0.25% trypsin-EDTA for 3 min, and then washed with DMEM containing 10% FBS. The cells were centrifuged at 150 × g for 2 min and re-suspended in PBS. Mature oocytes were enucleated with a glass pipette by aspirating the first polar body and MII plate of the partial cytoplasm in TL-HEPES containing 7.5 µg/ml cytochalasin B. The nuclei were stained with Hoechst-33342 and aspirated cytoplasm was viewed under ultraviolet light to confirm the removal of nuclei. A single cell was injected into the peri-vitelline space of the enucleated oocyte cytoplast. The reconstructed embryos were fused using a fusion chamber with two stainless steel electrodes (1 mm apart) in a fusion medium consisting of 0.3 M mannitol, 0.5 mM HEPES, 0.3% BSA, 0.1 mM CaCl2, and 0.1 mM MgCl2. A single direct current pulse of 1.6 kV/cm for 20 µs was applied using an Electro Cell Manipulator 2001 (BTX, San Diego, CA, USA). The fused embryos were activated using a modification of a previously described method [52]. Two hours after electrofusion, the fused embryos were activated with 5 µM ionomycin for 5 min; they were then treated with 2.5 mM 6-dimethyl-aminopurine (6-DMAP) in CR1-aa medium containing 0.3% BSA and incubated at 38.5°C under 5% CO_2 in air for 4 h. In vitro-fertilized reconstructed embryos were transferred to CR1-aa medium containing 0.3% BSA and incubated at 38.5°C under 5% CO_2 in air for 3 d. On day 3 of culture, the cleaved embryos were collected and the cleavage rate was evaluated. The embryos were then transferred to CR1-aa containing 10% FBS and cultured for an additional 4 days (evaluation of developmental rates at this time).

Preparation of Medium and Culture of MB-iSCNT Embryos

Some of the iSCNT embryos were cultured in CR1-aa containing 0.3% BSA and 10% FBS, whereas other reconstructed iSCNT embryos were cultured in the media indicated below. The following in vitro culture media were used only for MB-iSCNT embryo culture: IVC-1 and IVC-2 (IVF media series, InVitrocare Inc., Frederick, MD) supplemented with 10% human serum albumin; G-1 and G-2 media (Vitrolife AB, Kungsbacka, Sweden) containing human serum albumin; and complete early cleavage and complete blastocyst medium (Irvine Scientific, Santa Ana, CA) supplemented with 10% serum substitute supplement (SSS). Each medium was used in one of two steps in the culture process (early embryo development or blastocyst formation). We used the early development medium to support cleavage of MB-iSCNT embryos for 3 days and the blastocyst formation medium to support the already cleaved MB-iSCNT embryos for 4 days.

Analysis of Species-Specific mtDNA

Amplification of D-loop mtDNA was used to confirm the fusion of MB-iSCNT embryos. We analyzed each 8-cell MB-iSCNT embryo via PCR using specific

primers for monkey and bovine mtDNA. The primer sequences for each mtDNA
gene were: monkey, (GenBank™ accession number, AY612638), 5'-TAT TGC
ATA AGC TTC ATA AAT AAC TCT AGC-3' (sense), 5'-TTA TTT AAT AGA
TAT GTG CTA TGT CCG ATG-3' (antisense); bovine (GenBank™ accession
number, NC006853), 5'-AAA TGT AAA ACG ACG ACG GCC AGT AAT
CCC AAT AAC TCA ACA C-3' (sense), 5'-AAA CAG GAA ACA GCT ATG
ACC ACT CAT CTA GGC ATT TTC-3' (antisense). PCR was conducted with
an initial step of 94°C for 10 min and 30 cycles of 94°C for 40 s, 60°C for 40 s,
and 72°C for 45 s using the primer for the D-loop region of monkey mtDNA;
when using the primer for the D-loop region of bovine mtDNA, PCR was performed with an initial step of 94°C for 10 min and 30 cycles of 94°C for 40 s,
55°C for 40 s, and 72°C for 45 s. The final 326- and 944-bp products were detected by agarose gel electrophoresis.

Evaluation of Zygotic Gene Expression in 4- and 8-cell Embryos

We isolated the mRNA of 4- and 8-cell IVF, BB-SCNT, and MB-iSCNT embryos. Poly(A) mRNAs were extracted using a Dynabeads mRNA Direct kit
(DYNAL), according to the manufacturer's instructions. After thawing, the
samples were lysed in 300 μl of lysis/binding buffer (DYNAL) at room temperature for 10 min. Dynabeads oligo(dT) 25 (10 μl) were added to each sample.
The beads were hybridized for 5 min and separated from the binding buffer
using a Dynal magnetic bar. The poly(A) mRNAs and beads were washed in
buffers A and B (DYNAL) and separated by adding 11 μl of diethylpyrocarbonate (DEPC)-treated water. The poly(A) mRNAs were reverse-transcribed
in a total volume of 20 μl containing 500 μg/ml oligo(dT) primer, 10× PCR
buffer, 20 IU RNase inhibitor, 200 U SuperScript II Reverse Transcriptase
(Invitrogen, Madrid, Spain), 15 mM MgCl2, and 1 μl of dNTP mix (10 mM
each). The secondary RNA structure was denatured at 65°C for 5 min; then,
the cDNA was maintained at room temperature for 10 min and at 42°C for
60 min to allow reverse transcription. The reaction was terminated by heating
at 70°C for 15 min. The inactivated cDNA was used as a template for PCR
amplification. We used a previously described primer to detect monkey transcripts in the IVF, BB-SCNT, and MB-iSCNT embryos [18] and designed a
bovine primer using the Primer3 program http://bioinfo.ebc.ee/mprimer3.
An ABI 7500 Fast Real Time PCR System (Applied Biosystems, Inc., Foster
City, CA, USA) and SyberGreen PCR Core reagents (Applied Biosystems)
were used for RT-PCR. The primers are listed in Table 2. The Hprt1 gene level
was used as the endogenous reference for each group.

Table 2. Primer sequences for real-time PCR analysis

Gene	Species	Sequence	GenBank accession no.	Size (bp)
HSP70	B	F:CCAGAGGAGGTGTCATCCAT R:GGGTGCTGGAAGAGAGAGTG	NM_174345	490
	M	F:TATTGGAGCCAGGCCTACAC R:GTCCGTAAAGGCGACATAGC	AF352832	168
PGKI	B	F:CTGCTGTTCCAAGCATCAAA R:GCACAAGCCTTCTCCACTTC	BC102308	202
	M	F:GTTGCACAGCATCTCAGCTC R:TCACTTGGTTTTAACAGGCAAA	AB125189	140
PDHAI	B	F:ATCCTCTGTCGTCCCCTTCT R:CTTAGACTGCAAGGCGATCC	XM_581602	187
	M	F:TGTCACAACAGTGCTCACCA R:CAAGCTTCCTGACCATCACA	AB083322	147
HPRTI	B	F:TGGCTCGAGATGTGATGAAG R:ACACTTCGAGGGGTCCTTTT	NM_001034035	370
	M	F:TTATACCACCGTGTGTTAGAAAAG R:ACACTACTAAAATAATTCCAGGACAGA	M31642	100
NDN	B	F:TCGCCAAGAATAGTGTGCTG R:TGAGTGGAAGAGCTGTGGTG	BC146188	110
	M	F:GACGAGGACGACCCGAAG R:ACTGGAGAGGTGGAATGTG	AB172756	149
XIST	B	F:TGCCACGCCTACAGTTAGTG R:GGGTTTTTCCCAGGTTGATT	BC146188	168
	M	F:TTACAGCAGGGGGTACTTGG R:AGGGAAGTGAGTGGGGTCTT	NR_001564	200

Processing for TEM

Embryos at the 8- and 16-cell stages (84 h after insemination or after activation in the nuclear transfer group) were used for the TEM study. Cultured embryos were fixed with 3% glutaraldehyde in culture medium for 2 h at room temperature. They were then washed five times with 0.1 M cacodylate buffer containing 0.1% $CaCl_2$ at 4°C and post-fixed for 2 h at 4°C with 1% OsO_4 in 0.1 M cacodylate buffer (pH 7.2) containing 0.1% $CaCl_2$. The embryos were rinsed with cold distilled water, transferred to micro-centrifuge tubes at 4°C, collected by centrifugation, embedded in 1% ultra-low gelling temperature agarose (type IX), slowly dehydrated in a graded ethanol series and propylene oxide at 4°C, and then finally embedded in Spurr's epoxy resin [53]. The resin polymerized after 36 h at 70°C and serial sections were cut with a diamond knife on an ULTRACUT ultramicrotome (Leica, Austria) and mounted on formvar-coated slot grids. Sections were stained with 4% uranyl acetate for 10 min and lead citrate [54] for 7 min. They were observed with a Tecnai G2 Spirit Twin transmission electron microscope (FEI Company, USA) and a JEM ARM 1300S high-voltage electron microscope (JEOL, Japan).

Assessment of nucleolar developmental stages

Nucleolar structure was evaluated for developmental stages as described by Kopecny et al. [28], as follows: Stage 1, non-vacuolated NPB is an almost homogeneous fibrillar structure with densely packed fibrils; Stage 2, vacuolated NPB contains an eccentric center vacuole; Stage 3, nucleolus with secondary vacuoles;

Stage 4, fully reticulated nucleolus in which the NPB has been transformed into a functional, rRNA synthesizing nucleolus. In the NPB, primary and secondary vacuoles were displayed.

Assessment of Nucleolar Protein Expression by Immunocytochemistry

After assessing cleavage on day 3, the three types of embryos (8- to 16-cell stages) were fixed. The primary antibodies were: mouse monoclonal anti-UBTF (1:50, H00007343-M01; Abnova, Walnut, CA, USA), rabbit polyclonal anti-fibrillarin (1:50, sc-25397; Santa Cruz Biotechnology, Santa Cruz, CA, USA), mouse monoclonal anti-nucleophosmin (1:50, ab10530; Abcam, Cambridge, UK), rabbit polyclonal anti-nucleolin (1:50, ab16940; Abcam). The IVF, BB-SCNT, and MB-iSCNT embryos were fixed in 4% formaldehyde in PBS for 1 h at 4°C. The embryos and cells were washed for 30 min in PBS containing 0.1% Tween 20 (PBST) and then permeabilized for 1 h at room temperature in PBS-PVA containing 0.5% Triton X-100. The samples were treated with 2% BSA in PBST overnight at 4°C. The primary antibody was diluted 1:50 in PBST and co-incubated for 6 h at 4°C. After washing for 1 h, the samples were incubated for 30 min with Alexa 594 anti-rabbit IgG and FITC-conjugated anti-mouse IgM, and then washed for an additional hour. Samples were mounted on slides with mounting medium containing 1.5 µg/ml 4,6-diamidino-2-phenylindole (VECTASHIELD with DAPI; Vector Laboratories, Servion, Switzerland). The samples were viewed under a Zeiss AxioVert 200 M microscope with ApoTome (Zeiss, Oberkochen, Germany). The method to assess nucleolar protein expression was previously described [55]. We classified embryos as early stage (8–11 cells) or late stage (12–16 cells) and determined the proportion of nucleolar protein expressed in blastomeres to the total number of nuclei of embryos.

Statistical Analyses

All experiments were replicated more than three times. Data are presented as the mean ± standard error (SE) of the cultured oocytes. The data were analyzed via analysis of variance (ANOVA) followed by Duncan's multiple range test using the SAS software package (SAS Institute, Inc., Cary, NC, USA). $P < 0.05$ was considered statistically significant.

Authors' Contributions

BSS performed all experiments. SHL was responsible for analyzing TEM data. SUK, BSS, JSK, and JSP performed interspecies SCNT. CHK, KTC, YMH, and

KKL provided animal care services (specifically monkey breeding), provided rhesus monkey ear skin fibroblasts, performed bovine oocyte culture, and participated in manuscript preparation. DSL and DBK initiated and directed this study, participated in the analysis of nucleolar protein expression via immunocytochemistry, and wrote most of the manuscript. All authors read and approved the final manuscript.

Acknowledgements

This study was supported by grants (2006-04082 and R01-2008-000-21076-0) from KOSEF, Ministry of Science and Technology, Republic of Korea, and a grant (Code# 20070401034015) from the BioGreen21 Program, Rural Development Administration, Republic of Korea.

References

1. Illmensee K, Levanduski M, Zavos PM: Evaluation of the embryonic preimplantation potential of human adult somatic cells via an embryo interspecies bioassay using bovine oocytes. Fertil Steril 2006, 85(Suppl 1):1248–60.

2. Chen T, Zhang YL, Jiang Y, Liu JH, Schatten H, Chen DY, Sun QY: Interspecies nuclear transfer reveals that demethylation of specific repetitive sequences is determined by recipient ooplasm but not by donor intrinsic property in cloned embryos. Mol Reprod Dev 2006, 73:313–7.

3. Chang KH, Lim JM, Kang SK, Lee BC, Moon SY, Hwang WS: An optimized protocol of a human-to-cattle interspecies somatic cell nuclear transfer. Fertil Steril 2004, 82:960–2.

4. Zhao ZJ, Ouyang YC, Nan CL, Lei ZL, Song XF, Sun QY, Chen DY: Rabbit oocyte cytoplasm supports development of nuclear transfer embryos derived from the somatic cells of the camel and Tibetan antelope. J Reprod Dev 2006, 52:449–59.

5. Arat S, Rzucidlo SJ, Stice SL: Gene expression and in vitro development of inter-species nuclear transfer embryos. Mol Reprod Dev 2003, 66:334–42.

6. Dominko T, Mitalipova M, Haley B, Beyhan Z, Memili E, McKusick B, First NL: Bovine oocyte cytoplasm supports development of embryos produced by nuclear transfer of somatic cell nuclei from various mammalian species. Biol Reprod 1999, 60:1496–502.

7. Kitiyanant Y, Saikhun J, Chaisalee B, White KL, Pavasuthipaisit K: Somatic cell cloning in Buffalo (Bubalus bubalis): effects of interspecies cytoplasmic recipients and activation procedures. Cloning Stem Cells 2001, 3:97–104.

8. Lorthongpanich C, Laowtammathron C, Chan AW, Ketudat-Cairns M, Parnpai R: Development of Interspecies Cloned Monkey Embryos Reconstructed with Bovine Enucleated Oocytes. J Reprod Dev 2008, 54:306–13.

9. Murakami M, Otoi T, Wongsrikeao P, Agung B, Sambuu R, Suzuki T: Development of interspecies cloned embryos in yak and dog. Cloning Stem Cells 2005, 7:77–81.

10. Chang KH, Lim JM, Kang SK, Lee BC, Moon SY, Hwang WS: Blastocyst formation, karyotype, and mitochondrial DNA of interspecies embryos derived from nuclear transfer of human cord fibroblasts into enucleated bovine oocytes. Fertil Steril 2003, 80:1380–7.

11. Dindot SV, Farin PW, Farin CE, Romano J, Walker S, Long C, Piedrahita JA: Epigenetic and genomic imprinting analysis in nuclear transfer derived Bos gaurus/Bos taurus hybrid fetuses. Biol Reprod 2004, 71:470–8.

12. Park SH, Shin MR, Kim NH: Bovine oocyte cytoplasm supports nuclear remodeling but not reprogramming of murine fibroblast cells. Mol Reprod Dev 2004, 68:25–34.

13. Meirelles FV, Caetano AR, Watanabe YF, Ripamonte P, Carambula SF, Merighe GK, Garcia SM: Genome activation and developmental block in bovine embryos. Anim Reprod Sci 2004, 82–83:13–20.

14. Kidder GM, McLachlin JR: Timing of transcription and protein synthesis underlying morphogenesis in preimplantation mouse embryos. Dev Biol 1985, 112:265–75.

15. Tesarik J, Kopecny V, Plachot M, Mandelbaum J: High-resolution autoradiographic localization of DNA-containing sites and RNA synthesis in developing nucleoli of human preimplantation embryos: a new concept of embryonic nucleologenesis. Development 1987, 101:777–91.

16. Camous S, Kopecny V, Flechon JE: Autoradiographic detection of the earliest stage of [3H]-uridine incorporation into the cow embryo. Biol Cell 1986, 58:195–200.

17. Schramm RD, Bavister BD: Onset of nucleolar and extranucleolar transcription and expression of fibrillarin in macaque embryos developing in vitro. Biol Reprod 1999, 60:721–8.

18. Zheng P, Patel B, McMenamin M, Reddy SE, Paprocki AM, Schramm RD, Latham KE: The primate embryo gene expression resource: a novel resource to facilitate rapid analysis of gene expression patterns in non-human primate oocytes and preimplantation stage embryos. Biol Reprod 2004, 70:1411–8.

19. Wen DC, Bi CM, Xu Y, Yang CX, Zhu ZY, Sun QY, Chen DY: Hybrid embryos produced by transferring panda or cat somatic nuclei into rabbit MII

oocytes can develop to blastocyst in vitro. J Exp Zoolog A Comp Exp Biol 2005, 303:689–97.

20. Geuskens M, Alexandre H: Ultrastructural and autoradiographic studies of nucleolar development and rDNA transcription in preimplantation mouse embryos. Cell Differ 1984, 14:125–34.

21. Laurincik J, Schmoll F, Mahabir E, Schneider H, Stojkovic M, Zakhartchenko V, Prelle K, Hendrixen PJ, Voss PL, Moeszlacher GG, et al.: Nucleolar proteins and ultrastructure in bovine in vivo developed, in vitro produced, and parthenogenetic cleavage-stage embryos. Mol Reprod Dev 2003, 65:73–85.

22. Tomanek M, Kopecny V, Kanka J: Genome reactivation in developing early pig embryos: an ultrastructural and autoradiographic analysis. Anat Embryol (Berl) 1989, 180:309–16.

23. Schwarzacher HG, Wachtler F: The nucleolus. Anat Embryol (Berl) 1993, 188:515–36.

24. Ma N, Matsunaga S, Takata H, Ono-Maniwa R, Uchiyama S, Fukui K: Nucleolin functions in nucleolus formation and chromosome congression. J Cell Sci 2007, 120:2091–105.

25. Hamilton HM, Peura TT, Laurincik J, Walker SK, Maddocks S, Maddox-Hyttel P: Ovine ooplasm directs initial nucleolar assembly in embryos cloned from ovine, bovine, and porcine cells. Mol Reprod Dev 2004, 69:117–25.

26. Svarcova O, Dinnyes A, Polgar Z, Bodo S, Adorjan M, Meng Q, Maddox-Hyttel P: Nucleolar re-activation is delayed in mouse embryos cloned from two different cell lines. Mol Reprod Dev 2009, 76:132–41.

27. Niu Y, Yang S, Yu Y, Ding C, Yang J, Wang S, Ji S, He X, Xie Y, Tang X, et al.: Impairments in embryonic genome activation in rhesus monkey somatic cell nuclear transfer embryos. Cloning Stem Cells 2008, 10:25–36.

28. Kopecny V, Flechon JE, Camous S, Fulka J Jr: Nucleologenesis and the onset of transcription in the eight-cell bovine embryo: fine-structural autoradiographic study. Mol Reprod Dev 1989, 1:79–90.

29. Laurincik J, Thomsen PD, Hay-Schmidt A, Avery B, Greve T, Ochs RL, Hyttel P: Nucleolar proteins and nuclear ultrastructure in preimplantation bovine embryos produced in vitro. Biol Reprod 2000, 62:1024–32.

30. Lawrence FJ, McStay B, Matthews DA: Nucleolar protein upstream binding factor is sequestered into adenovirus DNA replication centres during infection without affecting RNA polymerase I location or ablating rRNA synthesis. J Cell Sci 2006, 119:2621–31.

31. Thongphakdee A, Kobayashi S, Imai K, Inaba Y, Tasai M, Tagami T, Nirasawa K, Nagai T, Saito N, Techakumphu M, et al.: Interspecies nuclear transfer

embryos reconstructed from cat somatic cells and bovine ooplasm. J Reprod Dev 2008, 54:142–7.

32. Schultz RM, Davis W Jr, Stein P, Svoboda P: Reprogramming of gene expression during preimplantation development. J Exp Zool 1999, 285:276–82.

33. Wee G, Koo DB, Song BS, Kim JS, Kang MJ, Moon SJ, Kang YK, Lee KK, Han YM: Inheritable histone H4 acetylation of somatic chromatins in cloned embryos. J Biol Chem 2006, 281:6048–57.

34. Tao Y, Cheng L, Zhang M, Li B, Ding J, Zhang Y, Fang F, Zhang X, Maddox-Hyttel P: Ultrastructural changes in goat interspecies and intraspecies reconstructed early embryos. Zygote 2008, 16:93–110.

35. Svarcova O, Laurincik J, Avery B, Mlyncek M, Niemann H, Maddox-Hyttel P: Nucleolar development and allocation of key nucleolar proteins require de novo transcription in bovine embryos. Mol Reprod Dev 2007, 74:1428–35.

36. Smith SD, Soloy E, Kanka J, Holm P, Callesen H: Influence of recipient cytoplasm cell stage on transcription in bovine nucleus transfer embryos. Mol Reprod Dev 1996, 45:444–50.

37. Laurincik J, Zakhartchenko V, Stojkovic M, Brem G, Wolf E, Muller M, Ochs RL, Maddox-Hyttel P: Nucleolar protein allocation and ultrastructure in bovine embryos produced by nuclear transfer from granulosa cells. Mol Reprod Dev 2002, 61:477–87.

38. Hernandez-Verdun D: Nucleolus: from structure to dynamics. Histochem Cell Biol 2006, 125:127–37.

39. Maeda Y, Hisatake K, Kondo T, Hanada K, Song CZ, Nishimura T, Muramatsu M: Mouse rRNA gene transcription factor mUBF requires both HMG-box1 and an acidic tail for nucleolar accumulation: molecular analysis of the nucleolar targeting mechanism. Embo J 1992, 11:3695–704.

40. Lafontaine DL, Tollervey D: Synthesis and assembly of the box C+D small nucleolar RNPs. Mol Cell Biol 2000, 20:2650–9.

41. Schimmang T, Tollervey D, Kern H, Frank R, Hurt EC: A yeast nucleolar protein related to mammalian fibrillarin is associated with small nucleolar RNA and is essential for viability. Embo J 1989, 8:4015–24.

42. Tollervey D, Lehtonen H, Jansen R, Kern H, Hurt EC: Temperature-sensitive mutations demonstrate roles for yeast fibrillarin in pre-rRNA processing, pre-rRNA methylation, and ribosome assembly. Cell 1993, 72:443–57.

43. Fomproix N, Gebrane-Younes J, Hernandez-Verdun D: Effects of anti-fibrillarin antibodies on building of functional nucleoli at the end of mitosis. J Cell Sci 1998, 111(Pt 3):359–72.

44. Newton K, Petfalski E, Tollervey D, Caceres JF: Fibrillarin is essential for early development and required for accumulation of an intron-encoded small nucleolar RNA in the mouse. Mol Cell Biol 2003, 23:8519–27.

45. Amin MA, Matsunaga S, Ma N, Takata H, Yokoyama M, Uchiyama S, Fukui K: Fibrillarin, a nucleolar protein, is required for normal nuclear morphology and cellular growth in HeLa cells. Biochem Biophys Res Commun 2007, 360:320–6.

46. Ghisolfi-Nieto L, Joseph G, Puvion-Dutilleul F, Amalric F, Bouvet P: Nucleolin is a sequence-specific RNA-binding protein: characterization of targets on pre-ribosomal RNA. J Mol Biol 1996, 260:34–53.

47. Biggiogera M, Burki K, Kaufmann SH, Shaper JH, Gas N, Amalric F, Fakan S: Nucleolar distribution of proteins B23 and nucleolin in mouse preimplantation embryos as visualized by immunoelectron microscopy. Development 1990, 110:1263–70.

48. Ginisty H, Sicard H, Roger B, Bouvet P: Structure and functions of nucleolin. J Cell Sci 1999, 112(Pt 6):761–72.

49. Li YP, Busch RK, Valdez BC, Busch H: C23 interacts with B23, a putative nucleolar-localization-signal-binding protein. Eur J Biochem 1996, 237:153–8.

50. Parrish JJ, Susko-Parrish JL, Leibfried-Rutledge ML, Critser ES, Eyestone WH, First NL: Bovine in vitro fertilization with frozen-thawed semen. Theriogenology 1986, 25:591–600.

51. Rosenkrans CF Jr, Zeng GQ, MCNamara GT, Schoff PK, First NL: Development of bovine embryos in vitro as affected by energy substrates. Biol Reprod 1993, 49:459–62.

52. Cibelli JB, Stice SL, Golueke PJ, Kane JJ, Jerry J, Blackwell C, Ponce de Leon FA, Robl JM: Cloned transgenic calves produced from nonquiescent fetal fibroblasts. Science 1998, 280:1256–8.

53. Spurr AR: A low-viscosity epoxy resin embedding medium for electron microscopy. J Ultrastruct Res 1969, 26:31–43.

54. Reynolds ES: The use of lead citrate at high pH as an electron-opaque stain in electron microscopy. J Cell Biol 1963, 17:208–12.

55. Schramm RD, Paprocki AM, Voort CA: Causes of developmental failure of in-vitro matured rhesus monkey oocytes: impairments in embryonic genome activation. Hum Reprod 2003, 18:826–33.

Expression of Transmembrane Carbonic Anhydrases, CAIX and CAXII, in Human Development

Shu-Yuan Liao, Michael I. Lerman and Eric J. Stanbridge

ABSTRACT

Background

Transmembrane CAIX and CAXII are members of the alpha carbonic anhydrase (CA) family. They play a crucial role in differentiation, proliferation, and pH regulation. Expression of CAIX and CAXII proteins in tumor tissues is primarily induced by hypoxia and this is particularly true for CAIX, which is regulated by the transcription factor, hypoxia inducible factor-1 (HIF-1). Their distributions in normal adult human tissues are restricted to highly specialized cells that are not always hypoxic. The human fetus exists in a relatively hypoxic environment. We examined expression of CAIX, CAXII and

HIF-1α in the developing human fetus and postnatal tissues to determine whether expression of CAIX and CAXII is exclusively regulated by HIF-1.

Results

The co-localization of CAIX and HIF-1α was limited to certain cell types in embryonic and early fetal tissues. Those cells comprised the primitive mesenchyma or involved chondrogenesis and skin development. Transient CAIX expression was limited to immature tissues of mesodermal origin and the skin and ependymal cells. The only tissues that persistently expressed CAIX protein were coelomic epithelium (mesothelium) and its remnants, the epithelium of the stomach and biliary tree, glands and crypt cells of duodenum and small intestine, and the cells located at those sites previously identified as harboring adult stem cells in, for example, the skin and large intestine. In many instances co-localization of CAIX and HIF-1α was not evident. CAXII expression is restricted to cells involved in secretion and water absorption such as parietal cells of the stomach, acinar cells of the salivary glands and pancreas, epithelium of the large intestine, and renal tubules. Co-localization of CAXII with CAIX or HIF-1α was not observed.

Conclusion

The study has showed that: 1) HIF-1α and CAIX expression co- localized in many, but not all, of the embryonic and early fetal tissues; 2) There is no evidence of co-localization of CAIX and CAXII; 3) CAIX and CAXII expression is closely related to cell origin and secretory activity involving proton transport, respectively. The intriguing finding of rare CAIX-expressing cells in those sites corresponding to stem cell niches requires further investigation.

Background

Carbonic anhydrases are metalloenzymes that are ubiquitous in nature, being found in prokaryotes and eukaryotes [1-3]. They are encoded by five distinct, evolutionarily related gene families (α, β, γ, δ and ζ). The complexity of the family membership increases with evolutionary progress and, in mammals, which contain 16 α-CA isozymes, the distribution of expression of family members differs enormously, varying from expression in virtually all cells (CAII) to very discrete temporal and tissue specific expression (CAIX, CAXII and CAXIV). Furthermore, the cellular localization of the CAs also varies, including cytoplasmic, mitochondrial, transmembrane and secreted [2-9].

The tra0nsmembrane CAIX and, to a lesser degree, transmembrane CAXII, have received a great deal of attention. This is because, in large measure, CAIX has been shown to be an excellent cellular biomarker of hypoxia [7,10,11] and a biomarker of certain malignancies, including renal clear cell carcinoma [12,13] and cervical dysplasia and carcinoma [14].

The association with hypoxia has been ascribed functionally to the fact that the CA9 promoter contains a hypoxia response element (HRE) that is the target for the hypoxia inducible factor 1 (HIF-1) transcription factor [10]. The alpha subunit of the heterodimeric HIF-1 (HIF-1α) is rapidly degraded under normoxic O2 levels but is stable under hypoxic conditions. Thus, HIF-1 is stable and transcriptionally active predominantly in hypoxic tissues. This close correlation between hypoxia and CAIX expression has led to studies that have identified CAIX expression in tumor cells that reside in hypoxic regions of solid tumor tissues, i.e. those most refractory to radiation treatment and certain drug modalities, as a potential therapeutic target [2,3,15].

Although it is clear that the CA9 gene is a transcriptional target of HIF-1 there have been several studies that have shown substantial but incomplete co-localization of CAIX expression and regions of hypoxia in solid tumors, with CAIX-positive areas extending beyond the region of hypoxia [7,10,11]. Other studies also showed that CA9 transcription is up-regulated by extracellular acidosis [7,16] and negatively controlled by an epigenetic mechanism that involves methylation of the CA9 promoter [17]. In addition to transcriptional regulation, control of CAIX expression may involve phosphatidyl inositol 3- kinase (P13-kinase) activity [18] or alternative splicing of the CA9 transcript [19].

Moreover, our previous studies also observed HIF-1α positive nuclear staining in squamous cells lining the oral cavity, esophagus and cervix but no CAIX expression was observed (data not shown). All of these studies would suggest the possibility that transcriptional factors, other than HIF-1, may modulate CAIX expression. Furthermore, although CA12 was initially thought to be a HIF-responsive gene [10], other evidence has questioned this correlation [7,20]. Our previous study of CAIX and CAXII expression in human tumor and normal adult tissues had also showed that co-expression of CAIX and CAXII in tumor tissue was mainly restricted to regions of hypoxia, but in normal adult tissues CAIX and CAXII expression did not appear to be induced only by hypoxia and seems to be related to the functional status and cell origin of the relevant tissue.

It has been well established that the mammalian embryo and early stages of fetal development are hypoxic [21-24]. As the process of vascularization occurs hypoxic tissues become progressively normoxic. Thus, the study of the distribution of CAIX, CAXII and HIF-1α expression in the developing human embryo and fetus would be an ideal in vivo system in which to determine whether the postulated coordination of hypoxia, HIF-1 activity and induction of expression of CAIX and CAXII is seen, and also to test our hypothesis that CAIX and CAXII expression may be related to cell origin and secretory activity involving proton transport.

Results and Discussion

During embryonic development, cytotrophoblasts are derived from the trophoblasts of the blastocyst. Later the cytotrophoblasts, together with the coelomic membrane and extraembryonic somatic mesoderm, constitute the chorion. The extraembryonic somatic mesoderm forms the connecting stalk (later becoming the umbilical cord). During the 3rd week (gastrulation), many cells of the epiblast detach themselves from the neighboring cells and form the mesoblast. The cells that remain in the epiblast eventually form the embryonic ectoderm. Some of migrating mesoblastic cells organize and form the intraembryonic mesoderm and the notochord process. Some mesoblastic cells form a loosely woven tissue of the embryonic mesenchyme. These mesenchymal cells have the potential to proliferate and differentiate into diverse types of cells (fibroblasts, chondroblasts, osteoblasts, myoblasts, etc.). At the end of the 3rd week, the intraembryonic coelom develops from the lateral and cardiogenic mesoderm and eventually gives rise to the body cavity [25]. It is at this stage that we were able to obtain our earliest embryonic tissue for analysis.

CAIX Expression During the Embryonic Period (4 to 8 Weeks)

As early as at the 3rd to 4th week of gestation, CAIX expression was already present in the cytotrophoblasts (Fig. 1A, arrow) and rare mesenchymal cells of the chorion and the connecting stalk. In the embryo there were rare cells along the external surface (ectoderm) (Fig. 1B and 1B2, short arrows) and in the primitive mesenchyme (mesoblastic cells) that also weakly expressed CAIX (Fig. 1B1 and 1B2, long arrows). As gestation progresses, increased levels of CAIX expression were seen in the basal layer of the skin, and in the undifferentiated mesenchymal cells of the embryo involved in chondrogenesis. At 7 to 8 weeks, high levels of CAIX expression were seen in the perichondrial undifferentiated mesenchymal cells of the skeletal system, such as facial bone (Fig. 1C), the tracheobronchial cartilage (Fig. 1D), limbs and pelvic bone (Fig. 1E, 1F), and vertebrae (data not shown). All epithelial cells (mesothelial cells) and underlying mesenchyma lining the body cavities showed varying degrees of CAIX expression (Fig. 1F, 1G). As mentioned above, at the 3rd to 4th week of gestation there were only a relatively few weakly positive CAIX expressing cells identified in the embryo, and some of these cells appeared to locate near the neural plate. At 7 to 8 weeks of gestation, the only cells in the nervous system that expressed CAIX were the primitive ependymal cells, derived from the neural crest cells (Fig. 1H). It is worth noting that no other embryonic cells, besides the cells described above, expressed CAIX. For example, CAIX expression was not seen in the lung, heart (Fig. 1D), intestine (Fig. 1F), kidney or gonad (Fig. 1G).

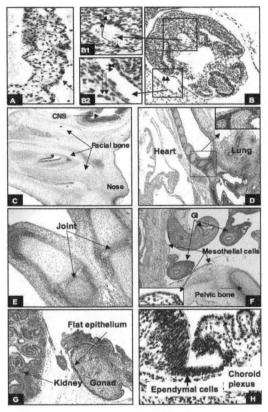

Figure 1. The embryonic period (4–8 weeks): CAIX expression was first seen in the cytotrophoblasts (A, arrow) and certain cells in the embryo (B), such as primitive mesenchyma (B1 long arrow), external surface of the embryo (B2, short double arrows), and mesenchymal cells of the chorion (B2, double long arrows). In the later embryonic stage, CAIX expression was primarily observed in the mesencyhmal cells involving chondrogenesis, as shown in the facial bone (C), bronchial tree (D), the limb (E), and the pelvic bone (F). All of the epithelial cells lining the body cavity were also positive for CAIX, e.g. the peritoneum (F), and the surface of the gonad (G). The only CAIX-positive cells in the CNS are ependymal cells (H). Original magnifications: A, B and H (20×); B1 and B2 (40×); C, D and F (4×); E and G (10×).

CAIX Expression During Fetal Development (9 to 40 Weeks)

During the fetal stage, CAIX expression was closely related to the cell origin (coelomic derivatives), state of cell differentiation, microenvironmental pH status and, to a large degree, the condition of hypoxia. Those tissues expressing CAIX during the embryonic period continued to retain high levels of CAIX expression during the first to 2nd trimester. However, with a few exceptions (body cavity lining cells, coelomic derived remnants, and the gastrointestinal tract), the levels of CAIX expression progressively diminished after 29–30 weeks of gestation and eventually vanished during the postnatal period. After the age of one year, CAIX expression is similar to that observed in normal adult tissues [7]. The details are as follows:

Placenta

High levels of CAIX expression in the mesenchymal cells of the connecting stalk/umbilical cord (Fig. 2A1) and the chorionic plate (Fig. 2A2) were persistent throughout fetal development to term pregnancy. No CAIX immunoreactivity was observed in the chorionic villi.

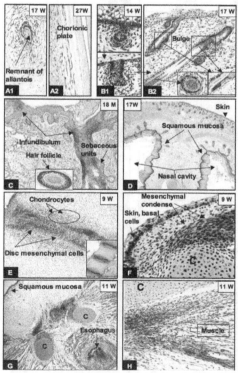

Figure 2. CAIX expression in the placenta, skin, squamous mucosa and skeletal system. The mesenchymal cells of the chorionic plate and umbilical cord retained their CAIX expression until birth (A1, A2). In the epidermis, CAIX expression was limited to the basal layer (B1, F), hair buds (B1) and sebaceous units, hair follicles and the bulges (B2, arrow and inserts). Between 18 to 24 months after birth, CAIX expression was restricted to the hair follicles, sebaceous units, and the infundibulum (C). In the skeletal system, persistent high levels of CAIX immunoreactivity was seen in the chondrocytes, mesenchymal condense (E, F, G) and tendoligamental tissues (G, H). The basal layer of squamous mucosa of the nose (D) and oral cavity (G) also expressed CAIX. W = gestational age in weeks; M = postnatal age in months. Original magnifications: A1, A2, B, C and H (20×); D and G (4×); E (10×) and F (40×).

The Body Cavity

The mesothelial cells lining the body cavities and the surfaces of visceral organ system showed high levels of CAIX expression throughout the developmental stages of the fetus and continued into adulthood [7]. Representative examples are shown in figs 1F, 1G, 3B, 3G1, and 3G2.

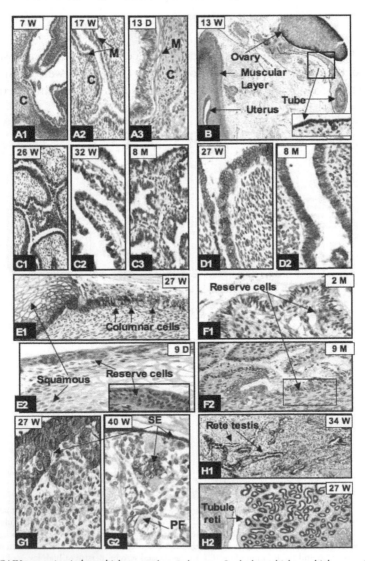

Figure 3. CAIX expression in bronchial trees and genital organs: In the bronchial trees high expression of CAIX was seen in the peribronchial immature mesenchymal cells and cartilage (designated C) (A1) but progressively diminished when the tissues became mature (A2, A3). CAIX positivity was persistently seen in peritoneal lining cells as shown in B (insert) and G1, G2 (surface epithelium [SE], arrow). As early as the 13th week of gestation,

CAIX positive cells were seen in the flat SE of the ovary, the outer muscular layer and rare epithelial cells of the uterus and fallopian tubes (B). Around 26 to 27 weeks, high levels of CAIX expression were transiently seen in the epithelium of fallopian tubes (C1), endometrium (D1), and the cervix (E1). After birth, there was either no CAIX expression (C3) or expression was limited to occasional endometrial cells (D2), reserve cells of the cervix (E2, and corresponding H&E stain in insert, and F1, F2). In the ovary, CAIX expression was observed in SE migrating into the stroma and forming the primordial follicle (PF) (G1, G2). Persistent expression of CAIX was observed in the coelomic remnants: rete testis (H1) and tubule reti (H2). W = gestational age in weeks; D, M = postnatal age in days and months, respectively. Original magnifications: A1, A2, A3, C1, C2, C3, E1, E2 and F2 (20×); B (4×); H1 and H2 (10×); D1, D2, F1, G1 and G2 (40×).

The Skin and Squamous Mucosa

During the fetal period the basal cells of the skin continuously proliferate and eventually form the hair follicle, and the sebaceous, apocrine and eccrine units. In the early fetal stage, increasing CAIX expression was seen along the basal layer and the hair bud; however, no expression of CAIX was seen in the hair bulb and condensation of mesenchyma (Fig. 2B1). Around 16 to 17 weeks of gestation, high levels of CAIX expression were seen in the basal cells, sebaceous unit, the bulge of the prospective site of attachment of the erector muscle, and the outer sheath of the hair follicle (Fig. 2B2). The expression was persistent throughout the fetal period, but progressively diminished during the postnatal period. 18 to 24 months after birth, CAIX immunoreactivity was similar to that seen in the adult skin and was concentrated in the outer sheath of hair follicles, the sebaceous unit and the infundibulum of the hair follicle (Fig. 2C). However, scattered CAIX positive cells were also present along the basal layer between the infundibulo-sebaceous-apocrine units (data not shown). Besides the skin, the dental lamina, enamel organ and the basal cells of the squamous mucosa of the nose, the oral cavity, esophagus and the genital organ system, including the cervix, vagina and anus, also showed variable degrees of CAIX expression, ranging from expression limited to a few basal cells (data not shown) to diffuse basal cell positivity, as shown in fig 2D (the nose at 17 weeks of gestation). After 26 weeks, the levels of CAIX expression progressively diminished. At one year of postnatal age, CAIX expression was restricted to a few basal cells. However, increased levels of CAIX expression in the reparative basal cells were noticed when there was tissue repair due to injury or inflammation (data not shown).

The Articular, Skeletal and Muscular System

There was progressively increased intensity of CAIX expression in the mesenchymal cells involved in chondrogenesis. By the 9th to 11th week of gestation, high levels of CAIX expression were seen in the primitive mesenchyme and chondroblasts in most parts of the skeletal system, especially in the articular disc of the vertebrae (Fig. 2E), the condensation of the mesenchyme around the immature cartilage of the limbs/facial bone (Fig. 2F, 2G), the primitive mesenchymal cells of tendoligamentous units (Fig. 2H) and the tracheobronchial trees (Fig. 3A1). These areas have been shown by us (data shown later) and others [24] to be predominantly hypoxic; thus, CAIX expression in these areas is most likely due to HIF-1 mediated gene expression. In the muscular system, CAIX positivity was observed around the 10th week of gestation. Although the levels of CAIX expression in the muscular system varied among the visceral organs it was often observed in peritracheal primitive muscle cells (Fig. 3A2), the external layer of the muscular propria of the genitourinary tract (Fig. 3B) and the gastrointestinal system, the aorta, and focally in cardiac muscle (data not shown). All of these tissues are derived from the splanchnic mesenchyma. In addition, in the heart there were specialized mesenchymal cells (coelomic origin) at the atrioventricular junction that also expressed CAIX (data not shown). After 20 weeks of gestation, CAIX expression progressively diminished and eventually disappeared when these mesenchymal cells differentiated into mature cartilage, joint capsule, ligament, synovium, and muscular propria of all visceral organs. An illustrative example of the progressive loss of CAIX expression during cell maturation is shown in Fig. 3A. Diffuse CAIX expression was seen in the peritracheobronchial primitive mesenchyma (Fig. 3A1). In the relatively well developed lung (after 17 weeks), CAIX expression was limited to the immature peritracheobronchial muscle and perichondrial mesenchymal cells (Fig. 3A2). After birth to one year of age, CAIX expression was no longer present in the cartilage or muscle (Fig. 3A3).

The Genital System

During fetal development, all of the genital organs derived from the coelom and paramesonephric ducts (Müllerian duct) showed variable degrees of CAIX expression. These levels of expression can basically be divided into three periods (13 to 24 weeks; 25 to 28 weeks; and 29 weeks to birth). As early as at the 13th week of gestation, CAIX expression was identified in the primitive muscular layer of the uterus and fallopian tubes (Fig. 3B), the flat surface epithelium (modified mesothelium) of all male and female genital organs (Fig. 3B insert), and in rare epithelial cells of the endometrium, cervix and fallopian tubes. By the end of the 2nd trimester (24 weeks), there was no longer any observable CAIX expression in the muscular propria. However, there were transient increased levels of CAIX expression in the epithelial cells of the fallopian tube (Fig. 3C1), endometrium (Fig. 3D1), cervix (Fig. 3E1) and vagina during 26 to 28 weeks of gestation. CAIX expression progressively diminished after 29 weeks but the levels of expression appear to vary from case to case. Usually near or after one year of age, CAIX expression was either extinguished, as seen in the fallopian tube (Fig. 3C3), or restricted to rare columnar cells of the endometrium (Fig. 3D2). In the cervix, high levels of CAIX expression were persistently observed in the reserve cells (Fig. 3E2) and, to a lesser degree, in columnar cells at 1–2 months postnatally. Between 2–12 months postnatally, CAIX expression was restricted to rare reserve cells (Fig. 3F1, 3F2). In contrast, throughout the fetal period CAIX was persistently expressed in the flat surface epithelium of all genital organs and the coelomic derived remnants, such as the rete ovarii, rete testis and tubuli reti, the hydatid of Morgagni, and the appendix of the testis. Interestingly, the efferent ductules that merged into the rete testis also showed diffuse immunoreactivity for CAIX.

Representative illustrations are shown in figs 3G1, 3G2 (ovarian surface and primordial follicles), and figs 3H1, 3H2 (the rete testis and tubule reti). With the exception of mature follicles of the ovary, these expression patterns are continuously seen throughout adult life [7]

The Urinary System

The urinary organs consist of the kidneys, ureters, urinary bladder, prostate and urethra. The kidneys and ureters are derived from the metanephrons, whereas the epithelium of the bladder, prostate and urethra are derived from the endoderm. With the exception of efferent ductules as described above, CAIX expression was limited to a few prostate terminal glands and rare primitive reserve cells of the urothelial epithelium (data not shown).

The Endocrine and Hematopoietic System

None of the endocrine organs and hematopoietic systems expressed CAIX, with the exception of the adrenal cortical cells and the thymus gland. The proliferation of the coelomic epithelium initially forms the fetal cortex of the adrenal glands. Around 8 weeks of gestation, more coelomic epithelial cells proliferate and constitute the permanent cortex. CAIX immunostaining in the adrenal cortical cells was first seen around 12–14 weeks of gestation and reached its highest levels of expression during gestational age of 20–28 weeks. The CAIX immunoreactivity was persistent until birth but the numbers of positive cortical cells progressively decreased. By the end of 2 years, no CAIX positive cells were identified in the adrenal glands. Representative illustrations of this temporal sequence of expression are shown in fig 4A. The developing thymus gland comprises the mesenchyme, hematopoietic stem cells and the epithelial cell cords derived from the third pair of endodermal pharyngeal pouches. At the 13th week of gestation, scattered CAIX positive cells were identified near the Hassall's corpuscles. After the 17th week of gestation, high levels of CAIX expression extended to the surface of the lobules (Fig. 4B, 16 weeks). The cells that expressed CAIX did not appear to co-express cytokeratin (data not shown). From this we speculate that these CAIX positive cells were probably immature mesenchymal cells rather than thymic epithelial cells. As the thymus progressively matured the numbers of CAIX positive cells decreased and, after the age of one year, no CAIX expression was seen in the thymus (Fig. 4B, 24 Months).

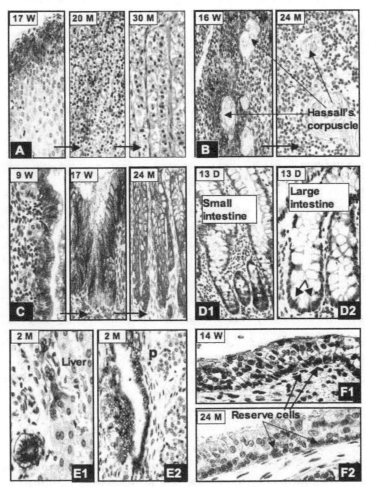

Figure 4. In the adrenal gland (A), strong CAIX immunoreactivity was seen in many cortical cells near the capsule (17W), but the level of expression progressively diminished after birth and, by the end of two years, no CAIX positive cells were seen. In the thymus gland (B), CAIX positive cells were primarily identified near the Hassall's corpuscles (16W) but disappeared by the first year after birth (24M). In the stomach (C), CAIX expression was first seen in surface columnar cells at the 9th week of gestation (9W). At 17 weeks, diffuse immunoreactivity was seen in the glandular and surface columnar cells (17W). This level of expression persisted after birth (24M). In the small intestine CAIX positive cells were restricted to the crypts (D1) and in the large intestine only rare positive cells were identified near the base of the crypts (D2). In the biliary trees, CAIX expression was seen in the epithelium of the entire ductal system and was persistent after birth. The liver (E1) and pancreas (E2) are shown as examples. During early fetal period, extensive CAIX expression was seen in the basal/reserve cells of the respiratory epithelium (F1, 14W) but after birth only rare reserve cells continued to express CAIX (F2, 24M). W = gestational age in weeks; D, M = postnatal age in days and months, respectively. Original magnifications: A1, A2, A3, B1 and C (20×); B2, D1, D2, E1, E2, F1 and F2 (40×).

The Epithelial Cells of the Respiratory and Gastrointestinal Systems (Endodermal Derivatives)

CAIX expression in the GI system was persistent throughout fetal development. CAIX immunoreactivity was first observed in surface columnar cells of the stomach (Fig. 4C), and the crypt cells of the duodenum at 9–12 weeks of gestation. After 16–17 weeks CAIX expression was restricted to certain epithelial cells along the GI system, including gastric glandular cells, neck mucous cells and fundic glandular cells (Fig. 4C), the pyloric and Brunner's glands; the crypt cells of the duodenum and small intestine (Fig. 4D1); and the appendix. The numbers of CAIX immunostained crypt cells were significantly decreased in the distal part of the small intestine. CAIX expression in the large intestine was limited to rare cells in the crypts and was not clearly identified until late in the third trimester. After birth, the numbers of positive cells in the large intestine varied from case to case, but were consistently located near the bottom of the crypts (Fig. 4D2). High levels of CAIX expression were seen in the ductal cells of the entire biliary system, including bile ducts of the liver (Fig. 4E1), common bile ducts, gallbladder, and pancreatic ducts (Fig. 4E2). In the esophagus and the respiratory system, CAIX expression was seen diffusely distributed in the basal/reserve cells of the squamous and respiratory mucosa during the early fetal period (Fig. 4F1) but, after birth, the expression progressively decreased. At the age of one year and thereafter, CAIX immunoreactivity was limited to a few basal cells of the squamous mucosa and primitive reserve cells of the respiratory epithelium (Fig. 4F2).

The Nervous System

During fetal development the only cells in the nervous system that expressed CAIX were the immature ependymal cells and leptomeningeal loose connective tissues of mesoderm derivatives. Its expression was persistent throughout the fetal period.

CAIX expression during the postnatal period (after birth to 8 years old)

During the postnatal period (especially after one year of age) and throughout adult life [7], CAIX expression was no longer present in the normal tissues, with the following exceptions: 1) persistent CAIX expression in mesothelial cells of the body cavity, the surface flat epithelium of all visceral organs, the rete ovarii, rete testis, tubuli reti, the hydatids of Morgagni, the appendix of the testis and efferent ductules. With the exception of efferent ductules, all of these CAIX positive organ systems are derived from coelomic epithelium; 2) persistent high levels of CAIX expression in the GI system, comprising the gastric surface epithelium, gastric glands, pyloric/Brunner's glands, crypt cells of the small intestine, and the biliary trees, including the gallbladder; 3) high levels of CAIX expression in the infundibulum and the outer sheath of hair follicles, and the sebaceous units of the skin; 4) variable degrees of CAIX expression in the primitive reserve cells or immature metaplastic squamous cells of the lining epithelium of all visceral organs and rare columnar cells in the crypts of the large intestine. Interestingly, in the reparative squamous or respiratory mucosa, not only do the numbers of basal/reserve cells expressing CAIX appear to increase, CAIX expression is also observed in rare columnar cells; and 5) the only mesenchymal tissues that retained CAIX expression after birth were the submesothelial stromal cells, the meniscus and the nucleus pulposus of the vertebrae. Schematic representations of the temporal and tissue-specific distribution of CAIX expressing cells, both pre- and postnatal, are shown in Figs 5 and 6. The patterns of expression clearly indicate that, during human development, all of the CAIX positive cells were derived from the mesoderm and were particularly related to the embryonic coelom and mesenchyme, with the exception of the skin, squamous mucosa, the upper gastrointestinal tract and the efferent ductules.

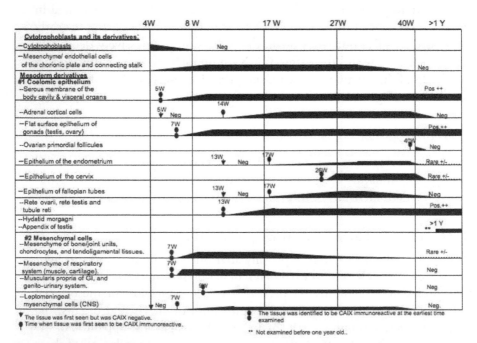

Figure 5. Schema of CAIX expression in the cytotrophoblast and extra-/intra-embryonic mesoderm derivatives during human development (embryonic, fetal and postnatal periods).

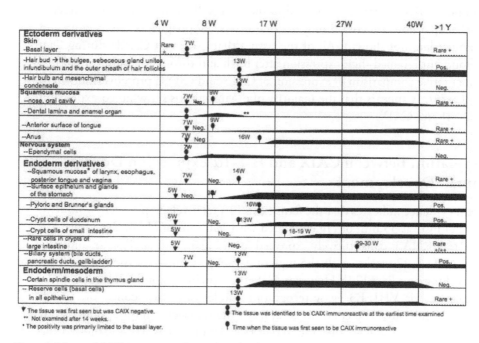

▼ The tissue was first seen but was CAIX negative.
** Not examined after 14 weeks.
* The positivity was primarily limited to the basal layer.

🔵 The tissue was identified to be CAIX immunoreactive at the earliest time examined
▼ Time when the tissue was first seen to be CAIX immunoreactive

Figure 6. Schema of CAIX expression in the ectodermal and endodermal derivatives during human development (embryonic, fetal and postnatal periods).

CAXII Expression During Human Development

In contrast to CAIX, the expression of CAXII was limited to very few organ systems. The numbers of cells stained and the intensity of positive staining were consistent throughout the embryonic and fetal periods and into the entire adult life. Its expression was observed predominantly in cells derived from the mesonephric duct or those cells whose differentiated function involves secretion and proton pumping. The coelomic cell origin, cell differentiation and degree of cellular hypoxia (the latter playing a significant role in regulation of CAIX expression), probably play no significant role in regulation of CAXII expression. CAXII expression was first observed in syncytiotrophoblasts during the embryonic period. Between 7–12 weeks of gestation, high levels of CAXII expression were already seen in those tissues involved in secretion or pH regulation. For example, the taste buds and underlying glands of the tongue (Fig. 7A), pancreatic acinar cells (Fig. 7B) and acinar cells of the minor/major salivary glands, the parietal cells of the stomach (Fig. 7C1), the epithelium of the large intestine (Fig. 7D1), renal tubules (Fig. 7E1), and choroid plexus (Fig. 7F1). CAXII expression was also observed in the remnants of mesonephric ducts of the testis and the ovary (Fig. 7G), and the epithelium of the inner ear (data not shown). Variable degrees of CAXII immunopositivity were also observed in the basal cells of the respiratory mucosa and the squamous mucosa lining the oral cavity, esophagus, cervix, vagina and the anus (data not shown). All of the tissues that expressed CAXII during fetal development retained their expression after birth and throughout adult life [7]. A representative comparison of expression of CAIX and CAXII proteins is shown in figs 7C–7H. It appears that the majority of organ systems show no co-expression of CAXII and CAIX.

Correlation Between CAIX and HIF-1α Expression During Human Development

Recent evidence indicates that the fetus exists in a relatively hypoxic environment. The fetal PO2 is 35 mm of Hg compared with 100 mm of Hg in the adult [23]. HIF-1α mRNA is constitutively expressed in all organs and at all stages of gestation from 14–22 weeks, with the highest expression noted in the brain, heart, kidney, lung and liver [23]. However, HIF-1α protein is rapidly degraded in the proteasome, via VHL E3 ligase recognition, under normoxic conditions and is stabilized under hypoxic conditions [2,3]. In the transgenic mouse model study conducted by Provot et al., they found that HIF-1α protein is stably expressed and transcriptionally active in limb bud mesenchyme and in mesenchyme condensations during fetal development [24]. Thus, HIF-1 appears to play an important role in early chondrogenesis and joint formation. We performed an immunohistochemical study for HIF-1α expression on the fetal tissues, representing all stages of human development from 5–38 weeks of gestation, and encompassing both embryonic and fetal development. We found that as early as 5–9 weeks of gestation, diffuse HIF-1α nuclear immunoreactivity was seen in most of the embryonic and fetal tissues, particularly in the endothelial cells, body mesenchymal cells and nervous system (Fig. 8A), primitive gastrointestinal tract (Fig. 8B), and the chondrocytes and mesenchyme of vertebral disc/limb bud (Fig. 8D, 8E). In the skin, HIF-1α expression was consistently observed during fetal development (Fig. 8E, 8H). Interestingly, the immature chorionic villi during the 1st trimester also showed diffuse HIF-1α nuclear staining (Fig. 8C) and this observation is consistent with that previously published [26]. Variable degrees of HIF-1α expression were also observed in the visceral organs, especially during 17–25 weeks of gestation. The best examples are shown in fig 8F (kidney) and 8G (rectum). Although HIF-1α expression was persistent throughout the fetal period, the numbers of expressing cells progressively diminished. During the last trimester of fetal development (27–40 weeks) stable HIF-1α expression was limited to chondrocytes, endothelial cells, skin, and epithelial cells of the squamous mucosa, such as the vagina and cervix (Fig. 8I, 8J), the oral cavity, esophagus and anus.

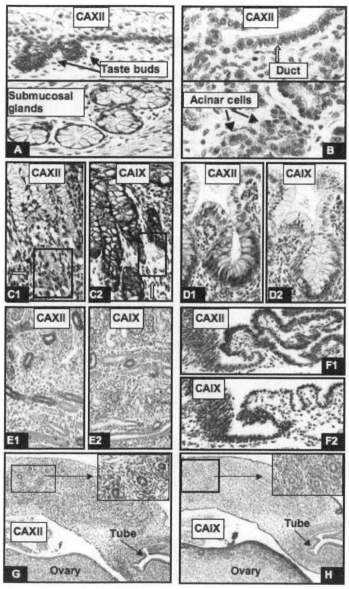

Figure 7. Comparison of CAIX and CAXII expression during human development. There is no correlation between CAXII and CAIX expression. CAXII positive cells were distributed in taste buds and the submucosal gland of the tongue (A, 17W), the acinar cells of the pancreas (B, lower panel, but not ductal cells [B, upper panel]), the parietal cells but not columnar cells of the stomach (C1, 27W), the large intestine (D1, 17W), the distal tubules and collecting ducts of the kidney (E1, 8W), the choroid plexus (F1, 8W) and the remnants of mesonephric ducts of the ovary (G, 13W). In contrast, CAIX positive cells were seen in the gastric columnar cells (C2, 27W), the ependymal cells (F2, 8W), and the surface epithelial cells of the ovary and peritoneum (H, 13W). However, CAIX expression was not seen in the large intestine (D2, 17W), kidney (E2, 8W) and the remnants of the ovarian mesonephric ducts (H, 13W). W = gestational age in weeks. Original magnifications: A, F1 and F2 (20×); B, C1, C2, D1 and D2 (40×); E1, E2, G and H (10×).

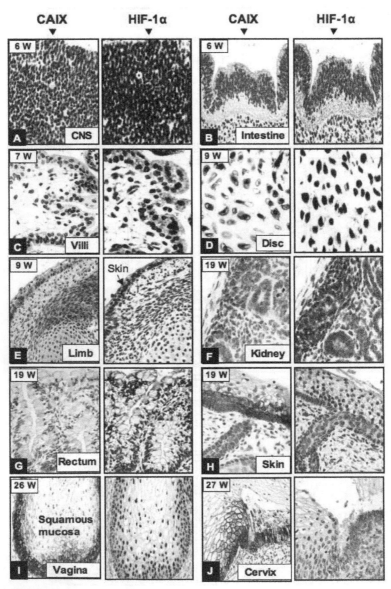

Figure 8. The correlation between CAIX and HIF-1α expression during human development. In the early stage (5–9 weeks of gestation) diffuse HIF-1α immunoreactivity was seen in the CNS (A), primitive intestine (B), chorionic villi (C) and chondrocytes of the disc (D), mesenchymal cells and chondrocytes of the limb (E), and the skin (E arrow). In contrast, no CAIX expression was seen in the CNS (A), primitive intestine (B), and chorionic villi (C). Co-expression of HIF-1α and CAIX was seen in the skin (E) and the cells involved in chondrogenesis (D, E). During later human development, around 19–20 weeks, HIF-1α expression was persistently observed in the kidney (F) and rectum (G) but no CAIX expressing cells were seen in these organs. However, a degree of overlap between CAIX and HIF-1α expression was seen in the skin (H), squamous mucosa of the vagina and the cervix (I and J). W = gestational age in weeks. Original magnifications: A, B, D-H (40×); C (left panel 20×; right panel 40×); I (20×); J (left panel 20×; right panel 40×).

It has been established that CAIX is an endogenous cellular hypoxia marker in solid tumor tissues, and that HIF-1α is an important factor in the regulation of CAIX expression [2-8]. This correspondence between HIF-1α and CAIX expression was also seen in this study in certain cell types during the embryonic period to early stage of fetal development, and between 17–25 weeks of gestational age. During the embryonic and early fetal periods, those cells comprising the primitive mesenchyme, or involving chondrogenesis and skin development, expressed high levels of CAIX. These CAIX positive primitive cells either co-expressed HIF-1α or were located adjacent to HIF-1α expressing cells. The best examples were the embryonic mesenchyme and chondrocytes of the vertebrae, limb buds, and basal cells of the skin (Fig. 8D, 8E, 8H). In the 17–28 week gestation period, there were transient high levels of CAIX expression in the epithelial cells derived from the coelom and the basal layer of the squamous mucosa. Some, but not all, of those CAIX-positive epithelial cells also showed nuclear immunostaining for HIF-1α; for example, vaginal squamous mucosa and the cervix (Fig. 8I, 8J). The findings appear to indicate that CAIX expression in the developing fetal tissues is partially under the influence of hypoxia. However, it is important to note that not all of the cells expressing CAIX protein were necessarily related to the hypoxic condition and, conversely, not all of the hypoxic cells (as indicated by stable HIF-1α expression) expressed CAIX. For example, the trophoblasts of the chorionic villi were hypoxic and showed diffuse nuclear immunoreactivity for HIF-1α, as shown by us and others, but were negative for CAIX expression (Fig. 8C).

The embryonic and fetal tissues during 5–12 weeks of gestation were uniformly hypoxic. By immunohistochemical study, high levels of HIF-1α expression was seen in most of the fetal tissues but no CAIX expression was present in the majority of visceral organs; for example, the primitive nervous system (Fig. 8A), kidney (Fig. 8F), gastrointestinal tract (Fig. 8B, 8G), lungs, cardiac muscle, and liver (data not shown). In contrast, during late fetal development there were epithelial cells of coelomic derivatives, stomach, biliary system, crypt cells of the intestine and squamous mucosa of the cervix, vagina that exhibited high levels of CAIX expression; however, no HIF-1α nuclear staining was seen in those CAIX positive tissues, with the exception of squamous mucosa of the cervix and vagina, as noted above.

It has been shown that CAIX expression in the gastrointestinal tract is not only involved in maintenance of tissue integrity and regulation of basolateral ion transport but is also closely related to cell proliferation and differentiation [27]. Thus, we also performed Ki67 immunostaining, a marker of cellular proliferation, in selected cases to determine whether the postulated coordination of proliferative activity and CAIX expression is seen. In this study, we found that the majority of CAIX positive cells were not in the proliferative stage (data not shown).

Conclusion

Although CAIX and CAXII expression during mouse embryonic development has been previously described [28], a systematic comparative study of CAIX, CAXII and HIF-1α expression in the embryonic, fetal and postnatal tissues during human development has not been reported. We find that the distributions of CAIX and CAXII immunoreactivity in the mouse tissues are similar but not identical to the immunoreactive patterns observed in the human organs. For example, staining for CAIX in the mouse embryos showed a relatively wide distribution pattern with moderate signals in the brain, lung, pancreas and liver, and weak expression in the kidney and stomach [28]. In human fetal tissues, CAIX expression in those organ systems was restricted to certain cell types, such as the primitive ependymal cells of the brain, immature mesenchymal cells of the bronchial tree system, and epithelial cells of the pancreatic ducts and stomach. High levels of CAXII expression were seen in the choroid plexus in both mouse and human tissues, but CAXII expression in the embryonic mouse kidneys and other organs was weak and became negative in the adult mice. In contrast, the intensity of CAXII expression in the human fetal tissues persists postnatally throughout the adult lifespan.

Several intriguing features concerning CAIX expression have emerged from the study of the developing human embryo and fetus. As expected, HIF-1α and CAIX expression colocalized in many, but not all, of the embryonic and early fetal tissues. As fetal development progressed, and vascularization increased, the lack of co-expression of HIF-1α and CAIX became more apparent in multiple tissues. In those instances where HIF-1α is not expressed but CAIX is, one must assume that its expression is regulated by other transcriptional factors. We, and others [16,29], have shown that SP1 is required.

Furthermore, transcription mediated indirectly by P13-kinase activity may also be a possibility [18]. Of particular interest is the situation where HIF-1α is stably expressed but no CAIX expression is found. Stable inactive HIF-1α expression has been noted in the presence of proteasome inhibitors [30]. One possibility is the expression of a co-repressor inhibiting the activity of HIF-1. Another important factor in the regulation of CAIX expression is the cell origin. CAIX expression was consistently observed in cells derived from the coelomic epithelium. The best examples are mesothelial cells, modified mesothelial cells (Müllerian epithelium), underlying mesenchymal cells lining the body cavity, and coelomic remnants, such as the rete ovarii, rete testis and tubuli reti, the hydatid of Morgagni and the appendix of the testis. Another feature of the regulation of CAIX expression is its pH control function. CAIX is a transmembrane carbonic anhydrase that possesses cell surface enzyme activity that catalyzes the conversion of CO_2 into bicarbonate and protons [31]. Thus, the CAIX expression observed in the gastrointestinal tract is more likely induced by cellular acidity and the requirement for proton transport.

Although CAXII expression was originally thought to be regulated by HIF-1 [10], this is clearly not the case. There is no obvious co-localization of HIF-1α and CAXII.

Furthermore, there is also no co-localization of CAIX and CAXII. The distribution of CAXII expression would suggest that it plays an important physiological role in secretory/absorptive cells in different organ systems, primarily involving ion transport and fluid concentration. Corroborating this suggestion is a previous study that found that CAXII is overexpressed in the ciliary epithelial cells of glaucomatous eyes and may be involved in aqueous humour production [32].

A particularly intriguing finding was the discrete distribution of CAIX expression in rare cells or niches in late stages of fetal development, and postnatally, that correspond to sites previously identified as harboring adult stem cells. For example, CAIX expression in the skin is restricted to the hair follicles, including the bulge, sebaceous gland, outer root sheath and infundibulum, plus rare cells in the inter-follicular zone. This corresponds to epidermal stem cell niches that have been described, using various stem cell biomarkers [33,34]. Also, the distribution of CAIX expression in the small and large intestine is very similar to that described for the identification of the intestinal stem cell niche, using the Lgr5 biomarker [35]. Finally, our observation of CAIX expression in Müllerian-type columnar cells and reserve cells of the cervix, during fetal development and postnatally, appears to be similar to that described in the study conducted by Martens and colleagues [36]. These authors claim that the Müllerian epithelial cells represent the stem cells for endocervical reserve cells and columnar cells. Thus, we hypothesize that the rare reserve cells expressing CAIX in the cervix and other organ systems may correspond to putative stem cell regions. However, at this time, such distributions and their relevance to stem cell identity are speculative and will require more comparison and functional analyses. These studies are in progress. If CAIX is found to be a stem cell marker for certain tissues it will be an attractive one. Its transmembrane location will be extremely useful for cell sorting analyses. It may also provide a caveat for cancer therapeutic regimens that target CAIX-expressing cells for destruction [37].

Methods

Tissue Specimens

The embryonic, fetal, and postnatal tissues were collected at St. Joseph Hospital (Orange, CA) from 1997 to 2005. The sources of these specimens included spontaneously abortive embryos, stillborn fetuses and organs of autopsy cases. Consent was obtained prior to any medical procedures being performed. The extra-tissue sections were prepared and labeled precisely according to the age distribution without any identification. The studied cases included 10 from the embryonic period (4 to 8 weeks), 42 from the fetal period (9 weeks to birth) and 26 from the postnatal period (one day to 8 years old). The cases from the fetal period were further divided into 9 to 12 weeks (5 cases), 13 to 16 weeks (9 cases), 17 to 20 weeks (7 cases), 21 to 25 weeks (4 cases), 26 to 40 weeks (17 cases). All of the tissues were fixed in 10% neutral-buffered formalin, paraffin-embedded, sectioned, and stained with hematoxylin and eosin (H&E).

Immunohistochemical Studies

The mouse monoclonal antibody (M75) used to detected the MN/CAIX protein and the rabbit polyclonal antibody to CAXII have been described previously and both antibodies have been shown to be specific for CAIX and CAXII, respectively [38,39].

Immunohistochemical staining of tissue sections with anti-CAIX and anti-CAXII antibodies was done using a peroxidase technique with a pressure cooking pretreatment.

The peroxidase method and the primary antibody dilution for CAIX (1:10,000) and CAXII (1:500) have been described previously [7,12,14]. For CAXII immunostaining, the avidin-biotin enzyme complex Kit (LSAB2, DAKO Corp., Carpinteria, CA) was used. Positive and negative controls were included in each run. Anti-HIF-1α (NeoMarkers INC. Fremont, CA, USA) and anti-Ki67 (Dako) mouse monoclonal antibodies at a 1:100 dilution, respectively, were also applied to selected cases. For HIF-1α, the sections were exposed to pressure cooking pretreatment in citrate buffer (pH 6.0) for 15 min and the DAKO Catalyzed Signal Amplification System (CSA) was used according to the manufacturer's instructions.

Abbreviations

CAIX: carbonic anhydrase IX; CAXII: carbonic anhydrase XII; HIF: hypoxia inducible factor.

Authors' Contributions

SYL and EJS conceived of the study and participated in its execution. SYL, EJS and MIL participated in its design and drafting of the manuscript. All authors read and approved the final manuscript.

Acknowledgements

This research was supported by Institutional Funds (EJS) and by the Intramural Research Program of NIH, National Cancer Institute, Center for Cancer Research and by Contract No. NO1-CO-56000 to Basic Research Program, SAIC-Frederick, Inc (MIL). The content of the publication does not necessarily reflect the views or policies of the Department of Health and Human Services, nor does mention of trade names, commercial products, or organizations imply endorsement by the U.S. Government.

References

1. Supuran CT: Carbonic anhydrases-an overview. Curr Pharm Des 2008, 14:603–14.

2. Pastorekova S, Parkkila S, Pastorek J, Supuran CT: Carbonic anhydrases:current state of the art, therapeutic applications and future prospects. J Enz Inhib Med Chem 2004, 19:199–229.

3. Potter C, Harris AL: Hypoxia inducible carbonic anhydrase IX, a marker of tumour hypoxia, survival pathway and therapy target. Cell Cycle 2004, 3:164–167.

4. Lehtonen J, Shen B, Vihinen M, Casini A, Scozzafava A, Supuran CT, Parkkila AK, Saarnio J, Kivela AJ, Waheed A, Sly WS, Parkkila S: Characterization of CAXIII, a novel member of the carbonic anhydrase isozyme family. J Biol Chem 2004, 279:2719–2727.

5. Tureci O, Sahin U, Vollmar E, Siemer S, Gottert E, Seitz G, Parkkila AK, Shah GN, Grubb JH, Pfreundschuh M, Sly WS: Human carbonic anhydrase XII: cDNA cloning, expression, and chromosomal localization of a carbonic anhydrase gene that is overexpressed in some renal cell cancers. Proc Natl Acad Sci USA 1998, 95:7608–7613.

6. Opavsky R, Pastorekova S, Zelnik V, Gibadulinova A, Stanbridge EJ, Zavada J, Kettmenn R, Pastorek J: Human MN/CA9 gene, a novel member of the carbonic anhydrase family: structure and exon to protein domain relationships. Genomics 1996, 33:480–487.

7. Ivanov S, Liao SY, Ivanova A, Danilkovitch-Miagkova A, Tarasova N, Weirich G, Merrill MJ, Proescholdt MA, Oldfield EH, Lee J, Zavada J, Waheed A, Sly W, Lerman MI, Stanbridge EJ: Expression of hypoxia-inducible cell-surface transmembrane carbonic anhydrases in human cancer. Am J Pathol 2001, 158:905–919.

8. Karhumaa P, Leinonen J, Parkkila S, Kaunisto K, Tapanainen J, Rajaniemi H: The identification of secreted carbonic anhydrase VI as a constitutive glycoprotein of human and rat milk. Proc Natl Acad Sci USA 2001, 98:11604–11608.

9. Pastorekova S, Parkkila S, Parkkila AK, Opavsky R, Zelnik V, Saarnio J, Pastorek J: Carbonic anhydrase IX, MN/CAIX: analysis of human and rat alimentary tracts. Gastroenterology 1997, 112:398–408.

10. Wykoff CC, Beasley NJ, Watson PH, Turner KJ, Pastorek J, Sibtain A, Wilson GD, Turley H, Talks KL, Maxwell PH, Pugh CW, Ratcliffe PJ, Harris AL: Hypoxia-inducible expression of tumor-associated carbonic anhydrases. Cancer Res 2000, 60:7075–7083.

11. Olive PL, Aquino-Parsons C, MacPhail SH, Liao SY, Raleigh JA, Lerman MI, Stanbridge EJ: Carbonic anhydrase 9 as an endogenous marker for hypoxic cells in cervical cancer. Cancer Res 2001, 61:8924–8929.

12. Liao SY, Aurelio ON, Jan K, Zavada J, Stanbridge EJ: Identification of the MN/CA9 protein as a reliable diagnostic biomarker of clear cell carcinoma of the kidney. Cancer Res 1997, 7:2827–2831.

13. Grabmaier K, Vissers JL, De Weijert MC, Oosterwijk-Wakka JC, VanBokhoven A, Brakenhoff RH, Noessner E, Mulders PA, Merkx G, Adema GJ, Oosterwijk E: Molecular cloning and immunogenicity of renal cell carcinomaassociated antigen 250. Int J Cancer 2000, 85:865–870.

14. Liao SY, Brewer C, Zavada J, Pastorek J, Pastrorekova S, Manetta A, Berman ML, DiSaia PJ, Stanbridge EJ: Identification of the MN antigen as a diagnostic biomarker of cervical intraepithelial squamous and glandular neoplasia and cervical carcinoma. Am J Pathol 1994, 145:598–609.

15. Thiry A, Dogne JM, Masareel B, Supuran CT: Targetingtumor-associated carbonic anhydrase IX in cancer therapy. Trends in Pharm Sciences 2006, 27:566–573.

16. Ihnatko R, Kubes M, Takacova M, Sedlakova O, Sedlak J, Pastorek J, Kopacek J, Pastorekova S: Extracellular acidosis elevates carbonic anhydrase IX in human glioblastoma cells via transcriptional modulation that does not depend on hypoxia. Int J Oncol 2006, 29:1925–1033.

17. Cho M, Uemura H, Kim SC, Kawada Y, Yoshida K, Hirao Y, Konishi N, Saga S, Yoshikawa K: Hypomethylation of the MN/CA9 promoter and upregulated MN/CA9 expression in human renal cell carcinoma. Br J Cancer 2001, 5:563–567.

18. Kaluz S, Kaluzová M, Chrastina A, Olive PL, Pastoreková S, Pastorek J, Lerman MI, Stanbridge EJ: Lowered oxygen tension induces expression of the

hypoxia marker MN/carbonic anhydrase IX in the absence of hypoxia-inducible factor 1α stabilization: a role for phosphatidylinositol 3' -kinase. Cancer Res 2002, 62:4469–4477.

19. Barathova M, Takacova M, Holotnocova T, Gibadulinova A, Ohradonova A, Zatovicova M, Hulikova A, Kopacek J, Parkkila S, Supuran CT, Pastorekova S, Pastorek J: Alternative splicing variant of the hypoxia marker carbonic anhydrase IX expressed independently of hypoxia and tumor phenotype. Br J Cancer 2008, 98:129–136.

20. KivelŠ A, Parkkila S, Saarnio J, Karttunen TJ, KivelŠ J, Parkkila H: Expression of a novel transmembrane carbonic anhydrase isozyme XII in normal human gut and colorectal tumors. Am J Pathol 2000, 156:577–584.

21. Chen E, Fujinaga M, Giaccia AJ: Hypoxia microenvironment within an embryo induced apoptosis and is essential for proper morphological development. Teratology 1999, 60:215–225.

22. Ramirez-Bergeron DL, Runge A, Cowden Dahl KD, Fehling HJ, Keller G, Simon MC: Hypoxia affects mesoderm and enhances hemangioblast specification during early development. Development 2004, 31:4623–4634.

23. Madan A, Varma S, Cohen HJ: Developmental stage-specific expression of the α and β subunits of the HIF-1 protein in the mouse and human fetus. Mol Genet Metabolism 2002, 75:244–249.

24. Provot S, Zinyk D, Gunes Y, Kathri R, Le Q, Kronenberg HM, Johnson RS, Longaker MT, Giaccia AJ, Schipani E: Hif-1α regulates differentiation of limb bud mesenchyme and joint development. J Cell Biol 2007, 177:451–464.

25. Moore KL, Persaud TVN, Torchia MG: The Developing Human: Clinically Oriented Embryology. 8th edition. WB Saunders/Elsevier, Philadelphia; 2008.

26. Caniggia I, Mostachfi H, Winter J, Gassman M, Lye SJ, Kuliszewski M, Post M: Hypoxia-inducible factor-1 mediates the biological effects of oxygen on human trophoblast differentiation through TGF-β3. J Clin Invest 2000, 105:577–587.

27. Saarnio J, Parkkila S, Parkkila AK, Waheed A, Casey MC, Zhou XY, Pastoreková S, Pastorek J, Karttunen T, Haukipuro K, Kairaluoma M, Sly WS: Immunohistochemistry of carbonic anhydrase isozyme IX (MN/CA IX) in human gut reveals polarized expression in the epithelial cells with the highest proliferative capacity. J Histochem Cytochem 1998, 46:497–504.

28. Kallio H, Pastorekova S, Pastorek J, Waheed A, Sly WS, Mannisto S, Heikinheiimo A, Parkkila S: Expression of carbonic anhydrases IX and XII during mouse embryonic development. BMC Development Biol 2006, 6:22–30.

29. Kaluzova M, Pastorekova S, Svastova E, Pastrorek J, Stanbridge EJ, Kaluz S: Characterization of the MN/CA9 promoter proximal region: a role for specificity protein (SP) and activator protein 1 (AP1 factors). Biochem J 2001, 359:669–677.

30. Kaluz S, Kaluzova M, Stanbridge EJ: Proteasomal inhibition attenuates transcriptional activity of hypoxia-inducible factor (HIF-1) via specific effect on the HIF-1 alpha C-terminal activation domain. Mol Cell Biol 2006, 25:5895–5907.

31. Swietach P, Vaughan-Jones RD, Harris AL: Regulation of tumor pH and the role of carbonic anhydrase 9. Cancer Metastasis Rev 2007, 26:299–310.

32. Liao SY, Ivanov S, Ivanova A, Ghosh S, Cote MA, Keefe K, Coca-Prados M, Stanbridge EJ, Lerman MI: Expression of cell surface transmembrane carbonic anhydrase genes CA9 and CA12 in the human eye: over-expression of CA12 (CAXII) in glaucoma. J Med Genet 2003, 40:257–261.

33. Moore AK, Lemischka R: Stem cells and their niches. Science 2006, 311:1880–1885.

34. Balanpain C, Horsley V, Fuchs E: Epithelial stem cells: turning over new leaves. Cell 2007, 128:445–458.

35. Barker N, van Es JH, Kuipers J, Kujala P, Born M, Cozjinsen M, Haegebarth A, Korving J, Begthel H, Peters PJ, Clevers H: Identification of stem cells in small intestine and colon by marker gene Lgr5. Nature 2007, 5:1003–1007.

36. Martens J, Smedts F, Muyden RV, Schoots C, Helmerhorst T, Hopman A, Ramaekers F, Arends J: Reserve cells in human uterine cervical epithelium are derived from Müllerian epithelium at midgestational age. Int J Gynecol Pathol 2000, 26:463–468.

37. Pastorekova S, Parkkila S, Pastorek J, Supuran CT: Carbonic anhydrases:current state of the art, therapeutic applications and future prospects. J Enzyme Inhib Med Chem 2004, 9:199–229.

38. Zavada J, Zavadova Z, Pastorekova S, Ciampor F, Pastorek J, Zelnik V: Expression of MaTu-MN protein in human tumor cultures and in clinical specimens. Int J Cancer 1993, 53:268–274.

39. Karhumaa P, Parkkila S, Tureci O, Waheed A, Grubb JH, Shah G, Parkkila A, Kaunisto K, Tapanainen J, Sly WS, Rajaniemi H: Identification of carbonic anhydrase XII as the membrane isozyme expressed in normal endometrial epithelium. Mol Hum Reprod 2000, 6:68–74.

Copyrights

Index

A

ABI 7500 fast real time PCR system, 270
actin protein, 51, 54
Affymetrix Fluidics Station, 99
Affymetrix GCOS imaging software, 39
Affymetrix GeneChip* chicken genome
  array, 141
Affymetrix GeneChip* Scanner 3000 7G,
  99
Affymetrix microarrays, 86, 122, 124,
  126, 144. *See also* embryonic stem cells,
  neural differentiation of
Affymetrix mouse 430 2.0 chip, 36, 39,
  40
af gene, 123
agarose-formaldehyde gel electrophoresis,
  98
Agilent 2100 bioanalyzer, 141
Agilent RNA 6000 nano reagents, 142
AlexaFluor-conjugated secondary antibodies, 96

Ambion MEGA Script T7 kit, 98
Ambion Purist Kit, 199
American Type Culture Collection
  (ATCC), 173
Amersham direct cDNA labeling kit, 199
analysis of variance (ANOVA), 52, 87, 99,
  145, 272
anterior cardinal vein (ACV), 231
antisense Hsp105 ODNs treatment, in
  embryos implantation, 58. *See also* rat
  uterus, Hsp 105 expression in
antisense oligodeoxy-nucleotides
  (A-ODNs), 51
  effect of, 58
  usage of, 56–57
apico-basal polarity, 77, 78, 79, 81
apoptosis, definition of, 49
AP substrate FastRed, 97
arsenic exposure, in zebrafish, 183
arteriovenous malformations (AVMs),
  235, 236
arteriovenous shunt (AVS), 234, 235

articular, skeletal and muscular system,
  CAIX expression in, 285–86. *See also*
  human development
ATCC. *See* American Type Culture
  Collection (ATCC)
autocrine induction mechanism, role of,
  78. *See also* embryonic stem cells, neural
  differentiation of
avidin-biotin enzyme complex kit, 298
AVMs. *See* arteriovenous malformations
  (AVMs)
AVS. *See* arteriovenous shunt (AVS)
Axon model 4000B dual-laser scanner,
  199

B

BABB. *See* benzyl alcohol and benzyl
  benzoate (BABB)
BabySentryPro, 114
B cell translocation gene 4 (btg4), 125,
  126, 129, 130, 134, 137
Beijing Municipal Science and Technology
  Commission, 49
BenchMark protein ladder, 218
benzyl alcohol and benzyl benzoate
  (BABB), 242
bESF. *See* bovine ear skin fibroblast
  (bESF)
biomarkers, 297
Bio-Plex
  buffer, usage of, 219
  cytokine assay, usage of, 218
  manager software, 219
  in sICAM-1 quantization, 209
  suspension array system, 219
  technology, usage of, 214
bmp15 gene, 133
body cavity, CAIX expression in, 284–85.
  See also human development
Bonferroni method, usage of, 145
bovine-bovine (BB)-SCNT embryos, 257
bovine ear skin fibroblast (bESF), 268
bovine oocyte, 255–57, 264, 266, 273

bovine ooplasm and rhesus monkey
  somatic cells
  interspecies cloned embryos between,
  263–67
    abnormal nucleolar morphology,
    259–60
    4- and 8-cell embryos, zygotic gene
    expression in, 270–71
    background of, 255–57
    donor cells, culture of, 268
    housekeeping and imprinting genes,
    down-expression of, 258–59
    IVF, BB-SCNT, and MB-iSCNT
    embryos, development of, 257–58
    IVM and IVF, 267–68
    MB-iSCNT embryos, medium and
    culture preparation of, 269
    nucleolar component proteins, in
    MB-iSCNT embryos, 261–63
    nucleolar protein expression, assess-
    ment of, 272
    SCNT and IVC, 268–69
    species-specific mtDNA, analysis of,
    269–70
    statistical analyses, 272
    TEM, processing for, 271–72
bovine serum albumin (BSA), 267
bronchial trees and genital organs, CAIX
  expression in, 284
BSA. *See* bovine serum albumin (BSA)
btg4. *See* B cell translocation gene 4 (btg4)

C

4CA. *See* 4-chloroaniline (4CA)
CA9 gene, 280
CAIX expression, in human development
  correlation between HIF-1α expression
  and, 292–95
  embryonic period, 281–82
  fetal development, 282–83
    articular, skeletal and muscular
    system, 285–86
    body cavity, 284–85

endocrine and hematopoietic system, 287–88

genital system, 286–87

placenta, 283

skin and squamous mucosa, 285

urinary system, 287

Cajal–Retzius neurons, 18, 21

calcium signalling, role of, 92. *See also* embryonic stem cells, neural differentiation of

Canadian Council on Animal Care, 242

candidate genes, in hen
expression of, 130
hierarchical clustering of, 132
mRNA, 135

CART analysis, usage of, 116. *See also* human embryo phenotypes, cohort-specific prognostic factors in

catalyzed signal amplification system (CSA), 298

CAXII expression, in human development, 291–92

cDNA microarray/chip technology, 162

central nervous system malformations, 34

cephalic plexus, development of, 230. *See also* mouse embryo, vascular development in

cervical intersomitic vessels, development of, 237

chaperone genes and toxicants response, 190. *See also* toxicogenomic responses, in zebrafish embryo

Chemiluminescence Imaging Geliance 600, usage of, 218

Chicken Affymetrix Microarrays, 124
usage of, 126–29

chicken vasa homolog protein (CVH), 124
gene, 133
protein, 138, 139

Chinese Academy of Sciences, 49

4-chloroaniline (4CA), 184

cleavage arrest rate, definition of, 115

Cluster 3.0 program, usage of, 144

COC. *See* cumulus-oocyte complexes (COC); cumulus ophorus complex (COC)

cohort-specific prognostic factors, in human embryo phenotypes, 106–7, 111–14
materials and methods
data collection, inclusion and exclusion criteria, 114
embryo development, assessment of, 114–15
patient, IVF cycle, and embryo parameters, 115
statistical analysis, 115–16
non-redundant prognostic variables, thresholds of, 110–11
outcomes, 107–8
prognostic significance and correlation of variables, 108–9

comparative embryology, 9

comparative mammalian brain collections, 19

confocal microscopy, usage of, 224

congenital malformations, 9, 27

cryoprotectant, usage of, 69. *See also* ovarian hyperstimulation syndrome (OHSS)

CSA. *See* catalyzed signal amplification system (CSA)

cumulus-oocyte complex (COC), 267

cumulus ophorus complex (COC), 217

CVH. *See* chicken vasa homolog protein (CVH)

CyberT, usage of, 39

Cy3-PECAM-1 signal, 226, 227, 232

cytochrome P4501A1, 190, 192

cytokine signalling, 194

D

DA. *See* dorsal aorta (DA)

DA-CCV arteriovenous shunts, 235

degenerated (DEG), 217

dense fibrillar components (DFCs), 256

DEPC. *See* diethylpyrocarbonate (DEPC)

descriptive embryology, 9

DFCs. *See* dense fibrillar components (DFCs)

diabetes-affected genes, classification of, 33. *See also* embryo development, maternal diabetes in

diabetic embryopathy. *See also* embryo development, maternal diabetes in
  animal model of, 28
  cause of, 27

diabetic pregnancies and embryo development, 35–38
  adverse outcomes, implications for prevention of, 38
  background of, 27–28
  diabetic embryopathy, animal model of, 28
  gene expression profiling, 29
  genes, functional roles of, 33–35
  methods
    animals, 38
    microarray analysis, 39–40
    NCBI gene expression omnibus repository, 40–41
    molecular classification, 31
    Q-RT-PCR, 29–31
    transcription regulation, 32–33

3,4-dichloroaniline, 183

diethylpyrocarbonate (DEPC), 270

6-dimethyl-aminopurine (6-DMAP), 269

dimethylsulfoxide (DMSO), 96, 198, 268

Dll1 gene, 82

6-DMAP. *See* 6-dimethyl-aminopurine (6-DMAP)

DM5000B microscope, usage of, 97

DMEM. *See* Dulbecco's modified Eagle's medium (DMEM)

DMSO. *See* dimethylsulfoxide (DMSO)

DNAseI treatment, 97

dorsal aorta (DA), 223, 227, 230, 231, 232

Down syndrome, 10

Dulbecco's modified Eagle's medium (DMEM), 268

Dynabeads mRNA Direct kit (DYNAL), 270

E

EB. *See* embryoid body (EB)

ectopic pregnancy
  chi-square testing, 154
  diagnosis of, 154
  occurrence of, 153
  rates, with day 3 *vs.* day 5 embryo transfer, 155–56
    background of, 153
    methods in, 154
    outcomes in, 154–55

EG. *See* embryonic germ (EG)

EGA. *See* embryonic genome activation (EGA)

elective embryo cryopreservation, 68. *See also* ovarian hyperstimulation syndrome (OHSS)

elective embryo freezing, usage of, 72

electro cell manipulator, 269

ELISA. *See* enzyme-linked immunosorbent assay (ELISA)

embryoblast, 10

embryo, definition of, 9–10

embryo development and oocyte maturation, in hen, 135–39
  background of, 122–24
  follicular maturation, gene expression during, 131–34
  gene expression, tissular pattern of, 129–31
  methods
    animals, 139–40
    bioinformatic analysis, 140–41
    hierarchical clustering, 144
    Inra-Urgv, Affymetrix equipment at, 141–42
    real time RT-PCR, 143–44
    RNA extraction and reverse transcription, 142
    RNA isolation and microarray analysis, 141
    in situ hybridization, 144
    statistical analysis, 144–45

tissues, oocytes and embryos, collection of, 140
murine oocyte genes, chicken homologs of, 124–25
oocyte and granulosa cells transcription profiles, comparing, 126–29
ovary, gene expression in, 134–35
embryo development, maternal diabetes in, 35–38
adverse outcomes, implications for prevention of, 38
background of, 27–28
diabetic embryopathy, animal model of, 28
gene expression profiling, 29
genes, functional roles of, 33–35
methods
animals, 38
microarray analysis, 39–40
NCBI gene expression omnibus repository, 40–41
molecular classification, 31
Q-RT-PCR, 29–31
transcription regulation, 32–33
embryoid body (EB), 78
embryo implantation, Hsp 105 expression in, 59–60
antisense Hsp105 ODNs treatment, 58
background of, 48–49
comparison of, 55–56
methods
animals, 49–50
FITC, tracking of, 51–52
immunohistochemistry, 50
microscopic assessment and statistical analysis, 52
oligodeoxynucleotides, design of, 51
western blot analysis, 50–51
in rat uterus during pseudo-pregnancy, 54–55
rat uterus, Hsp105 expression in, 53–54
suppression of, 56–57
in uterus during early pregnancy, 54

embryological study, types of, 9
embryology, definition of, 9
embryonic DarT assay, 195
embryonic genome activation (EGA), 256
fibrillarin proteins, 262
nucleophosmin and nucleolin proteins, 263
embryonic genome activation and nucleologenesis in
bovine ooplasm and rhesus monkey somatic cells, 263–67
abnormal nucleolar morphology, 259–60
4- and 8-cell embryos, zygotic gene expression in, 270–71
background of, 255–57
donor cells, culture of, 268
housekeeping and imprinting genes, down-expression of, 258–59
IVF, BB-SCNT, and MB-iSCNT embryos, development of, 257–58
IVM and IVF, 267–68
MB-iSCNT embryos, medium and culture preparation of, 269
nucleolar component proteins, in MB-iSCNT embryos, 261–63
nucleolar protein expression, assessment of, 272
SCNT and IVC, 268–69
species-specific mtDNA, analysis of, 269–70
statistical analyses, 272
TEM, processing for, 271–72
embryonic germ (EG), 161, 165
embryonic mouse vasculature, 3D visualization of, 226–30
embryonic period, 10, 283, 298
CAIX expression during, 281–82, 295
HIF-1α expression during, 295
syncytiotrophoblasts during, 291
embryonic stem (ES), 78, 161
cell lines, usage of, 95
embryonic stem cells, neural differentiation of, 77–79

materials and methods
  FACS analysis, 97
  immunocytochemistry, 96–97
  LY411575, treatment with, 96
  microarray sample preparation and data analysis, 98–99
  mouse ES cells, maintenance and differentiation of, 95–96
  RNA extraction and RT-PCR, 97–98
  in situ hybridization, 97
  time-lapse movie, 98
neural stem (NS) cells, 85
neuroepithelial progenitors (NPs)
  apico-basal polarity, 81–82
  improved generation of, 79–81
neuroepithelial rosettes, 89–91
  in vitro mammalian neural development, mechanisms of, 91–95
  neurogenic and gliogenic potential *in vitro*, 83–84
  rosette cultures, notch pathway in, 82–83
  in vitro neural commitment, transcriptional profiling of, 86–89
embryo transfer (ET), 72
  catheter, usage of, 70
endocrine and hematopoietic system, CAIX expression in, 287–88. *See also* human development
enzyme-linked immunosorbent assay (ELISA), 208, 218
ephrinB2, expression of, 239
epiblast stem cells (EpiSCs), 92
epidermal growth factor protein, role of, 124
EpiSCs. *See* epiblast stem cells (EpiSCs)
ES. *See* embryonic stem (ES)
ESGRO complete clonal grade medium, 95
ESTs. *See* expressed sequence tags (ESTs)
ET. *See* embryo transfer (ET)
Etnk1 gene, 34
evolutionary embryology, 9

experimental embryology, 9
expressed sequence tags (ESTs), 162
extraembryonic circulation, of mouse embryo, 231–34

F

FACS Aria cell sorter, 97
FACS Calibur cytometer, 97
false discovery rate (FDR), 39, 40, 99, 163, 186, 200
family wise error rate (FWER), 145
FBS. *See* fetal bovine serum (FBS)
FCs. *See* fibrillar centers (FCs)
FDR. *See* false discovery rate (FDR); frequency distance relationship (FDR)
fetal bovine serum (FBS), 95, 268
Fgf5, expression of, 81, 86, 92
FGF signalling, role of, 78
fibrillar centers (FCs), 256
fibrillarin protein, 266
FITC-labeled ODNs, tracking of, 51–52
fluorescence micrographs, of rat uteri, 57. *See also* rat uterus, Hsp 105 expression in
fourier transform (FT), 244
FOXO1, role of, 32
FOXO4, role of, 32
frequency distance relationship (FDR), 224, 243
  based deconvolution and filtering, 243–45
  deconvolution OPT technique, 229, 234
FT. *See* fourier transform (FT)
FunGenES European Consortium, 98
FWER. *See* family wise error rate (FWER)

G

GABAergic interneurons, 16, 17
GABA transporter SCL6 member, 196
GCOS. *See* GeneChip° Operating Software (GCOS)

GCs. See granular components (GCs);
    granulosa cells (GCs)
GD. See germinal disc (GD)
gdf9 gene, 125
GDR. See germinal disc region (GDR)
GeneChip° fluidics station, 141
GeneChip° IVT labelling kit, 141
GeneChip° Operating Software (GCOS),
    141
GeneChip° scanner 3000 7G, 141
gene expression omnibus (GEO), 40, 141,
    199
gene machines omnigrid 100, usage of,
    198
gene ontology (GO), 172, 196
classification of genes, 129
GenePix 6 software, usage of, 199
genes. See also embryo development, ma-
    ternal diabetes in
    expression profiling, in embryos, 29
    functional roles of, 33–35
    molecular classification of, 31
    in vivo function of, 34
GeneSpring7, usage of, 39
gene tools software, usage of, 218
genital system, CAIX expression in,
    286–87. See also human development
GEO. See gene expression omnibus
    (GEO)
germinal disc (GD), 123–24
germinal disc region (GDR), 126, 140
germinal vesicle (GV), 212, 215, 217
Glasgow Modified Eagles Medium
    (GMEM), 95
glutathione peroxidase 1 mRNA, 190
GMEM. See Glasgow Modified Eagles
    Medium (GMEM)
Gmfb gene, 34
GO. See gene ontology (GO)
GoTreeMachine, usage of, 189, 201
granular components (GCs), 256
granulosa cells (GCs), 126, 127, 136, 140,
    217
Γ-secretase inhibitor LY411575, treatment
    with, 96

γ-secretase inhibitor LY411575, usage of,
    82
GV. See germinal vesicle (GV)

H

Hamburger–Hamilton classification, 140
HAR1F, 18
H&E. See hematoxylin and eosin (H&E)
heat shock protein (Hsp) 105, expression
    of, 59–60
    antisense Hsp105 ODNs treatment, 58
    background of, 48–49
    comparison of, 55–56
    methods
        animals, 49–50
        FITC, tracking of, 51–52
        immunohistochemistry, 50
        microscopic assessment and statisti-
            cal analysis, 52
        oligodeoxynucleotides, design of, 51
        western blot analysis, 50–51
    in rat uterus during pseudo-pregnancy,
        54–55
    rat uterus, Hsp105 expression in,
        53–54
    suppression of, 56–57
    in uterus during early pregnancy, 54
hematopoietic stem (HS) cells, 162
hematoxylin and eosin (H&E), 298
hen, oocyte maturation and early embryo
    development in, 135–39
    background of, 122–24
    follicular maturation, gene expression
        during, 131–34
    gene expression, tissular pattern of,
        129–31
    methods
        animals, 139–40
        bioinformatic analysis, 140–41
        hierarchical clustering, 144
        Inra-Urgv, Affymetrix equipment at,
            141–42
        real time RT-PCR, 143–44

RNA extraction and reverse transcription, 142
RNA isolation and microarray analysis, 141
in situ hybridization, 144
statistical analysis, 144–45
tissues, oocytes and embryos, collection of, 140
murine oocyte genes, chicken homologs of, 124–25
oocyte and granulosa cells transcription profiles, comparing, 126–29
ovary, gene expression in, 134–35
hESC. *See* human embryonic stem cell (hESC)
hes5 gene, 82
hes6 gene, 82
hierarchical clustering, of candidate genes in hen, 132. *See also* hen, oocyte maturation and early embryo development in
HIF-1. *See* hypoxia inducible factor 1 (HIF-1)
HIF-1α. *See* hypoxia inducible factor 1α (HIF-1α)
high pure RNA isolation kit, usage of, 97, 98
histocompatibility leukocyte antigen-G (HLA-G), 208, 214
Hprt1 gene, 270
HRE. *See* hypoxia response element (HRE)
HS cells. *See* hematopoietic stem (HS) cells
Hsp70 gene, 194, 197, 256, 258, 264
human accelerated regions (HAR1), 18
Human Brain Project, 17
human development
transmembrane carbonic anhydrases, CAIX and CAXII expression in, 296–97
background of, 279–81
CAIX expression, in embryonic period, 281–82
CAXII expression during, 291–92
correlation between CAIX and HIF-1α expression, 292–95
fetal development, CAIX expression during, 282–88
methods, 297–98
respiratory and gastrointestinal systems, 289–91
human embryonic stem cell (hESC), 106, 255
human embryo phenotypes, cohort-specific prognostic factors in, 106–7, 111–14
materials and methods
data collection, inclusion and exclusion criteria, 114
embryo development, assessment of, 114–15
patient, IVF cycle, and embryo parameters, 115
statistical analysis, 115–16
non-redundant prognostic variables, thresholds of, 110–11
outcomes, 107–8
prognostic significance and correlation of variables, 108–9
human embryos, release of cytokines, chemokines and ICAM-1 by, 209
hypoxia inducible factor 1 (HIF-1), 280
role of, 32
hypoxia inducible factor 1α (HIF-1α)
and CAIX expression, correlation between, 292–95 (*see also* human development)
expression of, 31 (*see also* embryo development, maternal diabetes in)
hypoxia response element (HRE), 280

I

ICA. *See* internal carotid artery (ICA)
ICM. *See* inner cell mass (ICM)
ImageJ Cell Counter software, 97, 98
immunofluorescence confocal laser scanning microscopy, 257, 261

INM. *See* interkinetic nuclear movement (INM)

inner cell mass (ICM), 10, 70

in situ hybridization (ISH), 82, 97, 138, 140, 144, 197, 201
  for identification of tissue-specific genes, 190–92
  investigation of chicken genes, 137
  localization of gene expression in ovary by, 134–35

interkinetic nuclear movement (INM), 78

internal carotid artery (ICA), 227, 231

InterPro, usage of, 163. *See also* mouse stem cells, transcriptome analysis of

intersomitic vessels, usage of, 234

interspecies SCNT (iSCNT)
  application of, 255–56
  developmental capacity of embryos derived from, 257
  monkey and bovine mtDNA, 257

in vitro culture (IVC), 268–69

in vitro fertilisation (IVF), 67, 68, 70, 106, 107, 114, 256
  human embryos development by, 106–7, 111–14
    assessment of, 114–15
    data collection, inclusion and exclusion criteria, 114
    non-redundant prognostic variables, thresholds of, 110–11
    outcomes, 107–8
    patient, IVF cycle, and embryo parameters, 115
    prognostic significance and correlation of variables, 108–9
    statistical analysis, 115–16

in vitro fertilized (IVF) human embryos, 208–9, 215–16
  materials and methods
    Cyto/Chemokines and ICAM-1 profiles, 218–19
    patients, 216–18
    sICAM-1 levels, measurement of, 218

statistical analysis, 219
    western blotting, 218
  release of cytokines, chemokines and ICAM-1 by, 209
  sICAM-1
    levels and embryo grade, 213
    in mature and immature oocytes, 210–11
    in oocyte supernatants and pregnancy rate, 213–14
    production of, 212
    quantization of, 209–10
    and sHLA-G levels, comparison of, 214

in vitro mammalian neural development, mechanisms of, 91–95. *See also* embryonic stem cells, neural differentiation of

in vitro maturation (IVM), 209, 215, 267–68

In Vitro Maturation Protocol (IVM), 216

in vitro transcription (IVT), 141

iSCNT. *See* interspecies SCNT (iSCNT)

ISH. *See in situ* hybridization (ISH)

isosurface algorithm, usage of, 227. *See also* mouse embryo, vascular development in

Italian Law 40 on IVF, 217

IVC. *See in vitro* culture (IVC)

IVF. *See in vitro* fertilisation (IVF)

IVM. *See in vitro* maturation (IVM); In Vitro Maturation Protocol (IVM)

IVT. *See in vitro* transcription (IVT)

J

JEM ARM 1300S high-voltage electron microscope, 271

K

Ki67 immunostaining, usage of, 295

Korea Research Institute of Bioscience and Biotechnology (KRIBB), 268

ktfn gene, 133

L

laser-scanning densitometry, 52, 57
Leica DMIL microscope, 97
Leica MZFLIII stereozoom microscope, 242
leukaemia inhibitory factor (LIF), 92, 170
levamisole, 50
Lgr5 biomarker, 297
LIF. *See* leukaemia inhibitory factor (LIF)
LMP. *See* low melting point (LMP)
locally weighted regression smoother (LOESS), 200
logistic regression tests, importance of, 108. *See also* cohort-specific prognostic factors, in human embryo phenotypes
low melting point (LMP), 242

M

MAD. *See* median absolute deviation from median (MAD)
mark3 gene, 139
MART* method, usage of, 115
MasterPure™ RNA purification kit, 131
maternal antigen that embryos require (Mater), 123
maternal diabetes, in embryo development, 35–38
  adverse outcomes, implications for prevention of, 38
  background of, 27–28
  diabetic embryopathy, animal model of, 28
  gene expression profiling, 29
  genes, functional roles of, 33–35
  methods
    animals, 38
    microarray analysis, 39–40
    NCBI gene expression omnibus repository, 40–41
  molecular classification, 31
  Q-RT-PCR, 29–31
  transcription regulation, 32–33
maternal effect mutations, 123

Max-Delbrück-Centrum für Molekulare Medizin, 98
median absolute deviation from median (MAD), 200, 201
MeHg. *See* methylmercury chloride (MeHg)
mesenchymal stem (MS) cells, 162, 165
mESF. *See* monkey ear skin fibroblast (mESF)
metamorph software, role of, 98
metaphase I (MI), 217
methylmercury chloride (MeHg), 184, 185, 198
MI. *See* metaphase I (MI)
mitochondrial DNA (mtDNA), 257
  analysis of, 258, 269–70
M75 mouse monoclonal antibody, 298
MN/CAIX protein, 298
Moloney Murine Leukemia Virus, 142
monkey-bovine (MB)-iSCNT embryos, 257
  abnormal nucleolar morphology in, 259–60
  analysis of mtDNA in, 258
  expression and nucleolar component proteins in, 261–63
  expression of housekeeping and imprinting genes in, 258–59
  preparation of medium and culture of, 269
monkey ear skin fibroblast (mESF), 268
Mormyrid fishes, 19
Mos-MAPK pathway, role of, 137
mouse embryo, UV development in, 233
mouse embryo, vascular development in, 223–25
  DLAV establishment, 239–40
  embryonic and extraembryonic circulation, connection of, 231–34
  embryonic mouse vasculature, 3D visualization of, 226–30
  intersomitic and posterior cardinal vein, 237–38

materials and methods
  data visualization and segmentation, 245
  embryo collection and staining, 242
  embryos, OPT of, 242–43
  FDR-based deconvolution and filtering, 243–45
  point spread function acquisition, 243
  occipital region, intersomitic vessels of, 234–36
  outcomes of, 225
  PNVP, formation of, 240–41
  trunk region, intersomitic arteries of, 236–37
  vascular development between the 5 and 8 somite stage, 230–31
mouse ES cells, maintenance and differentiation of, 95–96
Mouse Genome 430 Version 2.0 array, 86, 98
mouse monoclonal antibody, usage of, 298
mouse monoclonal anti-nucleophosmin, 272
mouse monoclonal anti-UBTF, 272
mouse stem cells, transcriptome analysis of, 161–62, 173–74
  genes correlation with developmental potential of cells, 171–73
  materials and methods
    19 cDNA clones, analysis of, 175
    cDNA library construction, clone handling, and sequencing, 174
    clone frequencies, PCA of, 176
    differentially expressed genes, identification of, 175–76
    gene index, assembling of, 174
  novel genes, 162–63
  principal component analysis, 168–71
  signature genes, 163–68
mRNA, usage of, 37
MS cells. See mesenchymal stem (MS) cells
Msh4 gene, 139

mtDNA. See mitochondrial DNA (mtDNA)
Müllerian-type columnar cells, 297
murine oocyte genes, chicken homologs of, 124–25

N

National Center for Biotechnology Information (NCBI), 142, 163
National Institute on Aging (NIA), 163
N2B27-derived medium, usage of, 90
NCBI. See National Center for Biotechnology Information (NCBI)
NCBI gene expression omnibus, 199–201
  Q-RT-PCR, 40
  tissue expression, annotation for, 41
  transcription factor binding site prediction, 41
nebel gene, 123
neural differentiation, of embryonic stem cells, 77–79
  materials and methods
    FACS analysis, 97
    immunocytochemistry, 96–97
    LY411575, treatment with, 96
    microarray sample preparation and data analysis, 98–99
    mouse ES cells, maintenance and differentiation of, 95–96
    RNA extraction and RT-PCR, 97–98
    in situ hybridization, 97
    time-lapse movie, 98
  neural stem (NS) cells, 85
  neuroepithelial progenitors (NPs)
    apico-basal polarity, 81–82
    improved generation of, 79–81
  neuroepithelial rosettes, 89–91
    in vitro mammalian neural development, mechanisms of, 91–95
  neurogenic and gliogenic potential in vitro, 83–84
  rosette cultures, notch pathway in, 82–83

in vitro neural commitment, transcriptional profiling of, 86–89

neural induction, concept of, 89

neural stem (NS) cells, 96, 162, 170
    in monolayer cultures, 85

neural tube defects (NTD), 34, 36, 39

neuroepithelial progenitors (NPs), 77
    apico-basal polarity of, 81–82
    improved generation of, 79–81
    neurogenic and gliogenic potential *in vitro*, 83–84

neuroepithelial rosettes, 89–91
    in vitro mammalian neural development, mechanisms of, 91–95

neurogenic progenitors (nNPs), 88

neurons and glia, timing of production of, 84. *See also* neuroepithelial progenitors (NPs)

next stage of NP development (nNP), 93, 95

NIA. *See* National Institute on Aging (NIA)

NIA Mouse 15K cDNA Clone Set, 173

nNP. *See* next stage of NP development (nNP)

nNPs. *See* neurogenic progenitors (nNPs)

notch pathway, in rosette cultures, 82–83

NPB. *See* nucleolar precursor body (NPB)

NPs. *See* neuroepithelial progenitors (NPs)

NS cells. *See* neural stem (NS) cells

NT. *See* nuclear transfer (NT)

NTD. *See* neural tube defects (NTD)

nuclear transfer (NT), 114, 257, 260, 264, 268
    somatic cell, 268–69

nucleolar precursor body (NPB), 256

nucleolar protein expression, assessment of, 272

nucleolin
    expression of, 261
    protein, 266

nucleologenesis and embryonic genome, activation in
    bovine ooplasm and rhesus monkey somatic cells, 263–67

abnormal nucleolar morphology, 259–60

4- and 8-cell embryos, zygotic gene expression in, 270–71

background of, 255–57

donor cells, culture of, 268

housekeeping and imprinting genes, down-expression of, 258–59

IVF, BB-SCNT, and MB-iSCNT embryos, development of, 257–58

IVM and IVF, 267–68

MB-iSCNT embryos, medium and culture preparation of, 269

nucleolar component proteins, in MB-iSCNT embryos, 261–63

nucleolar protein expression, assessment of, 272

SCNT and IVC, 268–69

species-specific mtDNA, analysis of, 269–70

statistical analyses, 272

TEM, processing for, 271–72

nucleophosmin
    expression of, 261
    role of, 266

nucleospin RNA L kit, 199

O

OA. *See* omphalomesenteric artery (OA)

occipital intersomitic vessels, development of, 235

ODN. *See* oligodeoxynucleotides (ODN)

OET. *See* oocyte to embryo transition (OET)

OHSS. *See* ovarian hyperstimulation syndrome (OHSS)

oligodeoxynucleotides (ODN)
    design of, 51
    Hsp105 antisense, 49, 59

oligonucleotide microarrays, 28, 173

omphalomesenteric artery (OA), 231

omphalomesenteric veins (OV), 231

omphalomesenteric vessels, role of, 231

oncomodulin A, 192

one-cycle cDNA synthesis kit, usage of, 141
oocyte maturation and early embryo development, in hen, 135–39
  background of, 122–24
  follicular maturation, gene expression during, 131–34
  gene expression, tissular pattern of, 129–31
  methods
    animals, 139–40
    bioinformatic analysis, 140–41
    hierarchical clustering, 144
    Inra-Urgv, Affymetrix equipment at, 141–42
    real time RT-PCR, 143–44
    RNA extraction and reverse transcription, 142
    RNA isolation and microarray analysis, 141
    in situ hybridization, 144
    statistical analysis, 144–45
    tissues, oocytes and embryos, collection of, 140
  murine oocyte genes, chicken homologs of, 124–25
  oocyte and granulosa cells transcription profiles, comparison of, 126–29
  ovary, gene expression in, 134–35
oocytes, sICAM-1 in, 208–9, 215–16
  levels and embryo grade, 213
  materials and methods
    Cyto/Chemokines and ICAM-1 profiles, 218–19
    patients, 216–18
    sICAM-1 levels, measurement of, 218
    statistical analysis, 219
    western blotting, 218
  in mature and immature oocytes, 210–11
  in oocyte supernatants and pregnancy rate, 213–14
  production of, 212
  quantization of, 209–10
  release of cytokines, chemokines and Icam-1 by, 209
  and sHLA-G levels, comparison of, 214
oocyte to embryo transition (OET), 122
open reading frames (ORFs), 163
optical projection tomography (OPT), 224
  of embryos, 242–43
ORFs. *See* open reading frames (ORFs)
OV. *See* omphalomesenteric veins (OV)
Ov. *See* ovulated oocytes (Ov)
ovarian hyperstimulation syndrome (OHSS), 67–68, 71–73
  materials and methods
    embryo cryopreservation sequence, 69
    non-transferred blastocysts, secondary freeze for, 70
    patient selection and study design, 68–69
    statistical analysis, 70
    thaw, culture and transfer protocols, 70
  outcomes, 70–71
ovulated oocytes (Ov), 126, 136, 140–42
oxidative stress genes and toxicants response, 189. *See also* toxicogenomic responses, in zebrafish embryo

P

PA. *See* parthenogenetically activated (PA)
PAGE. *See* polyacrylamide gel electrophoresis (PAGE)
parthenogenetically activated (PA), 264
Pax3 gene, role of, 27, 31
PBS. *See* phosphate buffered saline (PBS)
PC. *See* principal component (PC)
PCA. *See* principal component analysis (PCA)
PCO. *See* polycystic ovarian syndrome (PCO)

PCR. *See* polymerase chain reaction (PCR)

Pdgfrα. *See* platelet derived growth factor receptor α (Pdgfrα)

PE. *See* primitive ectoderm (PE)

Pearson correlation coefficient, 115, 187, 201

PECAM-1, expression of, 230

penicillamine, hypotaurine and epinephrine (PHE), 268

peri-implantation period, in rat, 49. *See also* rat uterus, Hsp 105 expression in

perineural vascular plexus (PNVP), 226 formation of, 240–41

Perkin–Elmer nucleotide, 98

PGCs. *See* primordial germ cells (PGCs)

PHE. *See* penicillamine, hypotaurine and epinephrine (PHE)

phosphate buffered saline (PBS), 96, 140, 268

phosphatidyl inositol 3-kinase (P13-kinase), 280, 296

PHV. *See* primary head vein (PHV)

P13-kinase. *See* phosphatidyl inositol 3-kinase (P13-kinase)

placenta, CAIX expression in, 283. *See also* human development

platelet derived growth factor receptor α (Pdgfrα), 38

pluripotency genes, role of, 87

pluripotency markers, 91

PMA. *See* primitive maxillary artery (PMA)

PMSG. *See* pregnant mare's serum gonadotropin (PMSG)

PNVP. *See* perineural vascular plexus (PNVP)

Pole4. *See* polymerase epsilon 4 (Pole4)

polyacrylamide gel electrophoresis (PAGE), 51

polycystic ovarian syndrome (PCO), 216

polymerase chain reaction (PCR), 142, 257

polymerase epsilon 4 (Pole4), 40

pregnancy rate and oocyte supernatants, sICAM-1 in, 213–14

pregnant mare's serum gonadotropin (PMSG), 268

primary head vein (PHV), 231

primer3 program, usage of, 270

primitive ectoderm (PE), 81, 86

primitive maxillary artery (PMA), 231

primordial germ cells (PGCs), 171

principal component (PC), 168, 169, 201

principal component analysis (PCA), 168, 186

proneural genes, role of, 94

prophylactic human embryo cryopreservation, 67–68, 71–73
 materials and methods
  embryo cryopreservation sequence, 69
  non-transferred blastocysts, secondary freeze for, 70
  patient selection and study design, 68–69
  statistical analysis, 70
  thaw, culture & transfer protocols, 70
 outcomes, 70–71

protein-G HRP, usage of, 218

Q

quantitative real time PCR (Q-RT-PCR), 29–31

quantity one software, usage of, 51

R

rabbit polyclonal anti-fibrillarin, 272

rat uterus, Hsp 105 expression in, 59–60
 antisense Hsp105 ODNs treatment, 58
 background of, 48–49
 comparison of, 55–56
 methods
  animals, 49–50
  FITC, tracking of, 51–52
  immunohistochemistry, 50

microscopic assessment and statistical analysis, 52
oligodeoxynucleotides, design of, 51
western blot analysis, 50–51
in rat uterus during pseudo-pregnancy, 54–55
rat uterus, Hsp105 expression in, 53–54
suppression of, 56–57
in uterus during early pregnancy, 54
REACH. *See* registration, evaluation and authorization of chemicals (REACH)
real-time PCR analysis, of candidate genes, 130. *See also* hen, oocyte maturation and early embryo development in
registration, evaluation and authorization of chemicals (REACH), 182, 195
respiratory and gastrointestinal systems, epithelial cells of, 289–91. *See also* human development
Retiga Exi CCD camera, usage of, 243
reverse transcription (RT), 142
reverse transcription PCR (RT-PCR), 190
rhesus monkey somatic cells and bovine ooplasm
    interspecies cloned embryos between, 263–67
        abnormal nucleolar morphology, 259–60
        4- and 8-cell embryos, zygotic gene expression in, 270–71
        background of, 255–57
        donor cells, culture of, 268
        housekeeping and imprinting genes, down-expression of, 258–59
        IVF, BB-SCNT, and MB-iSCNT embryos, development of, 257–58
        IVM and IVF, 267–68
        MB-iSCNT embryos, medium and culture preparation of, 269
        nucleolar component proteins, in MB-iSCNT embryos, 261–63
        nucleolar protein expression, assessment of, 272

SCNT and IVC, 268–69
species-specific mtDNA, analysis of, 269–70
statistical analyses, 272
TEM, processing for, 271–72
RiboGreen® RNA quantification reagent, 141
ribosomal RNA (rRNA), 256
RNA extraction
    and reverse transcription, in hen, 142
    (*see also* hen, oocyte maturation and early embryo development in)
    and RT-PCR, 97–98
RNeasy Mini Kit, 131, 141
rosette cultures, notch pathway in, 82–83. *See also* embryonic stem cells, neural differentiation of
RQ1 DNase, usage of, 142
rRNA. *See* ribosomal RNA (rRNA)
RT. *See* reverse transcription (RT)
RT-PCR. *See* reverse transcription PCR (RT-PCR)

S

saline infusion sonogram, importance of, 68. *See also* ovarian hyperstimulation syndrome (OHSS)
SAS software package, usage of, 272
SCNT. *See* somatic cell nuclear transfer (SCNT)
SDS. *See* sodium dodecyl sulfate (SDS)
SE. *See* standard error (SE)
Sema3E, 240
sense oligodeoxynucleotides (S-ODNs), 51
sequence data analysis, flow chart of, 164. *See also* mouse stem cells, transcriptome analysis of
sequestosome1 gene, 194
serum-free medium N2B27, usage of, 79
serum hCG, usage of, 112
serum oestradiol, usage of, 73
serum substitute supplement (SSS), 154, 269

SGL. *See* subpial granular layer (SGL)

sICAM-1. *See* soluble Intercellular Adhesion Molecule 1 (sICAM-1)

signature genes, role of, 163–68

single lumen aspiration needle, in oocyte recovery, 217

singular value decomposition (SVD), 186

sinus venosus (SV), 223, 231, 232

skin and squamous mucosa, CAIX expression in, 285. *See also* human development

SLC. *See* solute carrier (SLC)

Smithsonian Institution, 14

sodium dodecyl sulfate (SDS), 50

S-ODNs. *See* sense oligodeoxynucleotides (S-ODNs)

soluble HLA-G (sHLA-G) and sICAM-1 levels, comparison, 214

soluble Intercellular Adhesion Molecule 1 (sICAM-1), 208–9, 215–16
  levels and embryo grade, 213
  materials and methods
    Cyto/Chemokines and ICAM-1 profiles, 218–19
    patients, 216–18
    sICAM-1 levels, measurement of, 218
    statistical analysis, 219
    western blotting, 218
  in mature and immature oocytes, 210–11
  in oocyte supernatants and pregnancy rate, 213–14
  production of, 212
  quantization of, 209–10
  release of cytokines, chemokines and Icam-1 by, 209
  and sHLA-G levels, comparison of, 214

solute carrier (SLC), 189, 196
  family genes and their response to toxicants, 190

somatic cell nuclear transfer (SCNT), 255, 268–69

Sox1 gene, 92

Sox2 gene, 85

Sox1-GFP analysis, usage of, 97. *See also* embryonic stem cells, neural differentiation of

Sox1-GFP knock-in ES cells, usage of, 79

SPSS. *See* statistical package for social science (SPSS)

Spurr's epoxy resin, 271

SSS. *See* serum substitute supplement (SSS)

standard error (SE), 272

statistical package for social science (SPSS), 52

stat view software package, usage of, 219

stem cell biomarkers, usage of, 297

STORM 840 phosphor screen imaging system, 142

Streptozotocin, 38

subpial granular layer (SGL), 21, 22

Superscript II Reverse Transcriptase, 97

SV. *See* sinus venosus (SV)

SVD. *See* singular value decomposition (SVD)

SyberGreen PCR Core reagents, 270

T

targeting-induced local lesions in genomes (TILLING), 197

tBHQ. *See* tert-butylhydroquinone (tBHQ)

TCDD. *See* 2,3,7,8-tetrachlorodibenzo-p-dioxin (TCDD)

Tecnai G2 Spirit Twin Transmission electron microscope, 271

Tefcat catheter, 154

TeleChem SMP3 pins, 198

TEM. *See* transmission electron microscopy (TEM)

tert-butylhydroquinone (tBHQ), 184, 198

2,3,7,8-tetrachlorodibenzo-p-dioxin (TCDD), 183, 184, 187, 188, 192, 196, 197, 198

TFBS. *See* transcription factor binding sites (TFBS)

thioredoxin-like mRNA, 192, 196

three-dimensional (3D) analysis, in mouse embryo, 223–25
  DLAV establishment, 239–40
  embryonic and extraembryonic circulation, connection of, 231–34
  embryonic mouse vasculature, 3D visualization of, 226–30
  intersomitic veins and the posterior cardinal vein, 237–38
  materials and methods
    data visualization and segmentation, 245
    embryo collection and staining, 242
    embryos, OPT of, 242–43
    FDR-based deconvolution and filtering, 243–45
    point spread function acquisition, 243
  occipital region, intersomitic vessels of, 234–36
  outcomes of, 225
  PNVP, formation of, 240–41
  trunk region, intersomitic arteries of, 236–37
  vascular development between the 5 and 8 somite stage, 230–31

TILLING. *See* targeting-induced local lesions in genomes (TILLING)

Tissue-Tek, 144

tNPs. *See* transient population of NPs (tNPs)

toxicogenomic responses, in zebrafish embryo, 187–90, 194–97
  background of, 182–83
  complex synergistic effects, 194
  genome in, 192–93
  materials and methods
    chemicals and embryo treatment, 198
    microarray analysis, 198–99
    Ncbi's gene expression omnibus, 199–201

model compounds in, 184–85
  stage-specific toxicogenomic responses, 186–87
  tissue-specific genes, identification of, 190–92

toxicogenomics, 183, 186, 187, 194, 196, 197

transcription factor binding sites (TFBS), 32

transcriptome analysis, of mouse stem cells, 161–62, 173–74
  genes correlation with developmental potential of cells, 171–73
  materials and methods
    19 cDNA clones, analysis of, 175
    cDNA library construction, clone handling, and sequencing, 174
    clone frequencies, PCA of, 176
    differentially expressed genes, identification of, 175–76
    gene index, assembling of, 174
  novel genes, 162–63
  principal component analysis, 168–71
  signature genes, 163–68

transferase dUTP nick-end labeling (TUNEL), 258

transient population of NPs (tNPs), 93, 95

transmembrane proteins, role of, 196

transmission electron microscopy (TEM), 257
  processing for, 271–72

transvaginal sonogram, 68

Tri-reagent, usage of, 142. *See also* hen, oocyte maturation and early embryo development in

Trizol', usage of, 39

trophoblast, 10, 49, 60, 162, 171, 281

trophoblast stem (TS) cells, 162

tumor suppressor genes, 10

TUNEL. *See* transferase dUTP nick-end labeling (TUNEL)

tyramide-Cy3 reagent, 242

U

UA. *See* umbilical artery (UA)
UBTF. *See* upstream binding transcription factor (UBTF)
U004160 gene, 163
U035352 gene, 163
ULTRACUT ultramicrotome, 271
umbilical artery (UA), 232, 234
umbilical vein (UV), 227, 232
    development in mouse embryo, 233
umbilical vessels, role of, 231
upstream binding transcription factor (UBTF), 256
    expression and localization of, 262
    localization of, 266
urinary system, CAIX expression in, 287.
    See also human development
UV. *See* umbilical vein (UV)
UV spectrophotometer, usage of, 50

V

valproic acid (VA), 184, 198
vascular development, in mouse embryo, 223–25
    DLAV establishment, 239–40
    embryonic and extraembryonic circulation, connection of, 231–34
    embryonic mouse vasculature, 3D visualization of, 226–30
    intersomitic veins and the posterior cardinal vein, 237–38
    materials and methods
        data visualization and segmentation, 245
        embryo collection and staining, 242
        embryos, OPT of, 242–43
        FDR-based deconvolution and filtering, 243–45
        point spread function acquisition, 243
    occipital region, intersomitic vessels of, 234–36
    outcomes of, 225

PNVP, formation of, 240–41
    trunk region, intersomitic arteries of, 236–37
    vascular development between the 5 and 8 somite stage, 230–31
vascular endothelial growth factor (VEGF), 72, 209, 219, 223, 240
Vecteur NTI software, usage of, 143
Vector Red AP substrates, 50
VEGF. *See* vascular endothelial growth factor (VEGF)
vertebral artery (VTA), 234, 235
virtual reality modeling language (VRML), 168
VTA. *See* vertebral artery (VTA)

W

wee gene, role of, 138
western blot analysis
    of Hsp105 expression, 54 (*see also* rat uterus, Hsp 105 expression in)
    in sICAM-1 analysis, 218 (*see also* in vitro fertilized (IVF) human embryos)
WHO. *See* World Health Organization (WHO)
whole genome rVISTA, usage of, 41
Wnt-pathway genes and maternal diabetes, 35. *See also* maternal diabetes, in embryo development
Wnt signaling, 27, 37, 139, 184
World Health Organization (WHO), 72

Z

zar1 gene, 133
zebrafish embryo, barcode-like toxicogenomic responses in, 187–90, 194–97
    background of, 182–83
    complex synergistic effects, 194
    genome in, 192–93
    materials and methods
        chemicals and embryo treatment, 198
        microarray analysis, 198–99

Ncbi's gene expression omnibus, 199–201

model compounds in, 184–85

stage-specific toxicogenomic responses, 186–87

tissue-specific genes, identification of, 190–92

Zeiss Axiovert 200 M microscope, usage of, 98, 272

zona pellucida (ZP), 138, 209

ZPC protein, role of, 138

zygotic gene expression, evaluation of, 270–71. *See also* bovine ooplasm and rhesus monkey somatic cells

Printed and bound by CPI Group (UK) Ltd, Croydon, CR0 4YY

23/10/2024

01777692-0002